PHOTOPOLYMERISATION AND PHOTOIMAGING SCIENCE AND TECHNOLOGY

PHOTOPOLYMERISATION AND PHOTOIMAGING SCIENCE AND TECHNOLOGY

Edited by

NORMAN S. ALLEN, BSc, PhD, DSc

Chemistry Department, Faculty of Science and Engineering,
Manchester Polytechnic, Manchester, UK

ELSEVIER APPLIED SCIENCE
LONDON and NEW YORK

ELSEVIER SCIENCE PUBLISHERS LTD
Crown House, Linton Road, Barking, Essex IG11 8JU, England

Sole distributor in the USA and Canada
ELSEVIER SCIENCE PUBLISHING CO., INC.
655 Avenue of the Americas, New York, NY 10010, USA

WITH 27 TABLES AND 78 ILLUSTRATIONS

British Library Cataloguing in Publication Data

Photopolymerisation and photoimaging
science and technology
1. Photopolymerisation
I. Allen, Norman S.
547'.28

ISBN-13: 978-94-010-6999-1 e-ISBN-13: 978-94-009-1127-7
DOI: 10.1007/978-94-009-1127-7

Library of Congress Cataloging-in-Publication Data

Photopolymerisation and photoimaging science and technology / edited by Norman S. Allen.
p. cm.
Bibliography: p.
Includes index.

1. Polymerization. 2. Photochemistry. 3. Photolithography.
I. Allen, Norman S.
QD381.9.P56P48 1989
547'.28—dc20

PREFACE

During the past decade the fields of photopolymerisation and photo-imaging science have grown from subjects of esoteric research specialities into major industrial developments and are now fields of central importance in polymer science and technology. Free-radical addition polymerisation, cationic polymerisation, photolithography, photocrosslinking, photografting, photochromic imaging and topo-chemical polymerisation all continue to gain important new grounds in industry. Much new research continues in these areas and today has given rise to many new applications in such fields as microelectronics, e.g. in resists, barrier coatings, encapsulants and printing board technologies. Without the use of photoresists it would not be possible to develop modern electronic and computer industries.

This book is meant to be a comprehensive and up-to-date text in the science and technology of all the above subjects with each chapter being written by a very prominent specialist or specialists working in the particular area concerned. The chapters have a strong international flavour and cover a very wide range of topics. The first, by Dr Hageman at Akzo Chemie in the Netherlands, deals with the most fundamental subject of them all, namely photoinitiated addition free-radical polymerisation. Here emphasis is on the mechanistic behaviour of both established and new carbonyl-containing systems. Cationic photopolymerisation has also established itself as a field of some importance and this is reviewed expertly by Dr Pappas from the North Dakota State University in the USA. Again the main emphasis here is on the mechanistic features of the individual types of photoinitiators. Chapter 3 has a more industrial flavour and deals with the chemistry of photolithography and photocrosslinking materials and processes; it is written by Drs Turner and Daly of the Eastman Kodak Company in the USA. Chapter 4 covers a somewhat different slant to photopolymerisation, that of photografting monomers onto polymer systems in order to improve a particular property requirement. This is dealt with effectively by Professor Bellobono from the University of Milan in Italy. Photochromic processes are only now just growing in importance and here the science and chemistry of the systems involved are reviewed in Chapter 5 by Dr Jacobson from the Central Polytechnic in London. The complex nature of solid-state photopolymerisation is dealt with in Chapter 6 by a well-known authority in the field, Professor Hasegawa, who is at the University of Tokyo, Japan. In Chapter 7 the excited-state properties of photoinitiators is seen through the use of lasers; this subject is well covered by Professor Fouassier at the University in Mulhouse, France, who is a noted expert in this area. Finally, Chapter 8 covers a subject which has hitherto often been omitted from books in this field. The post-cured stability of radiation-cured systems is now of some importance, especially in applications where long-term durability is required to both heat and light. Little has been published on this subject to date and this chapter details the extensive efforts of one of my Ph.D. students, Dr White, into this problem with emphasis on thermal and photooxidative stability and photoyellowing of radiation-curable resins.

This book should therefore appeal to all polymer scientists, physicists and technologists working in the fields of radiation curing and photoimaging science and technology as well as those with little knowledge of the subjects who wish to learn and develop some expertise.

NORMAN S. ALLEN

CONTENTS

LIST OF CONTRIBUTORS

N. S. Allen

*Department of Chemistry, Faculty of Science and Engineering, Manchester Polytechnic, Manchester M*1 *5GD, UK*

Ignazio Renato Bellobono

Università degli Studi di Milano, 20133 *Milano—Via C. Golgi* 19, *Italy*

Robert C. Daly

Electronics Research Laboratories, Eastman Kodak Company, Rochester, NY 14650, *USA*

J. P. Fouassier

Laboratoire de Photochimie Générale, Unité Associée au CNRS no 431, *Ecole Nationale Supérieure de Chimie,* 3 *rue A. Werner,* 68093 *Mulhouse Cedex, France*

H. J. Hageman

Akzo Research, Corporate Research Department, Postbus 60, 6800 *AB, Arnhem, The Netherlands*

Masaki Hasegawa

Department of Synthetic Chemistry, Faculty of Engineering, The University of Tokyo, Hongo, Bunkyo-ku, Tokyo 113, *Japan*

R. E. Jacobson

School of Communication, Polytechnic of Central London, 18/22 *Riding House Street, London W*1*P* 7*PP, UK*

S. Peter Pappas

Polymers and Coatings Department, North Dakota State University, Fargo, ND 58105, *USA*

P. J. Robinson

*Department of Chemistry, Faculty of Science and Engineering, Manchester Polytechnic, Manchester M*1 5*GD, UK*

Elena Selli

Università degli Studi di Milano, 20133 *Milano—Via C. Golgi* 19, *Italy*

R. Clancy

*Harcros Chemicals Limited, Silk Street, Eccles, Manchester M*30 0*BH, UK*

S. Richard Turner

Corporate Research Laboratories, Eastman Kodak Company, Rochester, NY 14650, *USA*

N. J. White

*Department of Chemistry, Faculty of Science and Engineering, Manchester Polytechnic, Manchester M*1 5*GD, UK*. Present address: *Banbury Laboratories, Alcan International Limited, Southam Road, Banbury OX*16 7*SP, Oxfordshire, UK*

Chapter 1

PHOTOINITIATORS AND PHOTOINITIATION MECHANISMS OF FREE-RADICAL POLYMERISATION PROCESSES

H. J. HAGEMAN

Akzo Corporate Research, Arnhem, The Netherlands

1 INTRODUCTION

Photoinitiated free-radical polymerisation, and crosslinking of multifunctional vinyl monomers and oligomers (prepolymers) in particular, is currently enjoying a lively commercial interest on account of its potential in a large number of applications. The reasons for the current developments are the particular attractive features of photoinitiated polymerisation processes, such as the concept of latency, high speed under mild conditions, reduced energy and space requirements, and last but not least considerable environmental advantages.

An essential ingredient of all UV-curable (crosslinkable) compositions is the photoinitiator, or more generally the photoinitiating system. Imperative requirements imposed on the photoinitiating system are:

(a) absorptivity preferably in the 300–400 nm range, with absorption coefficient (ε) high or low, depending on the specific application;
(b) efficient generation of free radicals capable of attacking the olefinic double bond of vinyl monomers;
(c) adequate solubility in unsaturated resin systems;
(d) should not adversely affect the viscosity of the overall system even at elevated temperatures, i.e. high storage stability in the dark;

(e) should not impart yellowing or unpleasant odours to the cured material;
(f) should be non-toxic itself, as should its irradiation products;
(g) it should preferably be liquid or low-melting for handling and blending convenience.

Over the last 15 years a large number of reviews have appeared on photoinitiators, photocrosslinkable resin systems and their applications,[1-36] illustrating the growing importance of UV curing science and technology. In this chapter an updating review is presented: without pretending to be exhaustive, it covers various aspects of photoinitiators and photoinitiated free-radical polymerisation, including general principles, mechanistic studies and recent developments.

2 STAGES OF PHOTOINITIATED FREE-RADICAL POLYMERISATION

The most important stages of photoinitiated free-radical polymerisation processes fall into three distinct categories

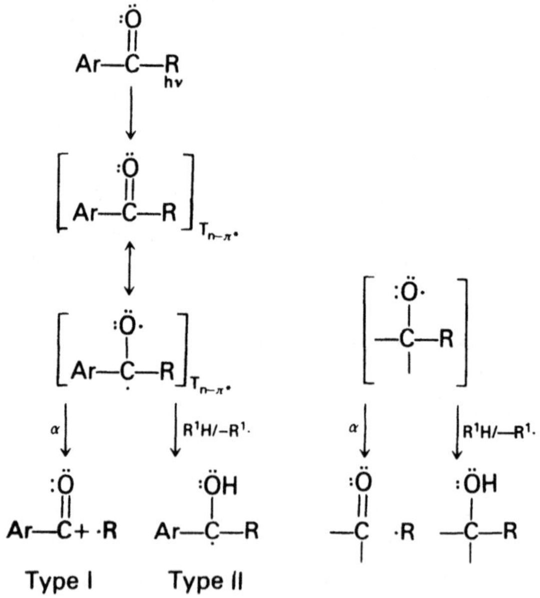

SCHEME 1. (n–π*) Triplet excited carbonyl reactions compared with reactions of an alkoxy radical.

2.1 Photophysical Processes[37]

The very first act in photoinitiated polymerisation processes is the absorption of UV light in the 300–400 nm range by the photoinitiator to populate an excited singlet manifold which may have π–π^* or n–π^* configuration. Through rapid internal conversion the lowest excited singlet state is then reached. In aromatic carbonyl compounds, to which category most photoinitiators belong, singlet–triplet energy splitting is usually small, and as a consequence intersystem crossing (i.s.c.) to the lowest excited triplet state (π–π^* or n–π^*) proceeds rapidly and efficiently (quantum yields between 0·9 and 1·0). Possible competing processes such as fluorescence and non-radiative decay of excited singlets to the ground state usually are of negligible importance.

It may be worth noting that in principle it is also possible to populate the lowest excited triplet state of the photoinitiator in a photosensitised process, i.e. absorption of the light by a suitable photosensitiser followed by energy transfer to the photoinitiator molecule.

2.2 Photochemical Processes[37]

Major chemical deactivation pathways open to the excited triplet state of these aromatic carbonyl compounds show an analogy with the major reactions of alkoxy radicals,[38,39] i.e. (i) fragmentation, and (ii) hydrogen-abstraction from a suitable hydrogen-donor, as is illustrated in Scheme 1.

On the basis of these two major reactions photoinitiators have been divided into two major types:

Type I photoinitiators, being those which undergo unimolecular fragmentation (Norrish type I, α-cleavage) into an aroyl and a second radical (R·); and
Type II photoinitiators, being those which only generate radicals through bimolecular hydrogen-abstraction from a suitable hydrogen-donor.

Which of these processes prevails depends primarily on the nature of the group R. When R is an alkyl group, preferably carrying substituents capable of stabilising an adjacent positive charge, α-cleavage is the almost exclusive process,[40] whereas when R is an aromatic group α-cleavage does not occur, but hydrogen-abstraction is the favoured process.[44]

Moreover, the configuration of the excited state involved also has implications for the rates at which the various processes take place: n–π^* triplets are far more reactive than their π–π^* counterparts in α-cleavage,[42,43] as well as in hydrogen-abstraction[44,45] (inter- and intra-molecular) reactions, except when tertiary amines are the hydrogen-donor involved.

Possible competing processes, such as phosphorescence, and non-radiative decay of the excited triplets to the ground state can safely be neglected. However, quenching of the excited triplet states by oxygen, vinyl monomers and other adventitious species does occur in certain Type I cases, and in Type II photoinitiators except when tertiary amines are used as the hydrogen-donor (*vide infra*).

2.3 Ground-State Free-Radical Processes

The primary radicals thus generated from the excited triplet state of the various photoinitiators can in principle undergo the usual set of reactions common to free radicals such as:

(i) addition to the double bond of a vinyl monomer present, i.e. the actual initiation step;
(ii) subsequent fragmentation generating a secondary radical that may or may not be reactive towards the double bond of vinyl monomers;
(iii) hydrogen-abstraction, creating another radical centre that again may or may not be reactive;
(iv) dimerisation, implying loss of radical activity; and
(v) coupling with other radicals, e.g. growing polymer radicals, implying primary radical termination.

Reaction (i) is of course the desired reaction, but reactions (ii) and (iii) may also contribute to the overall initiation of polymerisation.

In reactions (iv) and (v), however, potential initiating radicals are being destroyed and will therefore reduce the overall efficiency of the photoinitiation.

Which of these reactions actually prevails depends largely on the ratio of their rate constants. At the concentration normally used, i.e. low concentration of photoinitiator and relatively high concentration of vinyl functionalities, reactions (iv) and (v) will presumably be of minor importance. More important, however, are reactions of primary, secondary and growing polymer radicals with oxygen from the surrounding air atmosphere.

3 TECHNIQUES FOR MECHANISTIC STUDIES

The growing interest in commercial UV-curable systems has created a strong demand for more effective new photoinitiators. Their development now necessitated a more complete understanding of the photophysical and photochemical processes of the photoinitiator on the one hand and of the subsequent (ground-state) free-radical chemistry on the other. This section is devoted to a survey of the physical and chemical techniques used to study photoinitiator mechanisms in the widest sense.

3.1 Photolysis

The oldest method of studying the photochemical behaviour of organic compounds in general is simply photolysis in solution. The solvents to be used should preferably be inert (e.g. benzene, acetonitrile), or have a predictable effect such as a pronounced hydrogen-donating ability (e.g. 2-propanol). From the products formed conclusions can be drawn as to the most likely fragmentation pattern of the compounds under study. Quantum yields for the photodecomposition of starting compounds and for the formation of products can be obtained from kinetic analysis. Quenching and sensitisation studies can provide additional information on the nature and lifetime of the excited state involved.

3.2 Laser Flash Photolysis

More sophisticated is the more recently developed laser flash photolysis which gives some insight into the primary processes, i.e. the nature and lifetimes of excited states involved as well as the identity of the primary radicals formed. Unequivocal identification of the radicals from the transient optical absorption spectra, however, may present some difficulties. The decay kinetics of the radicals in the presence of vinyl monomers allow the determination of rate constants for the addition to the olefinic double bond of the vinyl monomer.

3.3 ESR Spectroscopy

A much more powerful technique for identifying radicals is electron spin resonance (ESR) spectroscopy, particularly when coupled with flash photolysis to produce relatively high concentrations of radicals within the ESR cavity. Kinetic ESR spectroscopy has been shown[46] to be a versatile tool to study radical reactions and obtain the corresponding rate constants. In order to detect and identify extremely short-lived radicals in reacting systems, the spin-trapping technique has been developed.[47] This technique is based on the addition reaction of such short-lived radicals to the double bond of certain nitrones and nitroso compounds to afford longer-lived secondary radicals.

An excellent review on typical applications of ESR spectroscopy to radical polymerisation has recently appeared.[48]

3.4 CIDNP and CIDEP

Chemically induced dynamic nuclear polarisation (CIDNP)[49] and the related chemically induced dynamic electron polarisation (CIDEP)[50] have the appearance of emission and enhanced absorption lines in the respective spectra taken during the course of radical reactions in common. In the CIDNP technique products are being looked at, whereas in the CIDEP technique transient radicals are being observed.

Both techniques, however, provide useful information on fragmentation patterns of the compounds studied and in particular on the multiplicity of the excited state(s) involved in photochemical reactions.

3.5 Radical Trapping

In contrast with the above-mentioned spectroscopic techniques, this method is based on the extremely fast reaction of the generated radicals with added stable nitroxyl radicals to give products that can be isolated and characterised.[51] This method has also been used for quantum yield determinations. Initiation mechanisms, however, can only be studied using initiators which generate radicals that do not couple with nitroxyl radicals, such as oxygen- and sulphur-centred radicals.

Most photoinitiators generate only carbon- and phosphorus-centred radicals, which are efficiently scavenged by the nitroxyl radicals before they can react with vinyl monomers.

3.6 Photoinitiated Polymerisation

Information on photoinitiator efficiency in polymerisations can be obtained from polymerisation studies of specific monomers in bulk or in solution. An extremely useful method recently developed is the solution-crosslinking of a polyfunctional 'monomer' followed by laser nephelometry.[52]

Information on the initiating species can in principle be obtained in a number of ways. A study of polymerisation kinetics using isotope-labelled photoinitiators[53] is a method of obtaining such information.

Laser flash photolysis of photoinitiators in the presence of vinyl monomers may provide information on the nature of the initiating species,[54] as may EST spectroscopy using spin-trapping.[55]

Isotope labelling of the initiator has also been used in end-group analysis by various NMR techniques (e.g. ^{13}C-, ^{15}N-, and ^{19}F-NMR).[56]

Photodecomposition of photoinitiators in the presence of non-polymerising model substrates for vinyl monomers leading to simple low molecular weight products is another useful method of identifying the initiating species.[57]

Photoinitiated crosslinking of multifunctional systems in thin films (UV curing) finally gives information on the efficiency of a given photoinitiator under more realistic conditions. Experiments have also been carried out in thin films of inert viscous solvents (model environment) to simulate the conditions in practical UV-curable systems.[58]

At the end of this section it should be emphasised that although each of the techniques mentioned above provides valuable information and is therefore useful in its own right, one should not rely on one technique alone if one seeks complete understanding of the mechanism of photoinitiators in photoinitiated polymerisation and crosslinking processes.

4 PHOTOINITIATORS

The various classes of photoinitiators known today based on the Type I and Type II classification will be discussed in some detail in this section.

4.1 Type I Photoinitiators (Unimolecular Radical Generation)

Aromatic carbonyl compounds carrying particular functional groups have been found to undergo direct photofragmentation into radicals. Depending on the nature of the functional group and its location in the molecule relative to the chromophoric carbonyl group, this fragmentation can take place at the bond adjacent to the carbonyl group (α-cleavage), at the bond in the β-position (β-cleavage) and in the case of a particularly weak bond elsewhere in the molecule, accordingly at this more remote position.

The photoinitiators undergoing direct photofragmentation are subdivided in different groups on account of their molecular structures.

4.1.1 Benzoin Derivatives

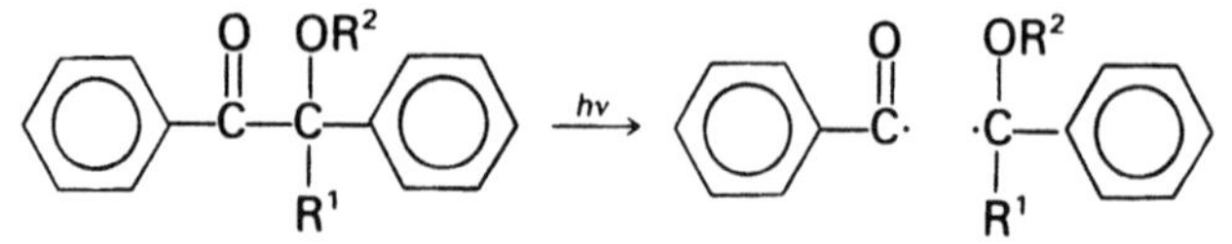

(1)

R^1 = H, Alk, substd Alk; R^2 = H, Alk, substd Alk

Benzoin and particularly its alkyl ethers have long been known for their ability to initiate free-radical polymerisation of the vinyl monomers upon irradiation with ultraviolet light in the 300–400 nm range. These compounds, with few exceptions are relatively low-melting (m.p. <100°C) colourless solids. In the most relevant spectral range (300–400 nm) their absorption spectra show a maximum at λ_{max} 320 nm (log ε 2·3).

The products found on photolysis of benzoin[59,60] and its alkyl ethers[40,61–63] in the absence of vinyl monomers have been explained as indicating α-cleavage into a benzoyl and an α-substituted benzyl radical:

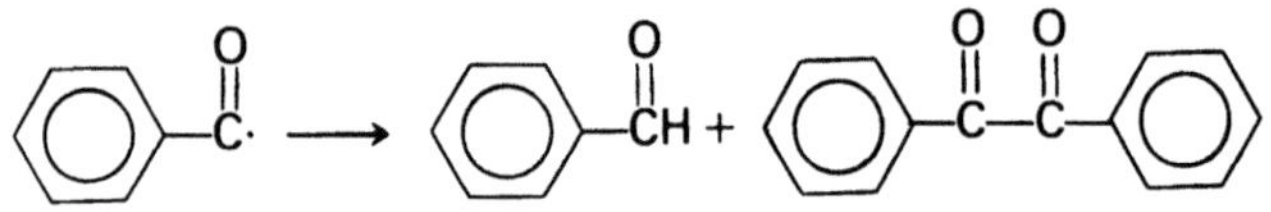

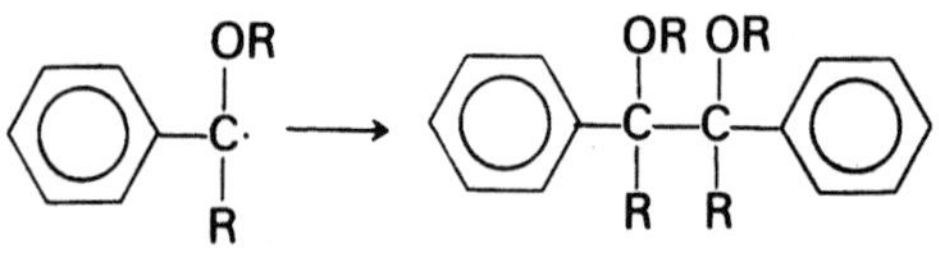

(*meso* + *DL*)

Thus benzaldehyde and benzil have been inferred to arise from the benzoyl radical through hydrogen-abstraction and dimerisation respectively [reaction (2)]. The pinacol derivative has been inferred to arise from the substituted benzyl radical through dimerisation [reaction (3)].

From photolysis in the presence of added thiol (RSH) in order to scavenge efficiently all non-cage benzoyl radicals and form benzaldehyde, quantum yields for α-cleavage have been determined to fall in the range 0·39–0·44.[40] This α-cleavage reaction of benzoin alkyl ethers cannot be quenched by the usual triplet quenchers,[40,61,62] which indicates that the α-cleavage is an exceptionally rapid triplet (?)-state process ($k_\alpha \geqslant 10^{10}\,s^{-1}$).

Confirmation is provided by laser flash photolysis studies.[54] The radical precursors (the triplet (?) excited states) of benzoin alkyl ethers could not be observed. In all cases a rapidly decaying (assigned to the benzoyl) and a more slowly decaying (assigned to the α-substituted benzyl) radical were observed. Electron spin resonance (ESR)[63,64] and ESR spin-trapping[65] studies have confirmed the assignments of the radicals generated:

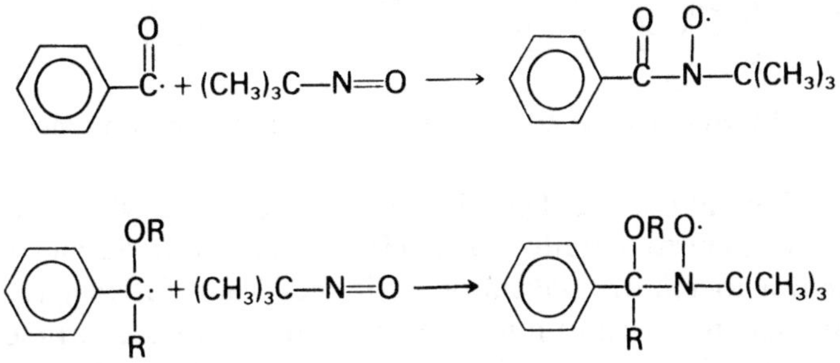

Additional confirmatory evidence comes from radical trapping experiments using 2,2,6,6-tetramethylpiperidin-1-oxyl (TMPO):[66]

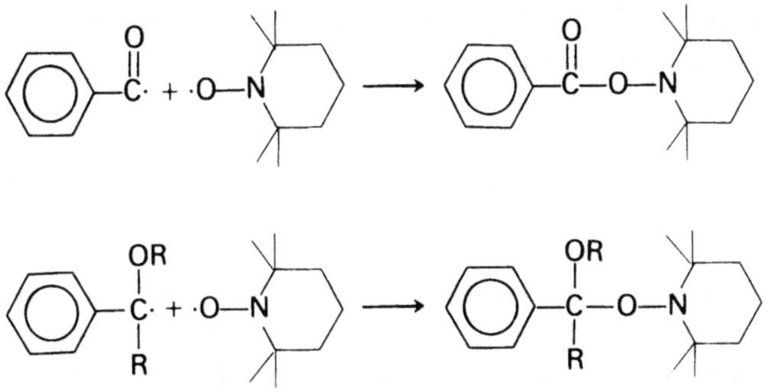

^{1}H-NMR–CIDNP[67–70] and ^{13}C-NMR–CIDNP[71] experiments finally confirmed not only the fragmentation pattern as concluded from all other techniques but in particular the involvement of the triplet excited state, which until then had been no more than a reasonable assumption.

In the first study of the photoinitiated polymerisation of methyl methacrylate (MMA) using ^{14}C-labelled benzoin methyl ether, the incorporation of 13–14 photoinitiator molecules (?) or radioactive fragments per polymer molecule was reported[72] without a satisfactory explanation being offered.

It has been generally accepted that the benzoyl radical adds to the olefinic double bond of vinyl monomers,[61,63,65,73] i.e. it initiates the polymerisation. The role played by the α-substituted benzyl radical, however, has aroused considerable dispute.

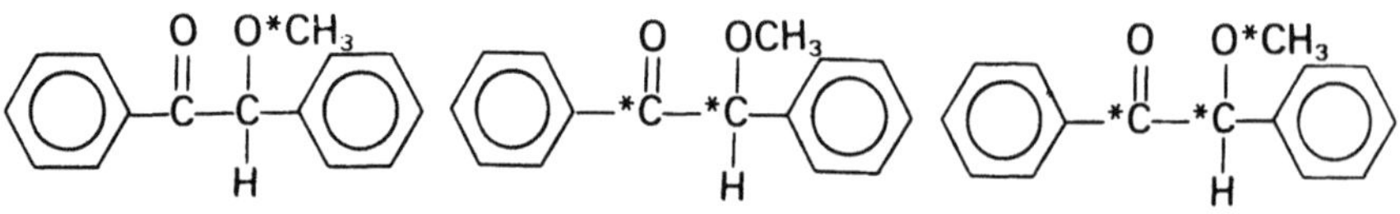

I Benzoin methyl ether: * indicates labelled carbon atoms

The photoinitiated polymerisation of methyl acrylate (MA) and MMA using benzoin methyl ether (**I**) ^{14}C-labelled in various positions was shown to follow conventional kinetics[53] and for that reason the benzoyl and the α-substituted benzyl radicals were concluded to be equally effective in initiating the polymerisation. Generation of α-methoxybenzyl radicals through hydrogen-abstraction from benzyl methyl ether by triplet excited benzophenone was also found (Ref. 74; Hageman, H. J. & Overeem, T., unpublished) to initiate the polymerisation of MMA [reaction (2)]. The initiating species, however, could not be identified.

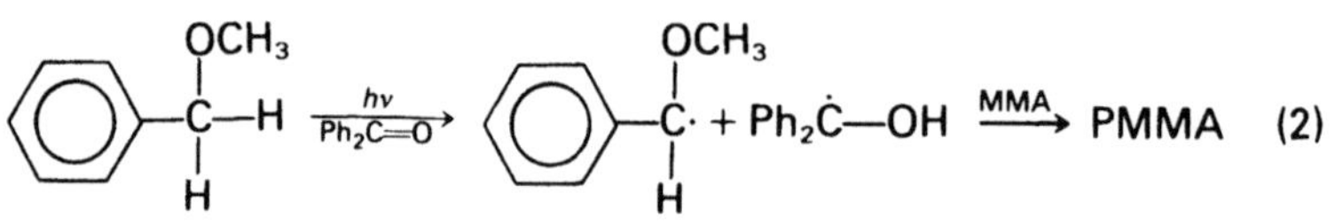

The photoinitiated polymerisation of styrene (St)[75] and MMA (Hageman, H. J., Janssen, P. C. G. M. & Maeden, F. B. D., unpublished) using relatively high concentrations of benzoin methyl ether was studied in order to identify the end groups of the resulting oligomer molecules (Scheme 2). Major products isolated and identified by gas chromatography/mass spectrometry (GC/MS) showed the benzoyl fragment located at the tail-end and the α-methoxylbenzyl fragment exclusively at the head-end of the respective monomer moieties.

This has been confirmed using 1,1-diphenylethylene[57a] and methyl α-methoxyacrylate[57b] as non-polymerising model substrates for styrene and MMA respectively (Scheme 3). These results have been explained as indicating the initiating propensity of the benzoyl radical and the lack of it in the α-methoxybenzyl radical. This conclusion is supported both by an earlier observation[61] that the dimerisation of the α-

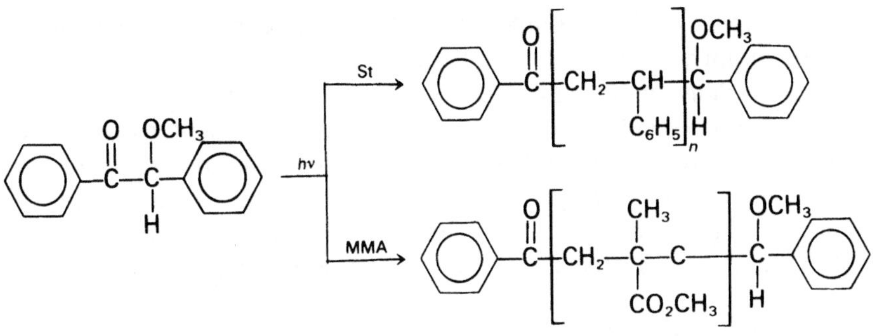

SCHEME 2

alkoxybenzyl radicals generated from a number of benzoin alkyl ethers is hardly affected by the presence of small amounts of vinyl monomers, and by recent flash photolysis experiments.[54]

Recently the kinetics of the polymerisation of styrene photoinitiated by benzoin isobutyl ether were reported.[76] The kinetic and molecular weight data suggest that at low temperatures, high light intensity and/or low monomer concentration, the benzoyl radical is the dominant initiating species, the α-isobutoxybenzyl radical being consumed mainly in the terminating step. At low light intensity, high temperatures and/or high monomer concentration, however, it appears that both benzoyl and α-isobutoxybenzyl radicals initiate the polymerisation. This then implies that the benzoyl radical has a higher rate constant for addition to the olefinic double bond of vinyl monomers than the α-alkoxybenzyl radicals.

In an earlier report[77] on the comparison of various benzoin alkyl ethers in the photoinitiated polymerisation of acrylates, it is stated that

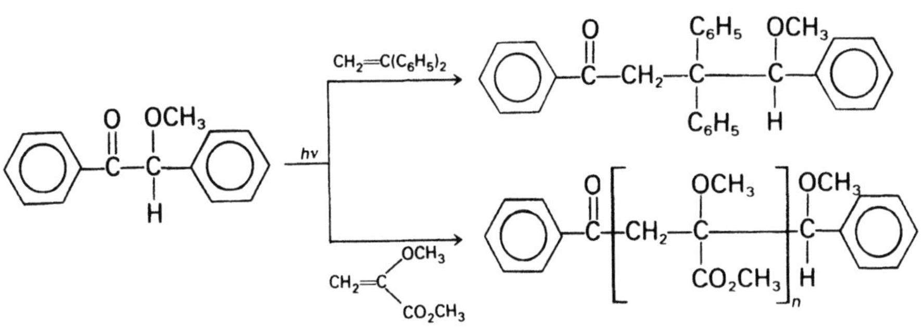

SCHEME 3

the effectivity of these compounds does not appear to be a function of the quantum yield for α-cleavage, e.g. the *t*-butyl ether has the highest rate of α-cleavage but nevertheless is the least effective photoinitiator. Corroborative evidence has been obtained from the crosslinking of trimethylolpropane triacrylate (TMPTA) in solution, monitored by laser nephelometry (Hageman, H. J. & Oosterhoff, P., unpublished). Since all these compounds generate the benzoyl radical, differences in photoinitiating effectivity must largely be ascribed to the role played by the different α-alkoxybenzyl radicals. Unfortunately, data on this particular aspect are not available.

Benzoin has been reported[78] readily to undergo base-catalysed *C*-alkylation followed by further *O*-alkylation to give a series of α-alkyl benzoin alkyl ethers (Scheme 4) which are expected to show a similar behaviour to the corresponding benzoin alkyl ethers.[57a]

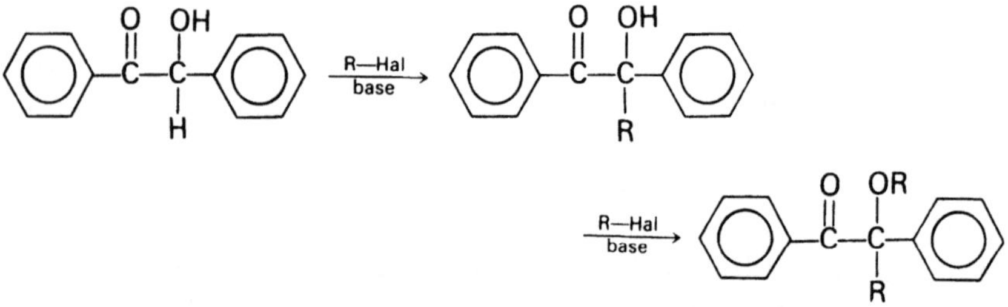

SCHEME 4

Benzoin ethers have been reported[79] to undergo Michael addition to several acrylic esters, acrylonitrile and acrylamide, leading to a series of different α-substituted benzoin ethers [reaction (3)]. No data are available, but these compounds will in all likelihood fall in the same category as the corresponding benzoin ethers.

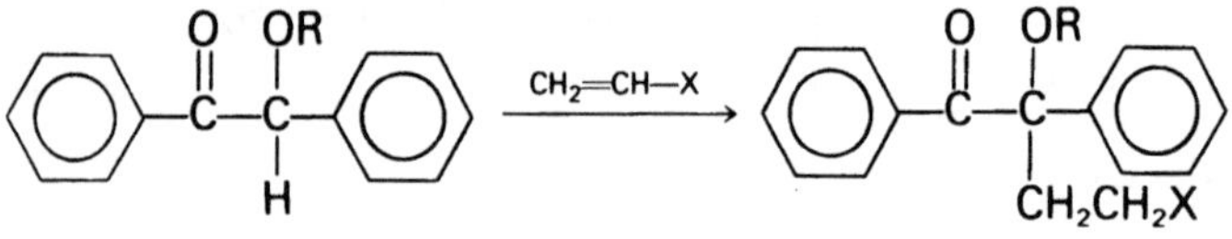

(3)

R = Alk, Ar; X = CO_2R^1, $CONH_2$, CN

Benzoin aryl ethers on irradiation (Scheme 5) undergo both α- (major) and β- (minor) cleavage from a triplet excited state of 'normal' lifetime (i.e. quenchable by naphthalene).[40,61] Whether the

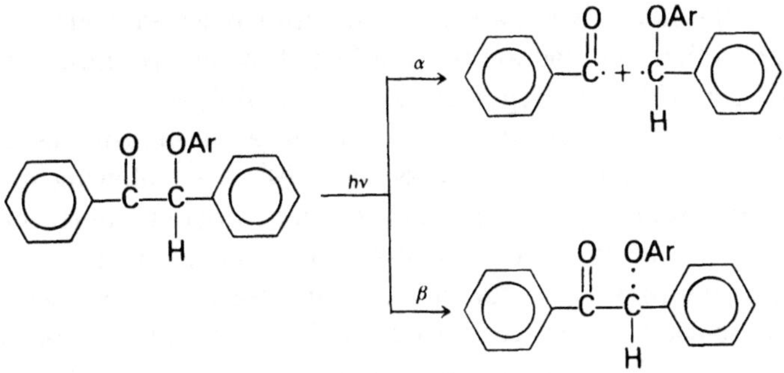

SCHEME 5

ratio of α- to β-cleavage is affected by the nature of the aryl ether group (e.g. the stability of the aryloxy radical that may be generated) is unknown at present.

The initiating efficiency of these compounds, for instance in unsaturated polyester (UP) resins is somewhat lower than with corresponding benzoin alkyl ethers,[4] presumably on account of slightly lower k_α values ($\sim 10^9\,s^{-1}$) and most certainly also the quenching by styrene (from the UP-resin) and atmospheric oxygen.

Benzoin esters (carboxylic and sulphonic) have been reported to undergo both α-cleavage (radical generation) and intramolecular cyclisation to a benzofuran derivative with concurrent heterolytic elimination of acid (Scheme 6).[80,81]

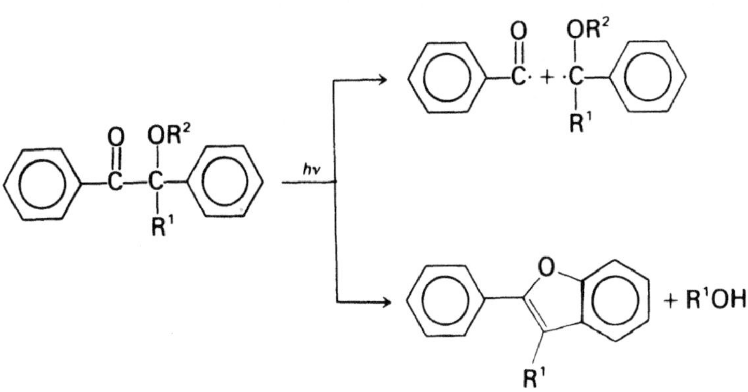

$R^1 = H$; $R^2 = -CO-R^3, -SO_2R^3$; $R^3 = $ Alk, Ar

SCHEME 6

Proof for the heterolytic nature of the latter reaction comes from the observation that the reaction is not affected by the presence of nitroxyl radicals (e.g. TMPO) (Hageman, H. J., unpublished).

Owing to the fact that the α-cleavage of benzoin esters is a relatively slow triplet-state process ($k_\alpha \sim 10^6\,s^{-1}$)[40] allowing for quenching by the usual triplet quenchers (oxygen, vinyl monomers) and the occurrence of a non-radical side-reaction, benzoin esters are less suitable as photoinitiators for radical polymerisation processes.

The base-catalysed reaction of formaldehyde with benzoin and its alkyl ethers yields the highly interesting α-hydroxymethylbenzoin and alkyl ethers having excellent thermal stability [reaction (4)]. Moreover, a host of derivatives, notably esters, useful as photoinitiators, have been obtained from these compounds through the primary hydroxy function.

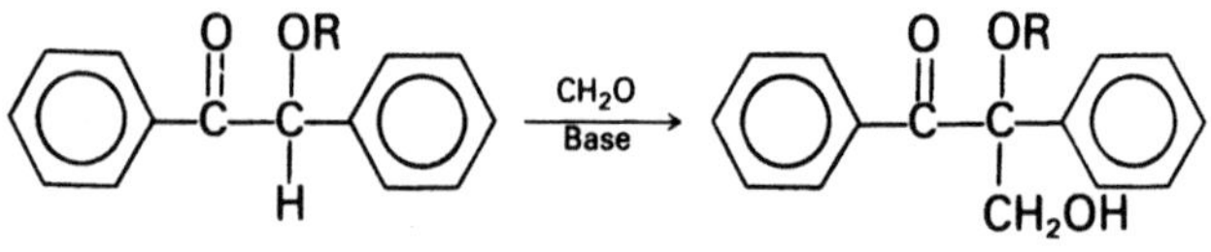

(4)

R = H, Alk

Sulphonic esters, originally developed as acid-releasing photocatalysts,[82] have now been shown to be excellent photoinitiators for free-radical polymerisation as well.[83,84] The alkanesulphonic esters, with one exception ($R = CH_3$), are low-melting (m.p. < 100°C) colourless solids having a $\lambda_{max} \sim 330$ nm ($\log \varepsilon \sim 2{\cdot}4$)

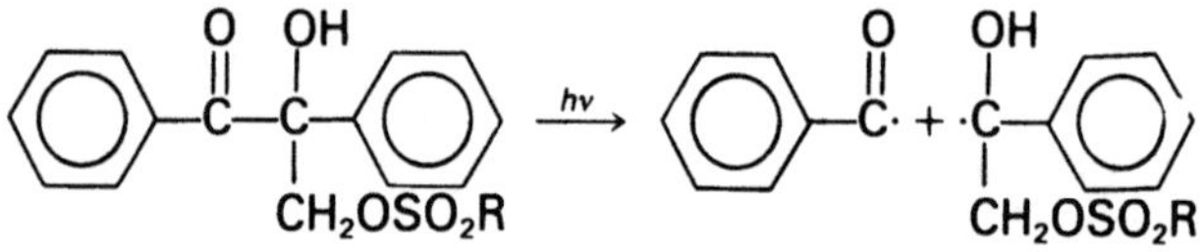

(5)

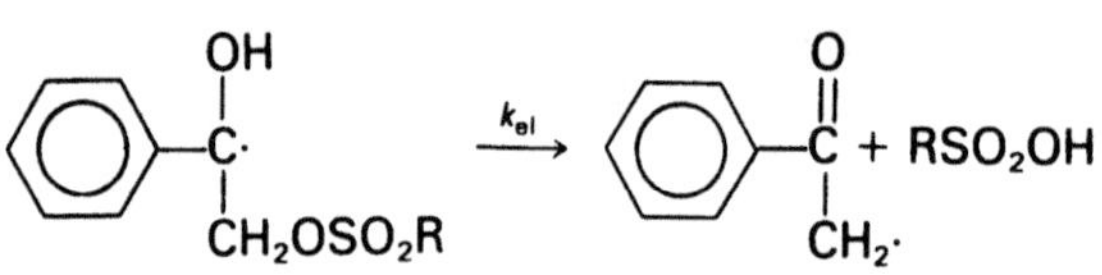

(6)

The primary photochemical reaction is, as expected, α-cleavage[85] ($k_\alpha > 10^8\,s^{-1}$) into a benzoyl and a β-sulphonyloxyketyl radical (reaction (5)]. ESR spectroscopy showed the presence of benzoyl and, a little surprisingly, benzoylmethyl radicals. The benzoylmethyl radical must have resulted from the β-sulphonyloxyketyl radical by elimination of sulphonic acid. ^{1}H-NMR–CIDNP experiments showed the excited state involved to be the (short-lived) triplet state. TMPO-trapping experiments showed the elimination of sulphonic acid [reaction (6)] to be an extremely fast process ($k_{el} > 10^8\ s^{-1}$).

Photolysis in the presence of 1,1-diphenylethylene as non-polymerising model substrate showed that the benzoyl and the benzoylmethyl radicals are equally reactive towards the olefinic double bond of vinyl monomers.

The corresponding sulphonic esters of α-hydroxymethyl benzoin alkyl ethers have been found to behave in a similar fashion,[86] generating benzoyl and α-alkoxy-β-sulphonyloxybenzyl radicals from a short-lived triplet excited state [reaction (7)]. The latter radicals, observable by ESR, and trapped by TMPO, eventually release sulphonic acid and generate benzoylmethyl radicals ($k_{el} < 10^8\ s^{-1}$) [reaction (8)]. The fate of the α-alkoxy moiety is still obscure.

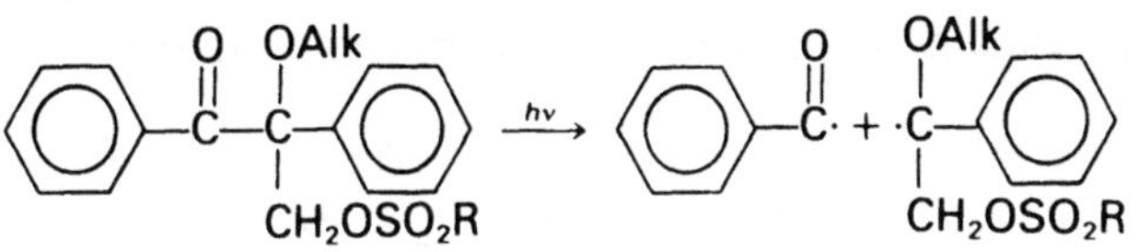

(7)

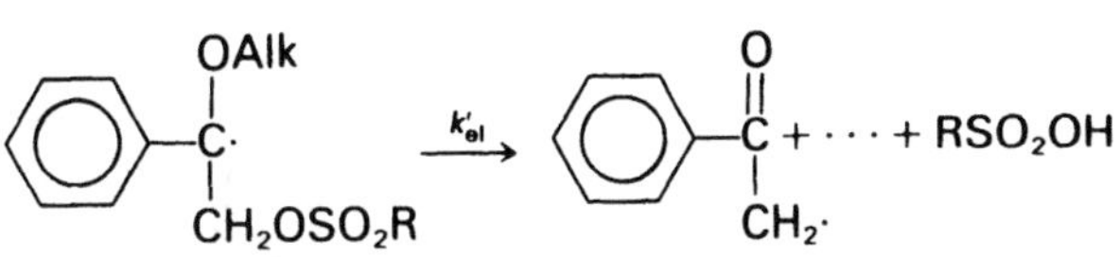

(8)

Carboxylic esters of α-hydroxymethylbenzoin claimed[87] as photoinitiators for free-radical polymerisation have also been studied in some detail (Scheme 7).[87–90] As expected, the primary photochemical reaction (9) again is α-cleavage into a benzoyl and a β-acyloxyketyl radical from a short-lived triplet excited state, as evidenced from product analysis,[88,90] ESR[90] and ^{1}H-NMR–CIDNP experiments.[90] The major secondary reaction of the β-acyloxyketyl radical (10) appears to be hydrogen-transfer to afford the corresponding phenacyl ester.[90] Elimination of carboxylic acid to give the benzoylmethyl radical does occur to some extent and at a much lower rate than earlier observed for the corresponding sulphonic esters.[85]

The homopolymer and copolymers of the acrylic ester of α-hydroxymethylbenzoin methylether have also been described.[89]

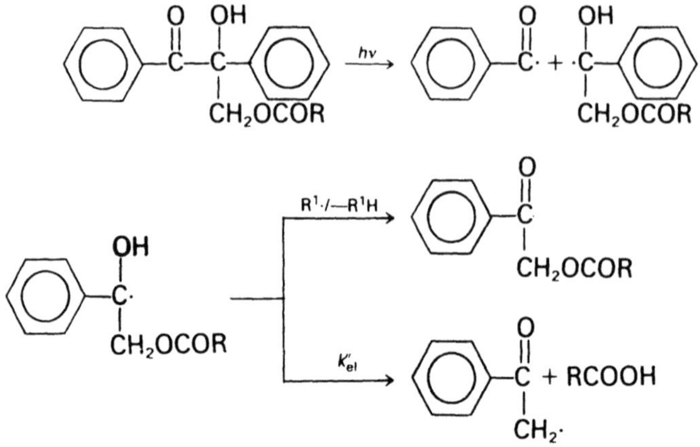

(9)

(10)

SCHEME 7

Benzoin and its alkyl ethers have also been reported[91] to undergo the Mannich reaction with a number of secondary amines to afford *α*-*N*,*N*-dialkylaminomethylbenzoins and benzoin ethers:

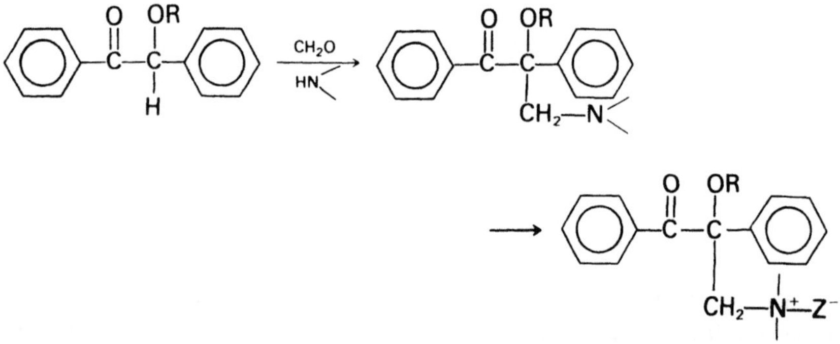

The tertiary amino group can be readily converted into quaternary ammonium salts, soluble in water and reportedly[91] useful in photo-initiated emulsion polymerisation of acrylates. Photochemical data are not available but these compounds can be safely assumed to undergo the usual *α*-cleavage upon irradiation.

α-Hydroxymethylbenzoin has been reported[92–94] to undergo acid-catalysed cyclisation with various carbonyl compounds to give 4-benzoyl-4-phenyl-1,3-dioxolanes [reaction (11)]. Reports on their photochemical behaviour are not available. From TMPO-trapping experiments (Hageman, H. J., unpublished), however, they are known to undergo the expected α-cleavage. The fate of the 1,3-dioxolan-4-yl radicals is unknown at present.

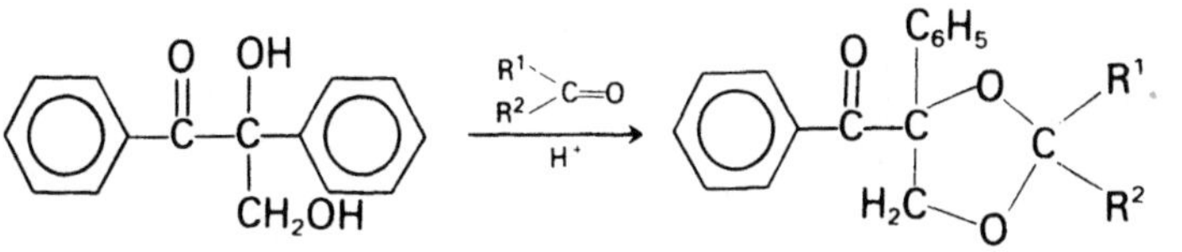

(11)

R^1 = H, Alk, Ar;
R^2 = H, Alk, Ar, OAlk

Benzoin-derived compounds such as desyl halides[64,80] and desyl aryl sulphides[95] reportedly undergo α-cleavage exclusively when irradiated:

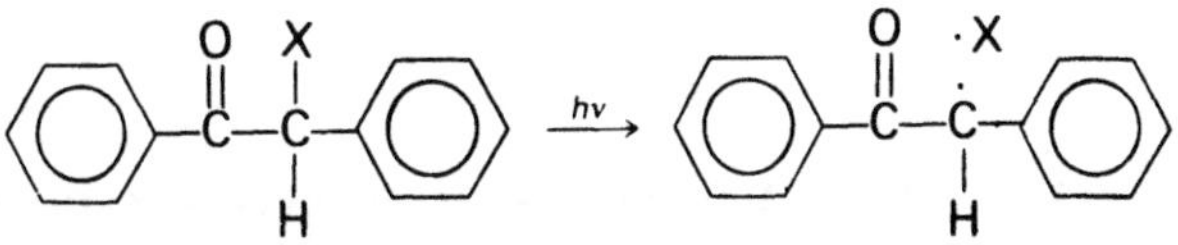

X = Cl, Br, SAr

Desyl aryl sulphides have been reported to photoinitiate the polymerisation of acrylates[96] and UP resins.[4]

The arylthiyl radical has been shown by ESR spin-trapping[97] to be capable of initiating the polymerisation of various vinyl monomers [reaction (12)]. Any contribution of the desyl radical to the initiation is unknown at present.

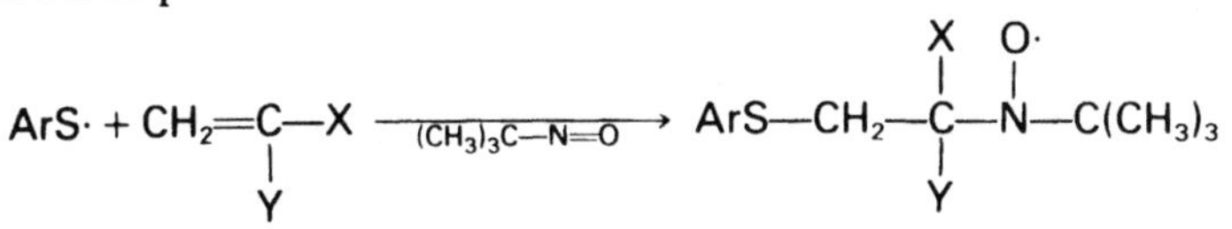

(12)

X = H, Alk; Y = COOAlk, Ar

4.1.2 Acetophenone Derivatives

α,α-Diethoxyacetophenone, **II** a colourless liquid with $\lambda_{max} \sim 330$ nm ($\log \varepsilon \simeq 1{\cdot}9$), the most prominent representative of this class of compounds, has been shown to be a useful photoinitiator for the radical polymerisations of vinyl monomers.[77,98,99] The initiation of the polymerisation was originally thought to proceed through a biradical following Norrish Type II photochemistry (intramolecular hydrogen-abstraction) from a short-lived triplet excited state (Scheme 8).

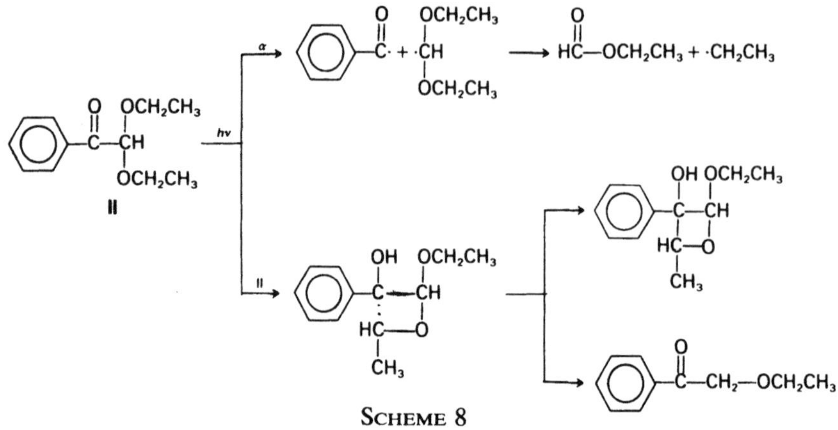

SCHEME 8

More recently the photoreactions of several α,α-dialkoxyacetophenones were studied[100] by ^{1}H-NMR–CIDNP both in the absence and in the presence of MMA. Both Norrish Type I (α-cleavage) and Type II (intramolecular hydrogen-abstraction) processes were observed and concluded to be of comparable importance. It was concluded that the initiation step predominantly involved radicals generated by the α-cleavage, i.e. benzoyl and alkyl radicals. These conclusions have been confirmed by trapping experiments both with TMPO and with 1,1-diphenylethylene as model substrate (Hageman, H. T. & Overeem, T., unpublished).

Several α,α-dialkoxyacetophenones were evaluated as photoinitiators in the curing of various acrylate systems.[101] The order of reactivity was discussed in terms of steric requirements of the different alkoxy groups and the implications for the rates of Type I and Type II reactions. Recently some of these α,α-dialkoxyacetophones were studied using both steady-state irradiation and time-resolved laser spectroscopy.[102] Several transients were observed, one of which was recognised as the benzoyl radical. No direct evidence was obtained for the occurrence of triplet excited states, which confirms their short-lived nature ($\tau \leqslant 10^{-7}$ s). The initiating species were found to be the benzoyl (major contributor) and the alkyl radicals, i.e. those arising from the α-cleavage. The slight differences in polymerisation rates obtained with the various α,α-dialkoxyacetophenones were ascribed to steric and/or electronic effects of the alkoxy groups favouring α-cleavage at the expense of the Norrish Type II process. α-

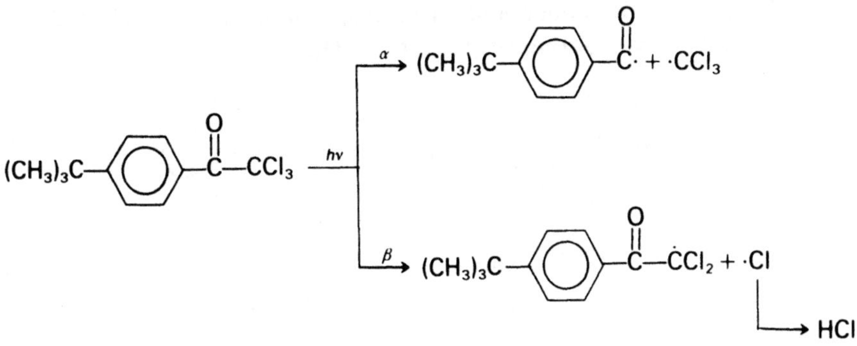

SCHEME 9

Halogenoacetophenones (mono-, di- and tri-) reportedly undergo photoinduced β-cleavage as observed for the corresponding desyl halides. The trichloro derivative shown in Scheme 9 has been studied in some detail.[13,14] ^{1}H-NMR–CIDNP experiments showed this compound to undergo α-cleavage (minor process) in addition to β-cleavage (major process).

The main application of these compounds is as photoinitiators for acid-catalysed curing processes on account of the hydrogen chloride generated through hydrogen-abstraction by the highly reactive chlorine radicals set free.

4.1.3 *Benzilketals*

Benzilketals (**III**) have been shown to be effective photoinitiators in many applications.[10,13–15,21,77,99,103] Benzildimethylketal ($R^1 = R^2 = CH_3$) a low-melting solid, $\lambda_{max} \sim 330$ nm ($\log \varepsilon \sim 2{\cdot}5$), the most

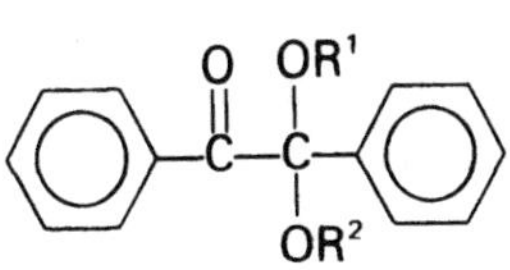

III

prominent representative, some 30 years ago was recognised for the first time to be sensitive to light,[110] benzoic acid being identified as a product. From photolysis in the absence of vinyl monomers[98] it was concluded that it underwent α-cleavage into a benzoyl and an α,α-dimethoxybenzyl radical [reaction (13)].

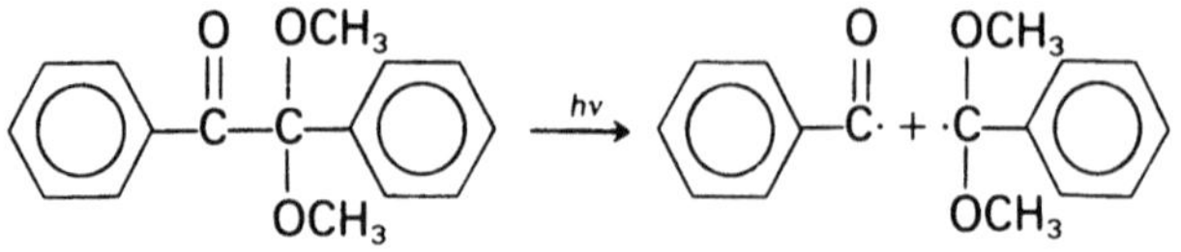

(13)

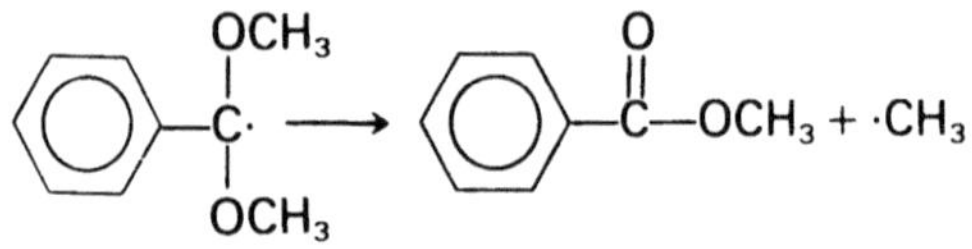

(14)

From the formation of methyl benzoate it was inferred that the α,α-dimethoxybenzyl radical undergoes sequential fragmentation, generating a methyl radical [reaction (14)]. Sensitisation and quenching studies suggested that the α-cleavage is an extremely rapid process ($k_\alpha \leqslant 10^{10}\,\mathrm{s}^{-1}$) and hence proceeds from a very short-lived triplet excited state.

^{1}H-NMR–CIDNP and ^{13}C-NMR–CIDNP studies[100,105] confirmed both the fragmentation pattern and the nature of the excited state involved. Laser flash photolysis[106] showed the formation of a short-lived transient ($\tau \sim 7{\cdot}5 \times 10^{-7}$ s), assigned to the benzoyl radical, and a relatively long-lived transient ($\sim 1{\cdot}7 \times 10^{-5}$ s), assigned to the α,α-dimethoxybenzyl radical. The precursor excited state could not be observed (too short-lived). ESR spectroscopy[57a] indicated the formation of benzoyl and methyl radicals.

Photolysis of benzildimethylketal in the presence of 1,1-diphenylethylene as a model substrate[57a] showed the benzoyl radical as the predominant initiating species. The major reaction of the α,α-dimethoxybenzyl radical proved to be sequential fragmentation generating a methyl radical which contributes to the initiation to some extent. Some of the α,α-dimethoxybenzyl radicals formed were found to participate in primary radical termination under the experimental conditions (low monomer concentration). They thus behave in much the same way as the α-alkoxybenzyl radicals generated from benzoin alkyl ethers, i.e. they have a substantially lower rate constant for addition to the olefinic double bond of vinyl monomers than the benzoyl radical.

The photodecomposition of benzildimethylketal has also been studied in thin films of two different non-polymerisable viscous model environments (dinonyl phthalate and 1,5-pentane diol respectively) to simulate typical UV-curable formulations.[107] In addition to the expected products such as benzaldehyde, benzoic acid, benzil and methyl benzoate, methyl 2-benzoylbenzoate and methyl 4-benzoylbenzoate were identified; their formation is exemplified in Scheme 10.

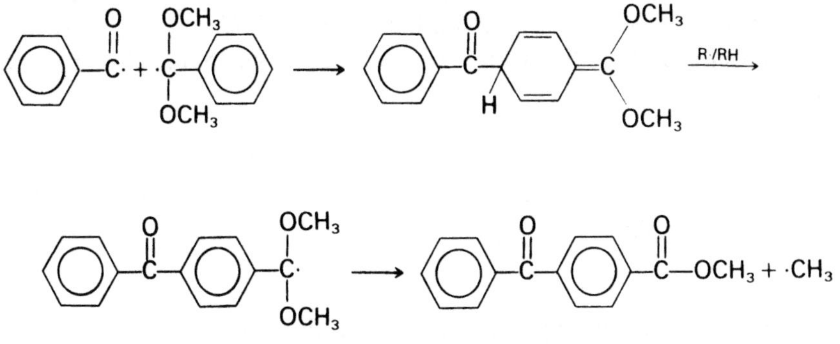

SCHEME 10

The identification of the latter two products provides beautiful confirmatory evidence for the correctness of the assignments of two cage-recombination products made in an earlier NMR–CIDNP study.[100]

4.1.4 α-Hydroxyalkylphenones and α-Aminoalkylphenones

A new group of photoinitiators is formed by the α-hydroxyalkylphenones (**IV**).[108,109] Most of these compounds are colourless liquids with $\lambda_{max} \sim 320$ nm ($\log \varepsilon \sim 2{\cdot}0$), with excellent solubility in the usual UV-curable formulations. They have been claimed[108,109] to photoinitiate the polymerisation of the usual vinyl monomers with an efficiency comparable with benzildimethylketal.[52] Particularly advantageous is the fact that these compounds do not impart yellowing to the cured composition, the formation of coloured cage-recombination products being impossible.

IV

From laser flash photolysis[110] relatively short-lived triplet excited states ($\tau_T \approx 5 \times 10^{-8}$ s) were observed and quantum yields for α-cleavage of 0·2–0·3 were estimated. ^{1}H-NMR–CIDNP experiments[111,112] and TMPO-trapping experiments (Hageman, H. J. & Overeem, T., unpublished) all indicated α-cleavage into benzoyl and ketyl radicals to take place [reaction (15)]. The benzoyl radical is known to be reactive as initiating species (*vide supra*) and the (aliphatic) ketyl radicals are likely[113] to contribute also to the initiation.

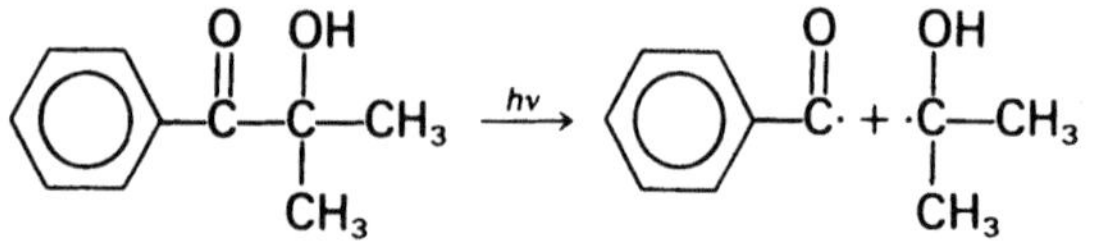

(15)

Conversion of the hydroxyl group of these compounds leads to derivatives showing similar characteristics to the corresponding derivatives in the benzoin series, i.e. alkyl ethers with good photoinitiator properties[112] and esters which apparently do not undergo α-cleavage.[114]

Polymers and copolymers having pendant α-hydroxyalkylphenone moieties (**V–VII**) have been synthesised as low-volatility and low-odour photoinitiators.[115,116]

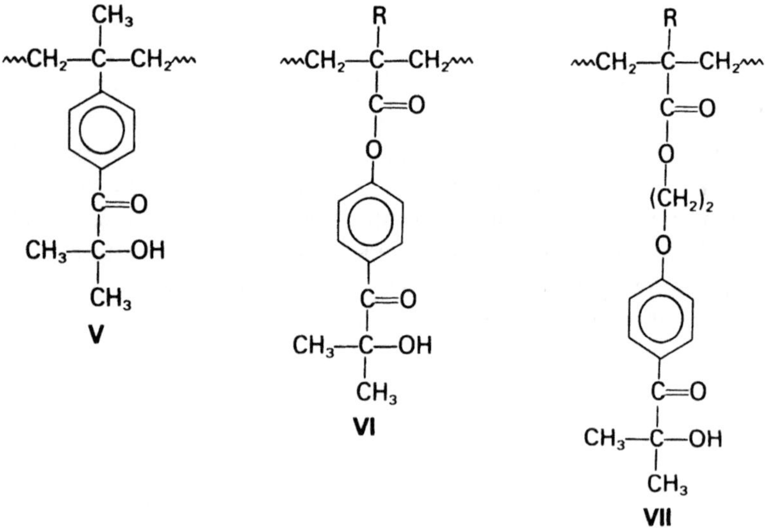

V **VI** **VII**

α-Hydroxyalkylphenones having polar substituents in the p-position of the aromatic ring, for example **VIII**, have been synthesised as

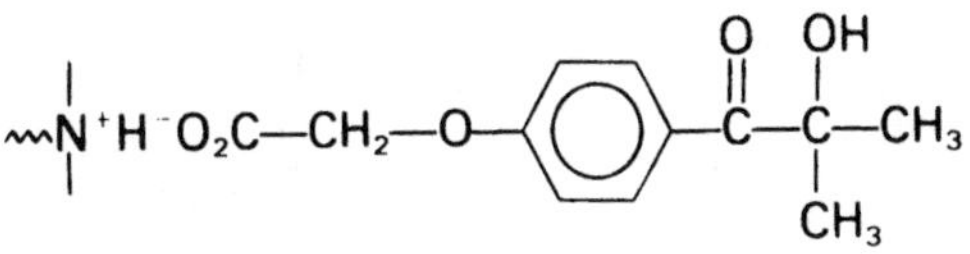

VIII

water-soluble photoinitiators[117] for use in emulsion polymerisation, reaction (16).

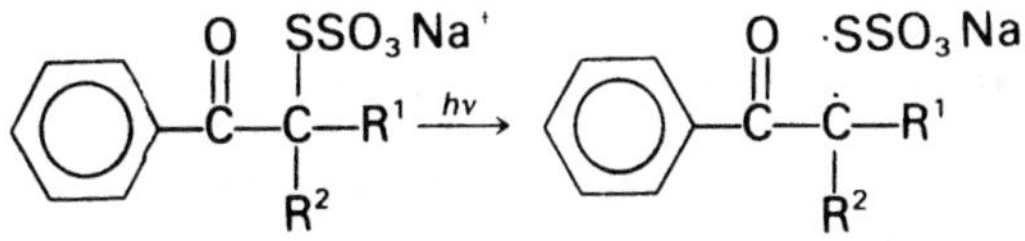

(16)

Related compounds, benzoylmethylthiosulphates, have also been claimed[118] as versatile water-soluble photoinitiators, reportedly undergoing photoinduced β-cleavage to release thiosulphate radicals.

Also recently some related α-aminoalkylphenones, notably the α-morpholino derivatives, have been developed particularly to accommodate the problem of curing white (TiO_2) pigmented coatings:[119]

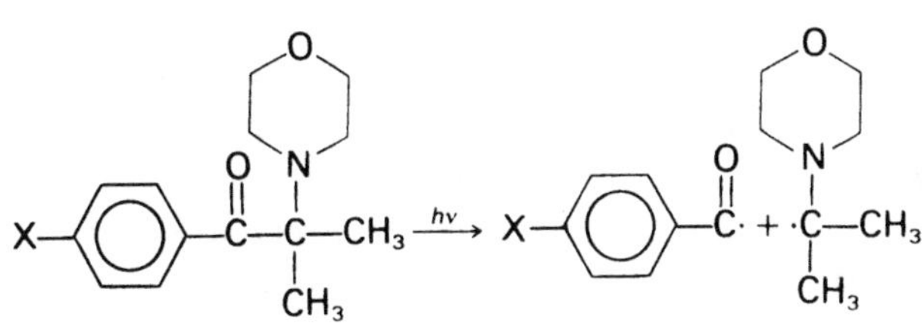

X = H, CH_3S

Both direct and thioxanthone-sensitised excitation were reported to result in α-cleavage as concluded from ^{1}H-NMR–CIDNP spectroscopy. Photolysis in the presence of methyl α-t-butylacrylate as non-polymerisable model substrate for acrylates[57b] revealed the benzoyl radical as the major initiating species.

4.1.5 *O-Acyl α-oximinoketones* (IX)

The sensitivity of O-acyl α-oximinoketones to light has in principle been known for quite some time. In 1904 the sunlight-induced decomposition of a product obtained from a Beckmann Type II rearrangement was described[120] which much later was shown[121] to be the [*E*] or anti-isomer of *O*-benzoyl 1,2-diphenyl-2-oximinoethanone.

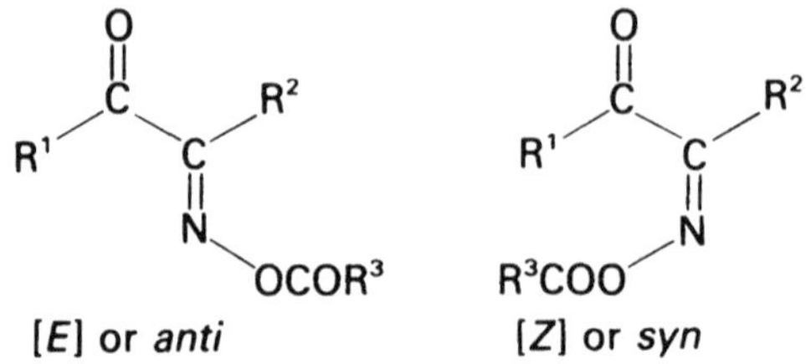

IX: R^1 = Ar; R^2 = Alk, Ar; R^3 = Alk, substd. Alk, Ar, OAlk

Again much later, *O*-acyl α-oximinoketones were shown to photoinitiate the polymerisation of vinyl monomers such as acrylamide,[122] various acrylates[13,15,122–124] and UP resins.[4]

Owing to isomerisation, about the carbon–nitrogen double bond these compounds exist in two isomeric forms, [*E*] or *anti* and [*Z*] or *syn*, as shown. Moreover, preferences for particular conformations have been shown to exist for the different isomers, depending on the substituents.[125] The synthetic routes commonly used for these compounds, however, nearly exclusively lead to the [*E*] isomers. The reported data on their use as photoinitiators for radical polymerisation can therefore, unless explicitly stated otherwise, be assumed to pertain to the [*E*] isomers. The [*Z*] isomers can be obtained through thermal or photo-isomerisation of the parent α-oximinoketones.[126]

The *O*-acyl α-oximinoketones, with few exceptions, are relatively low-melting (m.p. <100°C) colourless solids. Their absorption spectra show a maximum in the 300–400 nm range, $\lambda_{max} \sim 340$ nm, log $\varepsilon \sim 2{\cdot}3$.

Product studies[122b,127] in the absence of vinyl monomers supported by ESR spin-trapping experiments, have been reported. Photofragmentation through primary N–O bond cleavage into an α-ketoiminyl and an acyloxy radical followed by sequential fragmentation of both primary radicals (Scheme 11) has been proposed,[122b,127] and ever since has been assumed to occur.

The results of laser flash photolysis experiments[128] have been explained assuming a divergence in primary bond cleavage depending on the nature of R^3. Thus primary α-cleavage was assumed to occur for R^3 = Alk, and primary N–O bond cleavage was assumed to occur for R^3 = Ar and R^3 = OAlk on account of the stability of the acyloxy radical.

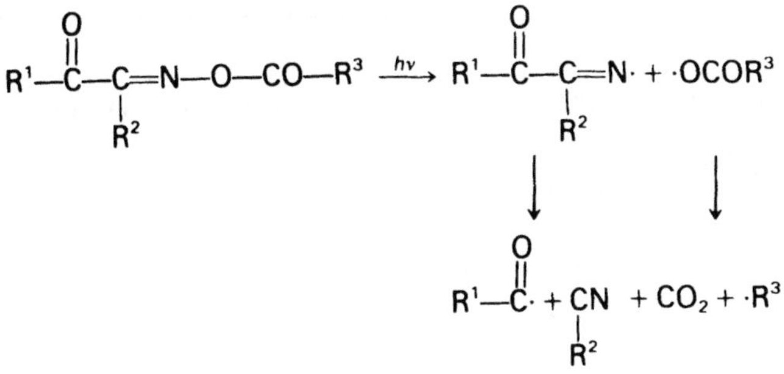

SCHEME 11

ESR spin-trapping experiments[129] with *O*-benzoyl 1,2-diphenyl 2-oximinoethanone ($R^1 = R^2 = R^3 = C_6H_5$) using *N*-*t*-butyl α-phenylnitrone (PBN) at room temperature showed the presence of trapped benzoyl, benzoyloxy and phenyl radicals. It proved impossible to decide on primary α-cleavage or N–O bond cleavage.

A detailed study[130] of a series of *O*-acyl α-oximinoketones, using ESR spin-trapping, low-temperature ESR, ^{1}H-NMR–CIDNP and radical trapping (TMPO, and thiol) provided the final answers. The ESR spin-trapping experiments run at room temperature only showed the trapped benzoyl, acyloxy and alkyl (from aliphatic esters) radicals. The low-temperature (−100°C) ESR spectra at long last showed the presence of the different α-ketoiminyl radicals ($R^1 = C_6H_5$, $R^2 = C_6H_5$ and CH_3), which on raising the temperature above −60°C disappeared, the benzoyl radical being generated concurrently. This should satisfactorily explain why α-ketoiminyl radicals so far (room temperature experiments) escaped our observations.

The ^{1}H-NMR–CIDNP experiments showed the excited state involved to be the triplet state, which must be short-lived as concluded from laser flash photolysis.[128] Moreover, the CIDNP spectra also showed the formation of cage-recombination products such as α-ketoimines: in fact this is confirmatory evidence for primary N–O bond cleavage (Scheme 12).

Product analysis of radical trapping experiments showed that $[E] = [Z]$ photoisomerisation is the predominant process for most *O*-acyl α-oximinoketones having R^2 = Ar, and photodecomposition is only a minor process (e.g. the quantum yield for trapped benzoyl radicals $\phi \simeq 0{\cdot}002$, reported[129] for *O*-benzoyl-1,2-diphenyl-2-oximinoethanone). *O*-Acyl α-oximinoketones having R^2 = Alk un-

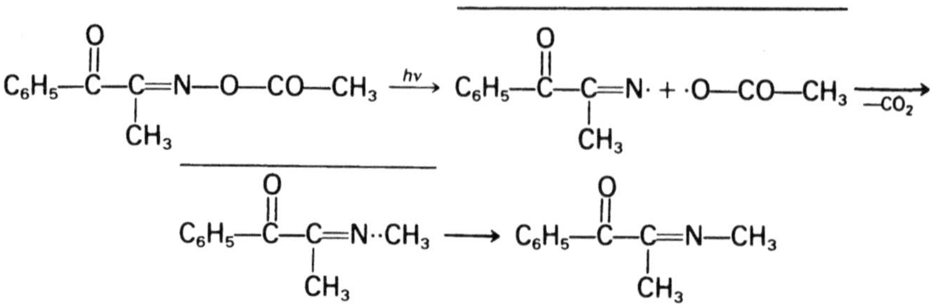

SCHEME 12

dergo photodecomposition almost exclusively (no noticeable photoisomerisation). These results explain nicely the differences in photoinitiating efficiency between two groups of *O*-acyl α-oximinoketones that have been reported.

One theoretically interesting question remains to be answered: does the photodecomposition of these compounds proceed equally well for both [*E*] and [*Z*] isomers or does it occur preferentially from one of these isomers? Unfortunately no data are available on this matter. However, from a study of cyclic *O*-methyl-α-oximinoketones (**X**)[131] it was concluded that, whereas direct excitation ($\lambda = 300$ and 350 nm, respectively) as well as triplet-sensitised excitation exclusively result in [*E*] = [*Z*] isomerisation ultimately leading to a photostationary state of the two isomers, direct excitation at $\lambda = 254$ nm results in photodecomposition (primary N–O bond cleavage) in addition to photoisomerisation, and this photodecomposition only occurs upon photoexcitation of the [*E*] isomers. *Ab initio* calculations[132] on a related model compound showed that N–O bond cleavage is a far more attractive process for the excited singlet π–π^* state of the [*E*] isomer than for the corresponding state of the [*Z*] isomer.

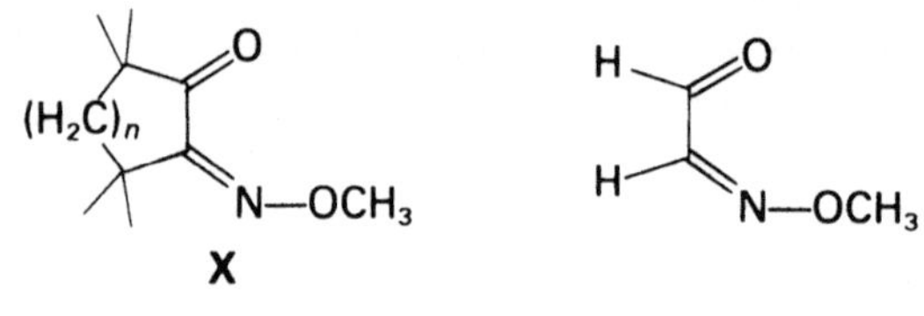

X

4.1.6 Acylphosphine Oxides and Acylphosphonates

Acylphosphine oxides (**XI**) and the related acylphosphonates (**XII**) were recently claimed[133] as useful photoinitiators for the radical

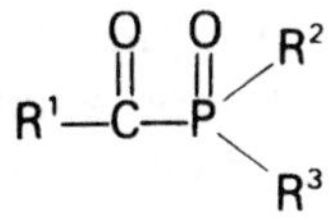

XI: R^1 = Ar, *t*-Bu; R^2, R^3 = C_6H_5

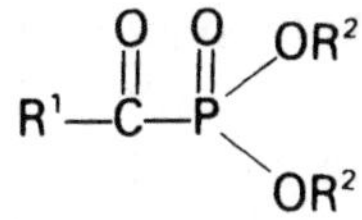

XII: R^1 = Ar, *t*-Bu; R^2, R^3 = Alk

polymerisation of acrylates and UP resins. These compounds are colourless solids (acylphosphine oxides) or liquids (acylphosphonates), having a particularly long-wavelength absorption ($\lambda_{max} \sim 380$ nm, $\log \varepsilon \sim 2,7$), which should broaden their scope compared with most other photoinitiators.

2,4,6-Trimethylbenzoyldiphenylphosphine oxide (TMBPO), the most prominent representative, has for that reason been studied in considerable detail. Thus photolysis of TMBPO in methanol in the presence of hydrogen peroxide afforded[1133d,134] trimethylbenzoic acid and diphenylphosphinic acid, suggesting α-cleavage [reaction (17)] to be the primary photoreaction. Laser flash photolysis[134] indicated the formation of a transient with a $\lambda_{max} \sim 330$ nm, which was assigned to the diphenylphosphinoyl radical. Excited states involved (singlet and/or triplet) were found to be very short-lived, the reason undoubtedly being the high rate constant for α-cleavage. Time-resolved ESR (CIDEP) spectroscopy[135] led to the unambiguous identification of the primary radicals resulting from a triplet excited state precursor.

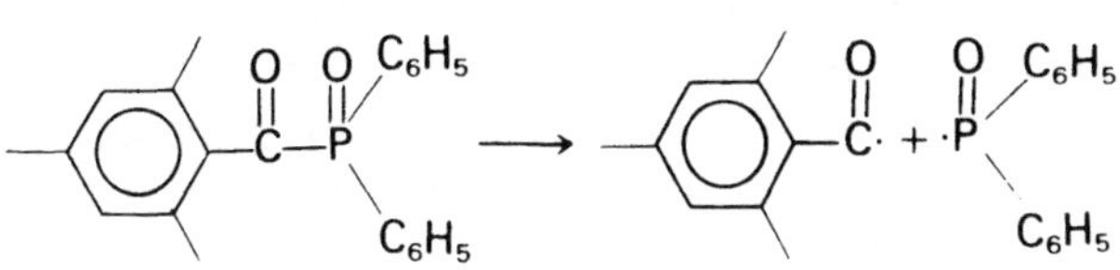

(17)

Further confirmatory evidence was provided by TMPO-trapping[137] [reaction (18)].

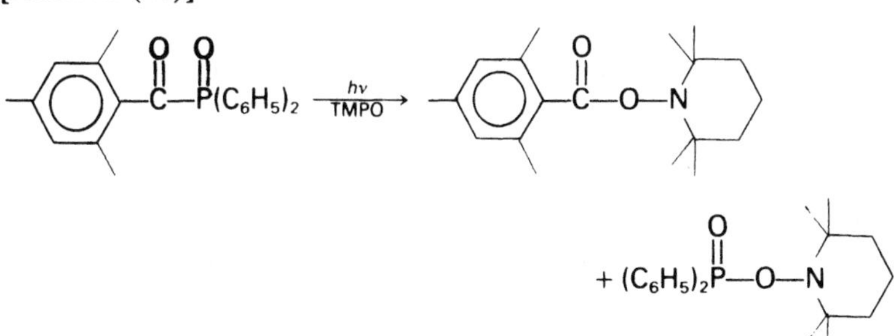

(18)

Laser flash photolysis of TMBPO in the presence of vinyl monomers[134] afforded (high) rate constants k_a for addition of the diphenylphosphinoyl radical to the olefinic double bond of the respective vinyl monomers.

The solution crosslinking of TMPA monitored by laser nephelometry[138] showed exceptionally short induction periods for TMBPO compared with benzoin isopropyl ether and benzildimethylketal, and a substantially higher rate of polymerisation as well.

In the curing of thin films of acrylates,[133a,139] UP resins[139] and pigmented acrylate systems,[140] however, satisfactory cure speeds could only be obtained upon the addition of tertiary amines.

Photolysis of TMBPO in the presence of 1,1-di(*p*-tolyl)ethylene as a model substrate in solution as well as in thin films of a non-polymerisable viscous solvent[141] indicated both the trimethylbenzoyl and the diphenylphosphinoyl radical as the initiating species, the latter being the more reactive. The influence of added triethylamine has also been studied (see Section 5).

Various acylphosphonates have been studied in some detail. From laser flash photolysis studies[134,142] it was concluded that the trimethylbenzoylphosphonates undergo α-cleavage less efficiently than acylphosphine oxides, and that benzoyldiethylphosphonate does not undergo α-cleavage at all, intramolecular hydrogen-abstraction being the major process, in agreement with previous reports.[143]

Trimethylbenzoylphosphonates proved to be less efficient photoinitiators than the acylphosphine oxides for bulk polymerisation of MMA, solution crosslinking of TMPTA, and UV curing of acrylates.[138] Benzoylphosphonates proved to be even less efficient.[138] The propensity of triplet excited benzoylphosphonates to abstract hydrogen rather than undergoing α-cleavage may also follow from the influence of added tertiary amines on the photoinitiated bulk polymerisation of MMA.[138]

The same order of reactivities was observed in the oxidative desulphurisation at pentacovalent phosphorus by photogenerated radicals from acylphosphine oxides and acylphosphonates respectively.[144] Confirmatory evidence is provided by radical-trapping experiments using TMPO as before.[145] 1,4,6-Trimethylbenzoylphosphonates clearly undergo α-cleavage, as is shown by the formation of 2,2,6,6-tetramethylpiperidino-2,4,6-trimethylbenzoate [reaction (19)]. The (unsubstituted) benzoylphosphonates, however, do not undergo α-

cleavage [absence of 2,2,6,6-tetramethylpiperidinobenzoate from reaction (20)].

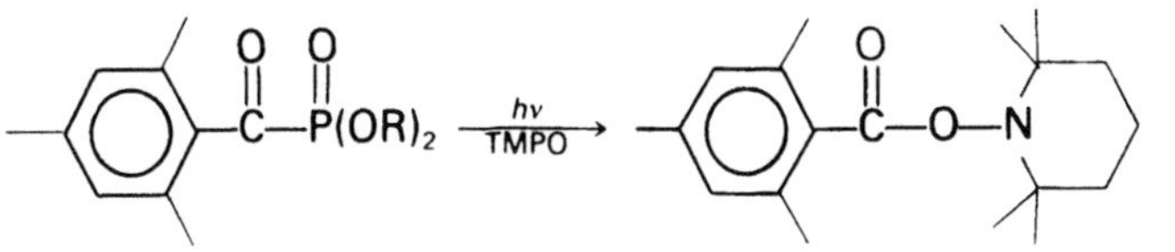
(19)

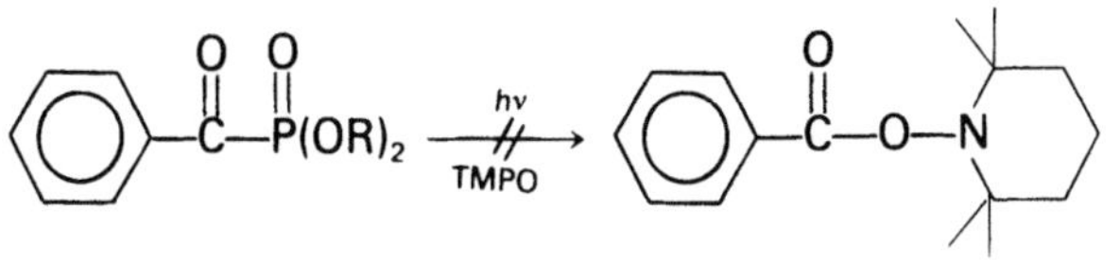
(20)

R = CH_3, CH_2CH_3

4.1.7 Miscellaneous Photoinitiators

In this section a few classes of compounds are brought together which have unimolecular radical formation in common, and are able to photoinitiate the free-radical polymerisation of vinyl monomers.

4.1.7.1 Peroxy compounds: Peroxy compounds, although primarily known as thermal initiators for the polymerisation of various unsaturated systems, can also be decomposed through absorption of UV light by the peroxide bond ($\lambda_{max} \sim 300$ nm, weak). Thus direct[146] but also singlet-sensitised[147] and triplet-sensitised[148] photolysis of diacyl peroxides has been reported.

Various benzophenone derivatives of dibenzoyl peroxide have been synthesised, for example **XIII**, and reported[149] to undergo photodecomposition (O–O bond cleavage) with quantum yields approaching unity, intramolecular energy transfer obviously proceeding far more efficiently than intermolecular energy transfer.

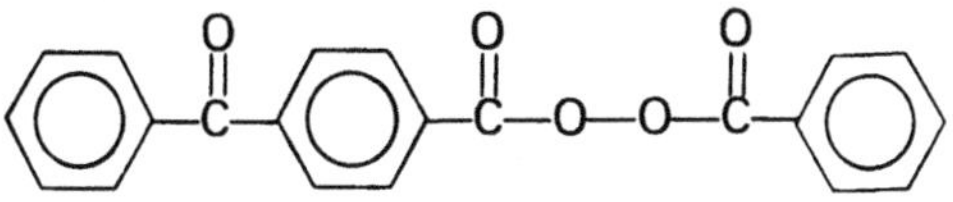
XIII

Based on the same principle some peresters have also been synthesised.[150] These compounds were reported to undergo photocleavage of the peroxide bond, neither inter- nor intra-molecular (Norrish Type II) hydrogen-abstraction by the triplet excited carbonyl moiety being observed (Scheme 13). When used as photoinitiators for

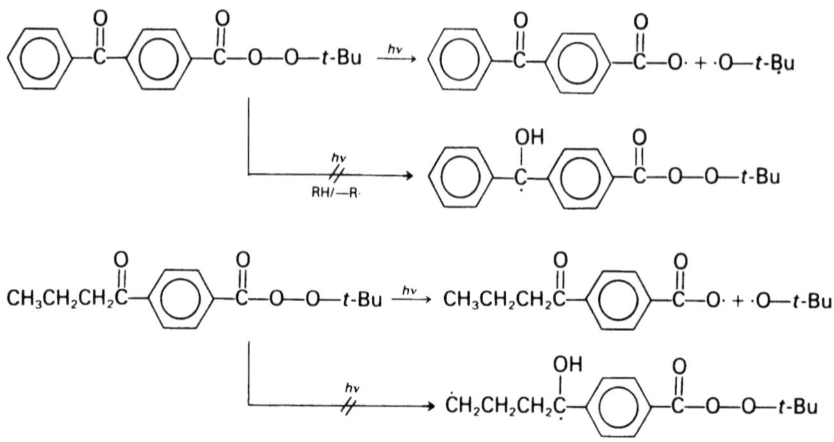

SCHEME 13

the polymerisation of methyl methacrylate and styrene, they showed higher rates of polymerisation than are obtained with the corresponding benzophenone–dibenzoyl peroxide system. However, in the case of styrene, quenching of the triplet excited carbonyl of the benzophenone moiety could still not be entirely avoided.

The initiating radicals generated from these compounds as determined from end group analysis (IR) proved to be almost exclusively benzophenone moieties attached to the polymer chain both through an ester linkage and through a carbon–carbon bond, the latter end groups (20–30%) indicating decarboxylation prior to initiation.

4.1.7.2 Thiobenzoic S-esters:

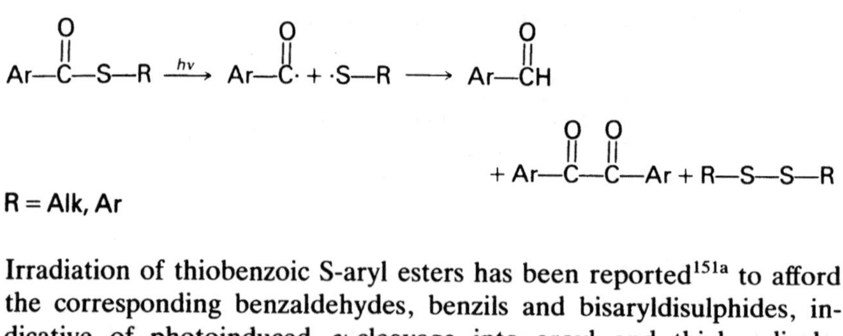

R = Alk, Ar

Irradiation of thiobenzoic S-aryl esters has been reported[151a] to afford the corresponding benzaldehydes, benzils and bisaryldisulphides, indicative of photoinduced α-cleavage into aroyl and thiyl radicals. Confirmation of this fragmentation mode has been provided by ^{1}H NMR–CIDNP experiments.[151b] These esters have been shown to photoinitiate the polymerisation of acrylates and unsaturated polyester resins.[151a] The cured compositions, however, turned out to be somewhat brittle, indicating a low degree of polymerisation, which was attributed to the involvement of thiyl radicals.

Thiobenzoic S-alkyl esters have also been reported[152] to undergo photoinduced α-cleavage.

4.2 Type II Photoinitiators

One of the first photoreactions studied at the beginning of the 20th century—'Bologna sunlight photochemistry', i.e. the photopinacolisation of benzophenone[41]—forms the basis of our present Type II photoinitiators. Fully aromatic ketones such as benzophenone are known to abstract hydrogen from suitable hydrogen-donors, often called coinitiators (*vide infra*), instead of undergoing α-cleavage, because the excitation energy acquired is insufficient for α-cleavage.

Because of the absence of relatively weak CO–C (or other) bonds, these compounds possess relatively long-lived triplet excited states perfectly suitable to enter into bimolecular reactions. Unfortunately, however, this very feature also allows for efficient (bimolecular) quenching by the usual triplet quenchers such as vinyl monomers, but above all by oxygen from the surrounding atmosphere.

This section has been arranged according to the various types of carbonyl and 1,2-dicarbonyl compounds displaying typical hydrogen-abstraction reactions and claimed and/or used in photoinitiated polymerisation and crosslinking processes.

4.2.1 Benzophenones

Many studies[153] have been devoted to the photoreduction of benzophenone and related compounds, the primary stages of which may be represented schematically as in Scheme 14.

$$\mathrm{Ar_2C{=}O_{S_0}} \xrightarrow{h\nu} \mathrm{Ar_2C{=}O^*_{S_1}}$$

$$\mathrm{AR_2C{=}O^*_{S_1}} \xrightarrow{\text{i.s.c.}} \mathrm{Ar_2C{=}O^*_{T_1}}$$

$$\mathrm{Ar_2C{=}O^*_{T_1} + RH} \longrightarrow \mathrm{Ar_2\dot{C}{-}OH + R\cdot}$$

$$\mathrm{Ar_2C{=}O^*_{T_1} + Q_{S_0}} \xrightarrow{k_q} \mathrm{Ar_2C{=}O_{S_0} + Q^*_{T_1}}$$

$$\mathrm{Ar_2C = O^*_{T_1} + O_2} \xrightarrow{k'_q} \mathrm{Ar_2C{=}O_{S_0} + O^*_2}$$

SCHEME 14

Major factors governing this hydrogen-abstraction process are:

(i) the triplet state configuration of the ketone, efficiencies decreasing in the order $T_{n-\pi^*} > T_{\pi-\pi^*} \gg T_{CT}$;[44,45]
(ii) the triplet energy of the ketone;
(iii) the bond strength of the carbon–hydrogen bond of the hydrogen-donor.

Laser flash photolysis has lately been used[154,155] to determine triplet lifetimes, rates of hydrogen-abstraction as well as quenching constants k_2 for a variety of vinyl monomers. General features that emerged are:

(i) alcohols and ethers are excellent hydrogen-donors;
(ii) vinyl monomers such as styrene and vinylpyrrolidone having lower triplet energies than benzophenone cannot be polymerised using this system (e.g. benzophenone–THF) because of quenching;
(iii) these systems are extremely susceptible to quenching by (atmospheric) oxygen.

Many studies[156–158] have also been devoted to the photoreduction of aromatic ketones by tertiary amines, for which Scheme 15 may be considered representative:

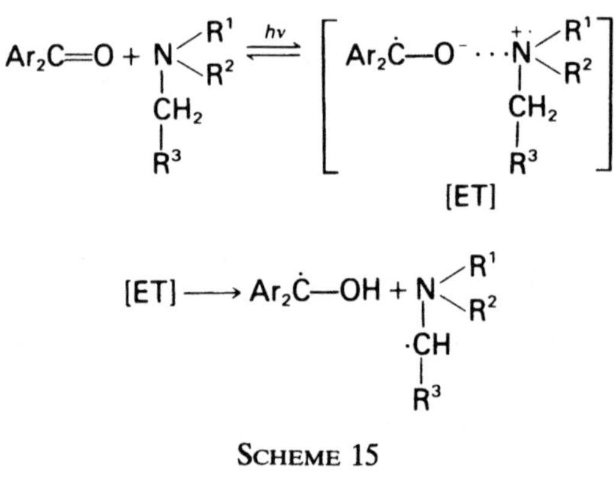

SCHEME 15

On excitation of the ketone an excited electron-transfer complex [ET] is efficiently formed. This [ET] exciplex may either revert back to starting materials through back-electron-transfer, or proceed through subsequent proton transfer to a ketyl radical and an α-aminoalkyl radical. The major factor governing the efficiency of this process is the ionisation potential of the tertiary amine.[153a]

Absorption spectra recorded for the systems benzophenone–aromatic tertiary amines,[159,163] and benzophenone–aliphatic tertiary amines,[164] showed, depending on the conditions, the ketyl radical-anion and/or the ketyl radical, and depending on the type of tert amine, the aminium radical-cation.

ESR spectroscopy using spin-trapping with 2-methyl-2-nitrosopropane (MNP) of the system benzophenone–triethylamine[165] identified the amine-derived radical:

$$(C_6H_5)_2C{=}O + N(CH_2CH_3)_3 \xrightarrow[MNP]{h\nu} (CH_3CH_2)_2N{-}\underset{\displaystyle CH_3}{\underset{|}{C}H}{-}\overset{\displaystyle O\cdot}{\overset{|}{N}}{-}C(CH_3)_3$$

The existence of α-aminoalkyl radicals and aminium radical-cations also followed from ^{1}H-NMR–CIDNP experiments.[166–168]

Laser flash photolysis has also extensively been used to study the aromatic ketone–tertiary amine system, e.g. the determination of quantum yield for photoreduction[169–171] and the determination of rate constants for decay of triplet excited ketones.[172,173]

Rates of polymerisation have been determined for various systems.[164,173–179]

Polymeric photoinitiators containing side-chain benzophenone chromophores have also been synthesised and their structure–activity relationship has been studied.[180–182] The results suggest that the activity strongly depends on conformational mobility of both pendant benzophenone and amine groups.

An important point concerning Type II photoinitiating systems is the nature of the initiating species: will it be the ketyl radical or the hydrogen-donor-derived radical, or both? Aromatic ketyl radicals are known for their tendency to enter into radical coupling reactions, e.g. pinacolisation. For this reason they cannot be expected to contribute to the initiation process to any large extent. This has been elegantly demonstrated in the case of the benzophenone ketyl radical [reaction (21)].[183,184]

$$(C_6H_5)_2C{=}O + (C_6H_5)_2\overset{\displaystyle H}{\overset{|}{C}}{-}OH \xrightarrow{h\nu} 2(C_6H_5)_2\dot{C}{-}OH \qquad (21)$$

The rate of polymerisation of MMA photoinitiated by benzophenone was retarded and the M_w of the resulting polymer was drastically reduced when benzhydrol was used as the hydrogen-donor, benzophenone ketyl radicals being the only primary radicals generated from both precursors. The initiation using a benzophenone–hydrogen-donor system therefore will have to be induced predominantly if not exclusively by the radicals derived from the hydrogen-donors. Chemical confirmation is provided by the attempted cyclodimerisation of atropic esters photosensitised by benzophenone in THF solution [reaction (22)].[185] The major product proved to be the 1:1 adduct of THF to the atropic esters, the benzophenone ketyl radical ending up as benzopinacol and cross-coupled to the THF radical. Similar results were obtained in 1,4-dioxan as solvent.

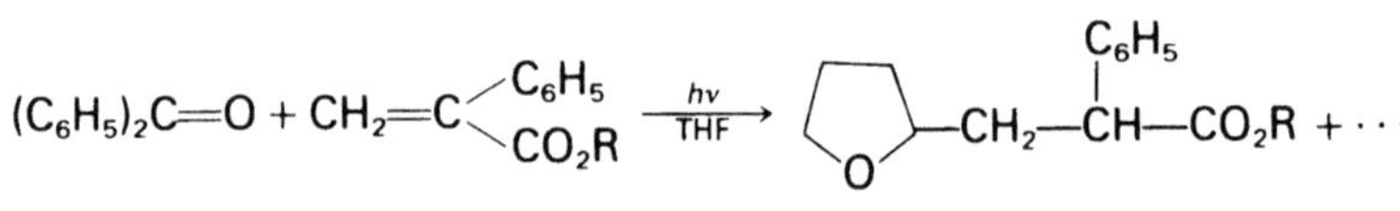

(22)

Likewise, in the benzophenone–tertiary amine system, the α-aminoalkyl radical may be expected to induce the polymerisation. Indirect evidence is provided by the observation that, for a particular ketone, the rate of polymerisation in bulk and in solution is strongly dependent on the structure of the α-aminoalkyl radical produced.[173,186,187] For the curing of thin films of acrylates the same dependence was observed.[173] Nitrogen analyses of poly-MMA resulting from polymerisation of MMA photoinitiated by benzophenone–triethylamine indicated[188] triethylamine to be incorporated in the polymer, presumably as an end group, implying initiation by a triethylamine-derived radical.

Only very recently[189] direct evidence for the involvement of the α-aminoalkyl radical in the initiation step was obtained. Irradiation of benzophenone–tertiary amine in the presence of 1,1-di(*p*-tolyl)ethylene as a non-polymerisable model substrate resulted in the 1:1 adduct of the tertiary amine and the model substrate as the major product [reaction (23)]. In addition substantial amounts of benzopinacol were isolated.

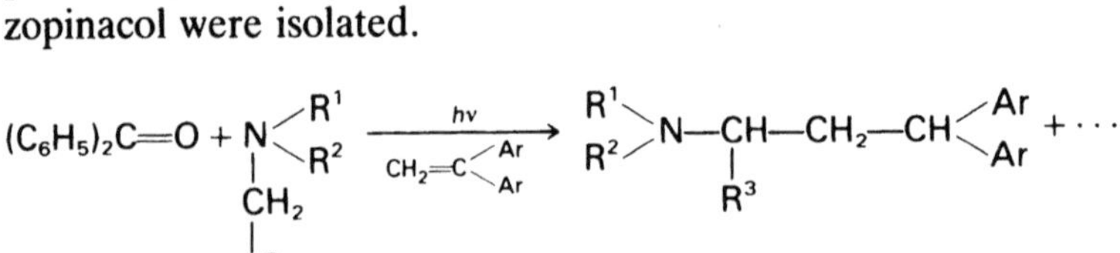

(23)

4.2.2 *Xanthones and Thioxanthones* (**XIV**)

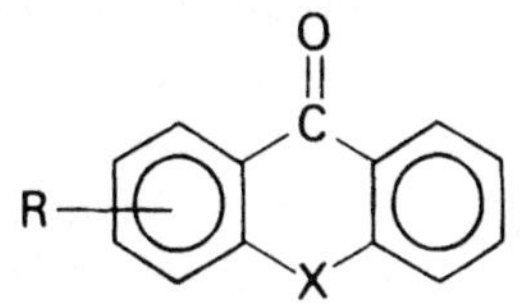

XIV: X = O, S; R = Alk, OAlk, OAlk—$SO_3^-M^+$, OAlk—N^+—Z

Photochemically these compounds behave in much the same way as the benzophenones discussed in Section 4.2.1, the essential difference being mainly their absorption spectra, λ_{max} ranging from 380 to 420 nm, with log ε 3·5–4·0,[128,186,187,190–196] which makes them particularly suitable for use in white (TiO_2) pigmented unsaturated systems.

Flash photolysis[190–196] and laser flash photolysis[128,186] studies of a variety of substituted thioxanthones, including water-soluble,[190] polymerisable[196] and polymeric thioxanthones[196] of the thioxanthone–tertiary amine systems have characterised the α-aminoalkyl radicals produced, which can be held responsible for the initiation of the polymerisation. Thus radicals **A–D** were found.

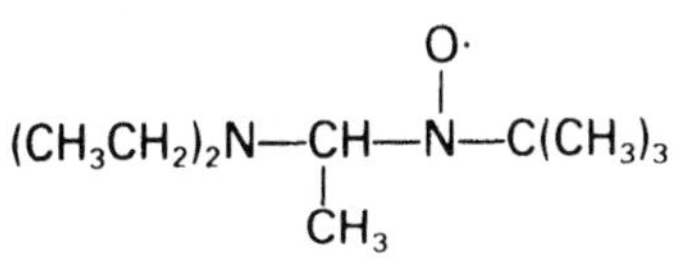

A From triethylamine

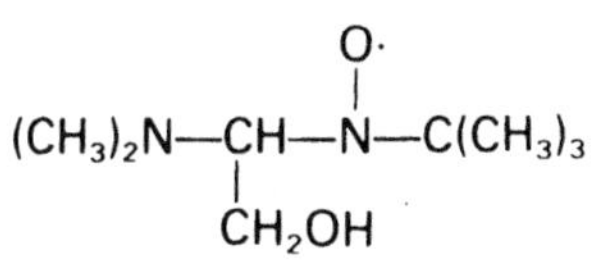

B From *N,N*-dimethylethanolamine

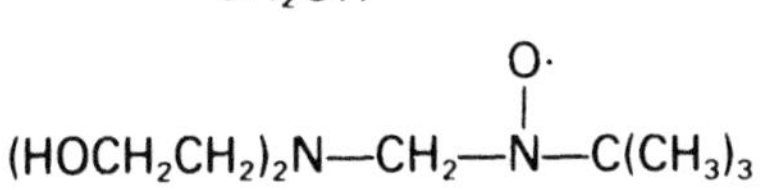

C From *N*-methyldiethanolamine

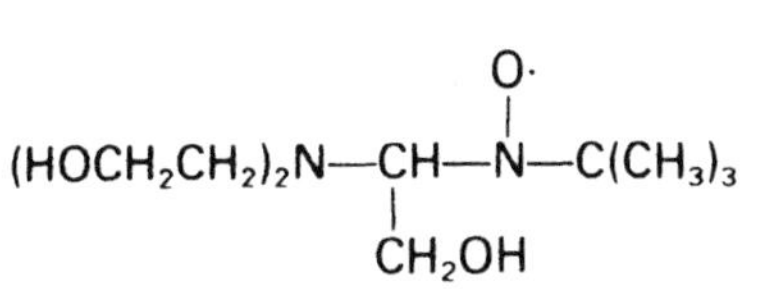

D From triethanolamine

Radical **A** from triethylamine had already been observed in the benzophenone triethylamine system.[165] The formation of the radical **C** from *N*-methyldiethanolamine is remarkable in the light of the formation of radical **B** from *N,N*-dimethylethanolamine. The formation of radical **D** from triethanolamine is again according to expectations. However, from triethanolamine also minor amounts of radical **C** were formed, which necessitates the fragmentation of some precursor radical prior to trapping by MNP. A fragmentation scheme based on the one observed[197] in the case of 2-(*N*-aryl)aminophenylethanols has been proposed to account for this observation.

4.2.3 α-Ketocoumarins

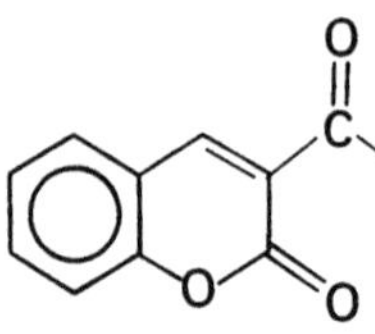

XV

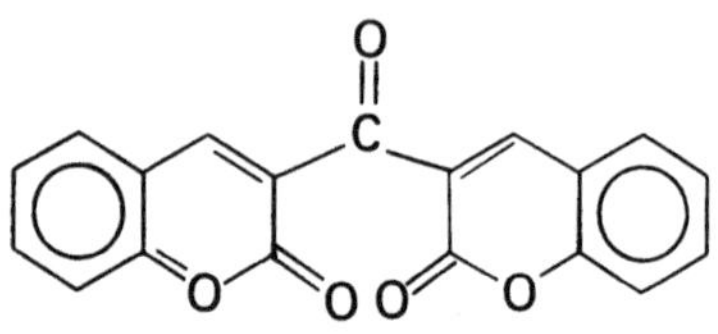

XVI

Recently several α-ketocoumarins (**XV, XVI**) were synthesised and were shown[198] to fulfill most of the photophysical criteria required for efficient triplet sensitisers. Thus these compounds show strong absorption maxima (λ_{max} 330–450 nm, log ε 4–5), singlet–triplet intersystem crossing efficiencies approaching unity in several cases, and triplet energies ranging from 48 to 60 kcal mol^{-1}.

As a logical extension these compounds were also evaluated[206] as Type II photoinitiators in conjunction with suitable coinitiators (hydrogen-donors) such as tertiary amines.

4.2.4 Aromatic 1,2-Diketones

Most if not all of the studies reported refer to benzil, being the simplest representative of this category.

In inert solvents such as benzene, benzil appears to be rather stable on irradiation. Thus, ESR spin-trapping experiments using MNP failed to detect trapped benzoyl radicals [reaction (24)].[65] Corroborating evidence is provided by corresponding experiments using TMPO as radical trapping agent [reaction (25)] (Hageman, H. J., unpublished). The conclusion is justified that benzil does not undergo α-cleavage in a primary photoreaction.

Photolysis of benzil in a hydrogen-donating solvent such as 2-propanol, however, leads to a number of products such as benzilpinacol, benzoin, benzoin benzoate and benzoic acid in approximately equal amounts [reaction (26)].[200]

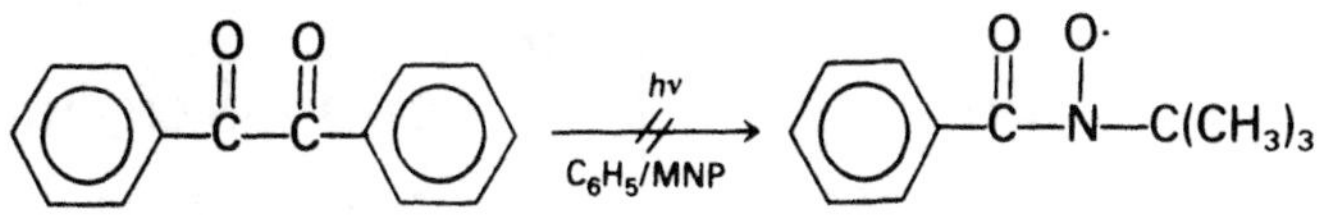

(24)

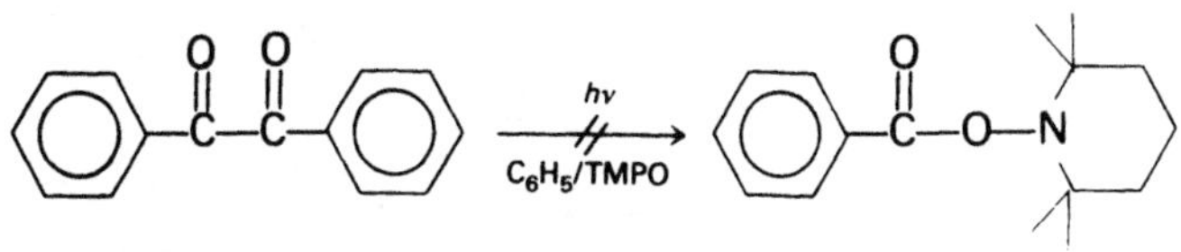

(25)

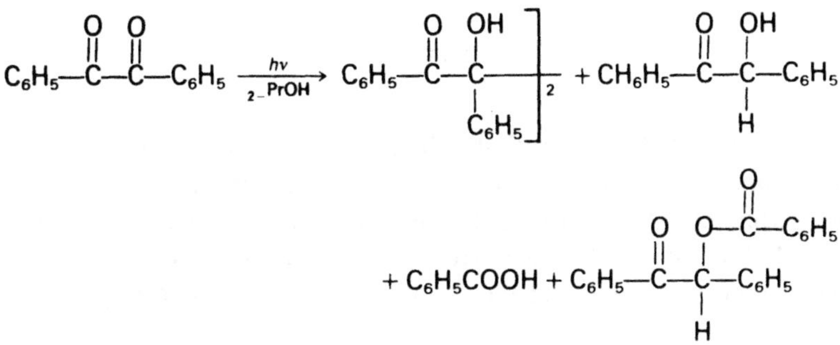

(26)

On addition of triethylamine, the formation of the pinacol is almost completely suppressed, benzoin and benzoin benzoate becoming the major products. On increasing the amine concentration, benzoin ultimately becomes the sole product. The formation of these products can be readily explained by a primary hydrogen-abstraction (or electron-transfer followed by proton-transfer in the case of tertiary amine being present) by the triplet excited benzil from the hydrogen-donor. On further irradiation the picture may well be obscured by subsequent photoreactions of the primary photoproducts such as α-cleavage of benzoin and of benzoin benzoate furnishing benzoyl radicals among others.

For this very reason it is not surprising that in conventional flash photolysis experiments (flash duration 10 μs) in 2-propanol, benzil was reported[201] to produce benzoyl radicals, as tentatively concluded from a transient absorption around 340 nm.

In the polymerisation of n-butyl methacrylate in solution in methyl ethyl ketone photoinitiated by benzil (among other compounds), it was reported[201] that benzil showed the greatest increase in rate of polymerisation upon the addition of a tertiary amine.

The polymerisation of methyl methacrylate (MMA) photoinitiated

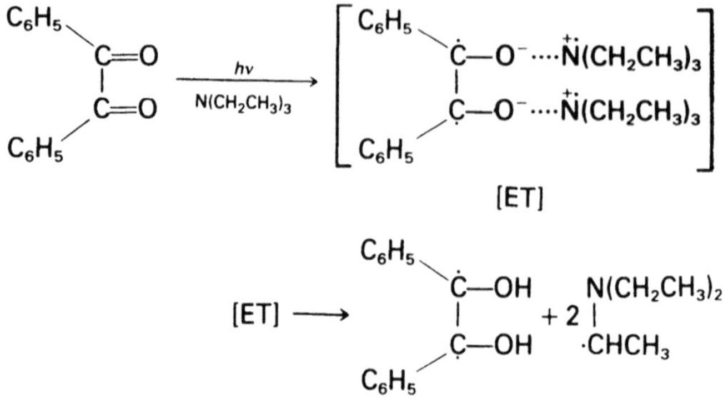

SCHEME 16

by the benzil–N,N-dimethylaniline[202] and by the benzil–triethylamine[203] system respectively has been reported.

The sequence of primary reactions shown in Scheme 16 has been proposed, the only difference with the general scheme for ketone–tertiary amine systems being the involvement of both carbonyl groups in this case.

Indications were obtained for some initiator-dependent termination process. End group analysis (UV absorbance) showed the presence of benzilketyl moieties. The presence of amino end groups was estimated by means of a dye partition technique.

4.2.5 Phenylglyoxylates

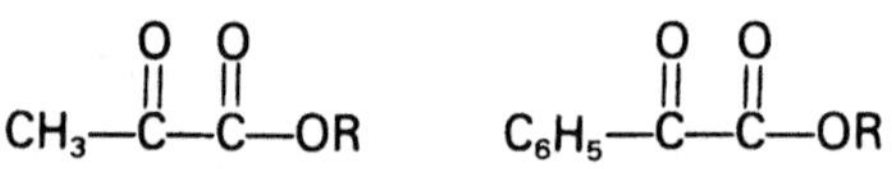

XVII **XVIII**

R = Alk

Esters of pyruvic acid (**XVII**) and phenylglyoxylic acid (**XVIII**) have been claimed as photoinitiators for free-radical polymerisation processes.[204] These compounds are light yellow liquids showing weak absorption maxima in the 320–375 nm range.

The photochemistry of the parent acids and of some esters has been studied under various conditions.

The irradiation of aqueous solutions of pyruvic acid and phenylglyoxylic acid was found[213] to lead to rapid decarboxylation, major products being acetoin and benzaldehyde respectively. In solvents like benzene, pyruvic acid is photochemically inert.[205] In hydrogen-donating solvents (alcohols) the predominant reaction of pyruvic acid is photoreduction of the α-carbonyl group, ultimately resulting in the dimerisation product of 2,3-dimethyltartaric acid.[205,206] The photo-decarboxylation and photoreduction respectively of pyruvic acid have been studied also by the ^{1}H-NMR–CIDNP technique.[207]

Phenylglyoxylic acid has also been reported[216] to undergo photo-reduction in alcohols to give 2,3-diphenyltartaric acid [reaction (27)]. Ketyl radicals have been observed using ESR spectroscopy, and rates of bimolecular radical decay have been reported.[209]

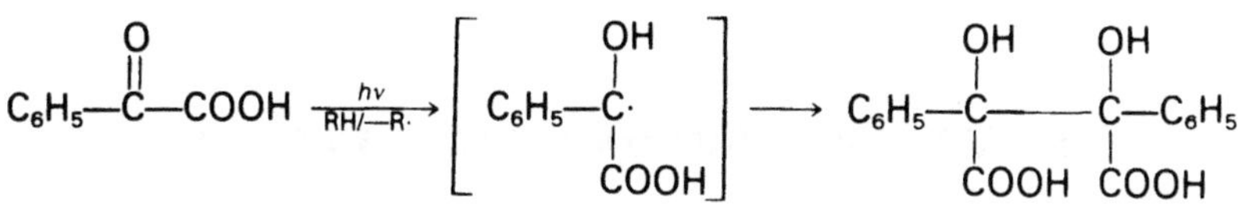

(27)

Pyruvic esters and phenylglyoxylates have been reported to undergo fragmentation in inert solvents such as benzene, acetonitrile and *n*-heptane:[210,211]

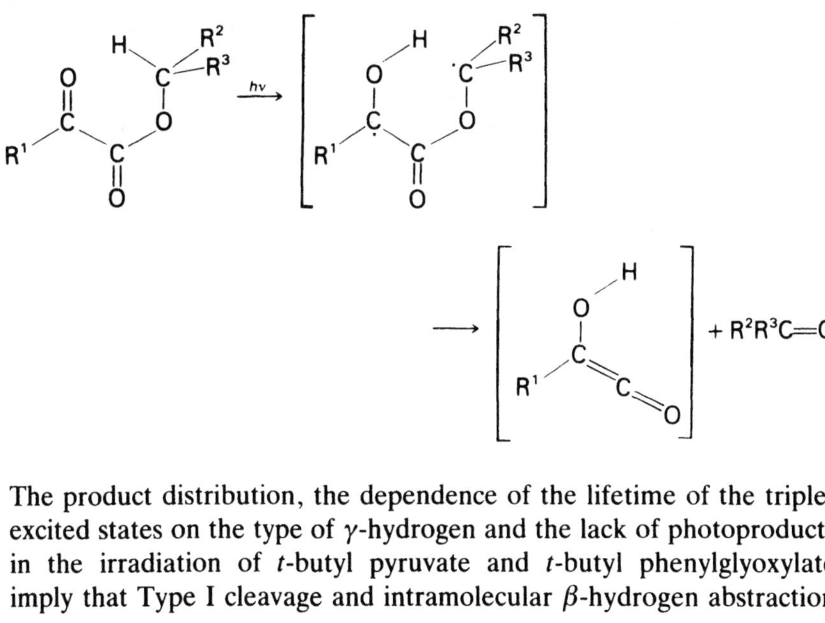

The product distribution, the dependence of the lifetime of the triplet excited states on the type of γ-hydrogen and the lack of photoproducts in the irradiation of *t*-butyl pyruvate and *t*-butyl phenylglyoxylate imply that Type I cleavage and intramolecular β-hydrogen abstraction are negligible.[211] The observation that irradiation of methyl phenylglyoxylate in benzene in the presence of TMPO does not lead to any product (Hageman, H. J., unpublished) seems to confirm these conclusions, although a triplet quenching process cannot be entirely excluded.

Pyruvic esters and phenylglyoxylates undergo efficient photoreduction when irradiated in alcohols.[212,213]

In the photoinitiated crosslinking of acrylate coatings, methyl phenylglyoxylate displays an efficiency comparable with that found for 2-hydroxy-2-methylpropiophenone (Hageman, H. J. & Jansen, L. C. J., unpublished). The mechanism of initiation (e.g. initiating species), however, is not known with certainty.

5 ATMOSPHERIC OXYGEN INHIBITION

In most if not in all practical applications, UV curing of coatings is carried out in an air atmosphere. Oxygen from the surrounding air may play a role in competing processes at several stages of the photoinitiated free-radical polymerisation, such as quenching of the triplet excited state of photoinitiators, and scavenging of carbon-centred (and some other) radicals both from the photoinitiating system and from propagating polymer radicals which occurs at high rate constants ($k > 10^9\,\mathrm{mol}^{-1}\,\mathrm{s}^{-1}$)[214] leading to less reactive peroxy radicals. This inevitably results in a slowing down of the overall rate of polymerisation (crosslinking), giving tacky (undercured) coatings, an effect commonly referred to as oxygen inhibition.

Oxygen inhibition has been well demonstrated by kinetic analyses,[225] differential scanning calorimetry[216] (DSC), and laser nephelometry[52] to result in:

(i) an induction period;
(ii) a reduced rate of polymerisation (crosslinking);
(iii) an incomplete consumption of unsaturated functionalities.

Most Type I photoinitiators used nowadays appear to undergo unimolecular fragmentation (most α-cleavage) into radicals at high rate constants ($k_\alpha > 10^9\,\mathrm{s}^{-1}$). As a consequence the triplet state precursors are extremely short-lived, too short-lived to be quenched by oxygen in a bimolecular process to any appreciable extent. In fact the photodecomposition of benzoin methyl ether has been reported[63,217] to proceed with considerably higher quantum yields in the presence of oxygen in a variety of solvents, in all likelihood because the primary radicals generated are prevented from recombining through scavenging by oxygen. This then leaves the scavenging of reactive carbon-centred (and some other) radicals by oxygen as the main cause for the oxygen inhibition observed.

Type II photoinitiating systems based on alcohols and ethers as hydrogen-donors are extremely susceptible to quenching by oxygen owing to the relatively long-lived triplet states of these ketones.

Type II photoinitiating systems based on tertiary amines as the hydrogen-donors, however, appear to form primary radicals within the [ET] exciplex without intervention by oxygen. In these cases one only has to deal with oxygen scavenging once more (Table 1).

Oxygen inhibition is particularly pronounced in the UV curing of coatings on account of the unfavourable surface-to-volume ratio, which creates optimal conditions for oxygen diffusion into the coating. From a kinetic study of oxygen inhibition in UV curing of acrylate coatings[218] it was concluded that the concentration of oxygen dissolved in the system has to drop by a factor of about 3×10^2 before polymerisation can actually start. In practice the effect of oxygen inhibition in the curing of polyester-acrylates photoinitiated by benzil dimethylketal has been demonstrated[219] by comparison with parallel experiments carried out under a transparent foil in order to exclude air. Parameters such as M_n (oligomer), vinyl functionality (oligomer and reactive diluent) and viscosity (Trommsdorf effect), which were found to affect the curing speed in air, became totally irrelevant when air was excluded. Similar effects were observed using acylphosphine oxides[138] and sulphonic esters of α-hydroxymethylbenzoin.[85]

Apart from excluding oxygen altogether by creating an inert atmosphere, e.g. by expensive nitrogen blanketing, the methods

TABLE 1
The Nature of Oxygen Inhibition in Different Photoinitiating Systems

Photoinitiators	*O_2 quenching*	*O_2 scavenging*
Type I	−	+ +
Type II/alcohols, ehters	+ +	(+ +)
Type II/tertiary amines	−	+

developed to overcome or at least reduce the effect of oxygen inhibition fall into the following two categories.

(i) Physical methods such as more effective light-sources,[218,220] and prevention of oxygen diffusion into the thin coating using paraffin waxes[221] or transparent foils[219] as oxygen-barriers.
(ii) Chemical methods aimed at in-situ generation of oxygen barriers, and depletion of oxygen present.

It has been found[83,84] that hexanesulphonic acid and its higher homologues as their metal salts, so efficiently generated from the corresponding alkanesulphonic esters of α-hydroxymethylbenzoin and its ethers, exhibit a pronounced surface activity, putting up an extremely effective oxygen barrier at the coating–air interface. Other photoinitiators generating sulphonic acids, such as *O*-sulphonyl-α-oximinoketones and phenacyl sulphonates, show the same effects (Hageman, H. J. & Jansen, L. G. J., unpublished).

Quite similar is the use of a mixture of one of the common photoinitiators and a small amount of a surface-active photoinitiator.[222] The latter is expected to create a relatively high radical concentration in the surface layers, ensuring a rapid cure locally, sealing off the coating surface for oxygen diffusion from the surrounding air.

Combinations of Type I photoinitiators and benzophenone (Type II) have also been claimed[223] for their superior performance. The Type I photoinitiator (not quenched by oxygen) is supposed to furnish the primary radicals which are sacrificed to deplete the oxygen present in the system. The Type II photoinitiator is then supposed to induce the polymerisation through both hydrogen-abstraction and photosensitised decomposition of hydroperoxides formed.

Depletion of oxygen can also be accomplished by the addition of readily oxidisable compounds such as phosphorus(III) compounds, thiols and amines. Thus combinations of photoinitiators (e.g. benzoin ethers) and phosphines and phosphites respectively have been claimed.[224] Phosphine oxides have been detected in such systems,[225] whereas phosphites are also known to be readily oxidised by hydroperoxides.[226]

Similarly, a polythiol such as pentaerythritol tetrakis-β-mercaptopropionate in combination with a photoinitiator appears to be effective in consuming oxygen.[238]

The beneficial effect of tertiary amines on the performance of Type I photoinitiators in photoinitiated crosslinking of various unsaturated systems has also been known for some time.[13,99,138–140,228] The effect has been ascribed to a suppression of the oxygen inhibition by preferential oxygen scavenging, since it is known that amines readily undergo photosensitised oxidation through initial hydrogen-abstraction.[229] Supporting evidence comes from the determination of oxygen consumption in photopolymerisable systems upon addition of tertiary amines,[227d] kinetic measurements of polymerisation in thin films of acrylates induced by UV and laser irradiation,[218] and measurements of polymerisation exotherms in the presence of aliphatic tertiary amines,[230] aromatic tertiary amines[231] and amine-functionalised methacrylates.[232]

The polymerisation of MMA in an air-saturated solution photoinitiated by benzil dimethylketal in the presence of triethylamine was recently reported[233] to show a five-fold decrease in quantum efficiency as well as a lower polymer yield compared with polymerisation under a nitrogen atmosphere. The conclusion that the addition of tertiary amines is not effective in relieving the oxygen inhibition is not justified, however, since the polymerisation in air-saturated solutions should have been studied also in the absence of triethylamine.

A recent product study[141] of the photodecomposition of an acylphosphine oxide in the presence of 1,1-di(*p*-tolyl)ethylene, both in solution and in thin films, clearly showed the adverse effects of oxygen, i.e. a dramatic increase in 'wastage products' and as a consequence a corresponding decrease in initiating efficiency. The addition of triethylamine was clearly shown to counteract the oxygen inhibition, however, only partly restoring the original situation in the absence of oxygen.

Since triethylamine appeared to have no effect in the absence of oxygen, its effect in the presence of oxygen was ascribed to oxygen scavenging by α-aminoalkyl radicals generated through hydrogen-abstraction, presumably by peroxy radicals. A contribution of α-aminoalkyl radicals to the initiation was not observed. This may well imply that the beneficial effect of tertiary amines in photoinitiated polymerisation and crosslinking processes is due to

(i) exclusive oxygen scavenging in combination with Type I photoinitiators;
(ii) initiation and in all likelihood also oxygen scavenging in combination with Type II photoinitiators.

An entirely different approach to overcome oxygen inhibition is a system based on the photosensitised conversion of oxygen (triplet ground state) into singlet oxygen.[234] Thus the system consists of a dye sensitiser, a trapping agent for singlet oxygen, e.g. 1,3-diphenylisobenzofuran, and last but not least a common photoinitiator.

6 CONCLUDING REMARKS

An attempt has been made to review the field of photoinitiated free-radical polymerisation processes, particularly emphasising mechanistic aspects but without pretending to have been exhaustive in coverage. It should also be emphasised that in the light of the present rate of progress this chapter has the character of a short-lived transient rather than of a stable situation.

Future trends may well be the design of tailor-made photoinitiators or photoinitiating systems for applications, the study of photoinitiators under less common conditions, e.g. in highly ordered media,[235] and last but not least new instrumental principles such as two-photon laser excitation.[236]

REFERENCES

1. Heine, H. G., Rosenkranz, H. J. & Rudolph, H., *Angew. Chem.*, **84** (1972) 1032.
2. Pappas, S. P., *Progr. Org. Coat.*, **2** (1973/74) 333.
3. Hutchison, J. & Ledwith, A., *Adv. Polym. Sci.*, **14** (1974) 49.
4. Heine, H. G. & Traencker, H. J. *Progr. Org. Coat.*, **3** (1975) 115.
5. Rudolph, H. J., Rosenkranz, H. J. & Heine, H. G., *Appl. Polym. Symp.*, **26** (1975) 157.
6. Young, S. E., *Progr. Org. Coat.*, **4** (1976) 225.
7. Ledwith, A., *J. Oil Col. Chem. Assoc.*, **59** (1976) 157.
8. Pryce, A., *J. Oil Col. Chem. Assoc.*, **59** (1976) 166.
9. Vrancken, A., *Farbe Lack*, **83** (1977) 171.
10. Ledwith, A., *Pure Appl. Chem.*, **49** (1977) 431.
11. Pappas, S. P. & McGinniss, V. D., *UV-Curing: Science and Technology*, Vol. 1. Technol. Mark. Corp., Norwalk, CT, 1978.
12. Papas, S. P., Chattopadhyay, A. & Carlblom, L. H., In *Ultraviolet Light Induced Reactions in Polymers*, ed. S. S. Labana, ACS Symp. Series, **25** (1976) 12.
13. Berner, G., Kirchmayr, R. & Rist, G., *J. Oil Col. Chem. Assoc.*, **61** (1978) 105.
14. Berner, G., Kirchmayr, R. & Rist, G., *Proc. 3rd Int. Conf. Org. Coat. Sci. Technol., Athens, Greece*, **1** (1979) 104.

15. Berner, G., Puglisi, J., Kirchmayr, R. & Rist, G., *J. Radiat. Curing*, **6**(6) (1979) 2.
16. McGinniss, V. D., *Photogr. Sci. Eng.*, **23** (1979) 124.
17. Green, G. E., Stark, B. P. & Zahir, S. A., *J. Macromol. Sci., Rev. Macromol. Chem.*, **C21** (1981/82) 187.
18. Roffey, C. G., *Photopolymerization of Surface Coatings*. J. Wiley, New York, 1982.
19. Dowbenko, R., Friendlander, C., Gruber, G., Prucnal, P. & Wismer, M., *Progr. Org. Coat.*, **11** (1983) 71.
20. Baumann, H., Timpe, H. J. & Böttcher, H., *Z. Chem.*, **28** (1983) 197.
21. Gatechair, L. R. & Wostratzky, D., *J. Radiat. Curing*, **10**(3) (1983) 4.
22. Senich, G. A. & Florin, R. E., *J. Macromol. Sci., Rev. Macromol. Chem.*, **C24** (1984) 239.
23. Holman, R., *UV and EB Curing Formulations for Printing Inks, Coatings and Paints*, Sita Technology, 1984.
24. Pappas, S. P. In *UV-Curing: Science and Technology*, Vol. II. Technol. Mark. Corp., Norwalk, CT, 1985.
25. Hageman, H. J., *Progr. Org. Coat.*, **13** (1985) 123.
26. Green, P. N., *Polym. Paint Colour J.*, **175**(4141) (1985) 246.
27. Fouassier, J. P., *Double Liaison Chim. Peint.*, **32**(356/357) (1985) 173.
28. Pappas, S. P., *Radiat. Phys. Chem.*, **25**(4/6) (1985) 633.
29. Pappas, S. P. In *New Trends in the Photochemistry of Polymers*, ed. N. S. Allen & J. F. Rabek. Elsevier Applied Science, London, 1985, Ch. 6.
30. LiBassi, G., *Double Liaison Chim. Peint.*, **32**(361) (1985) 382.
31. O'Hara, K., *Polym. Paint Colour J.*, **175**(4141) (1985) 254.
32. Paul, S., *Surface Coating Science and Technology*. Wiley Interscience, New York, 1986.
33. Vesley, G. F., *J. Radiat. Curing*, **13**(1) (1986) 4.
34. Pappas, S. P., *J. Radiat. Curing*, **14**(3) (1987) 6.
35. LiBassi, G., *J. Radiat. Curing*, **14**(3) (1987) 18.
36. Kloosterboer, J. G., *Adv. Polym. Sci.*, **84** (1988) 1.
37. Turro, N. J. *Modern Molecular Photochemistry*. Benjamin/Cummings, Menlo Park, 1978.
38. Kochi, J. K. In *Free Radicals*, Vol. II, ed. J. K. Kochi. Wiley, New York, 1973, p. 665.
39. Walling, C. & Gibian, M., *J. Am. Chem. Soc.*, **86** (1964) 3902.
40. Lewis, F. D., Lauterbach, R. T., Heine, H. G., Hartmann, W. & Rudolph, H., *J. Am. Chem. Soc.*, **97** (1975) 1519.
41. (a) Ciamician, G. & Silber, P., *Ber. Dtsch. Chem. Ges.*, **33** (1900) 2911; (b) Ciamician, G. & Silber, P., *Ber. Dtsch. Chem. Ges.*, **34** (1901) 1530.
42. Lewis, F. D., & Magyar, J. G., *J. Org. Chem.*, **37** (1972) 2102.
43. Baum, A. A., *J. Am. Chem. Soc.*, **94** (1972) 6866.
44. Baum, E. J., Wan, J. K. S. & Pitts, J. N., *J. Am. Chem. Soc.*, **88** (1966) 2652.
45. Wagner, P. J., Kemppainen, A. E. & Schott, H. N., *J. Am. Chem. Soc.*, **95** (1973) 5604.
46. Fischer, H. & Paul, H., *Acc. Chem. Res.*, **20** (1987) 200.
47. Janzen, E. G., *Acc. Chem. Res.*, **4** (1971) 31.

48. Kamachi, M., *Adv. Polym. Sci.*, **82** (1987) 207.
49. Kaptein, R., *Adv. Free Radical Chem.*, **5** (1975) 381.
50. (a) Hore, P. J., Joslin, C. G. & McLauchlan, K. A., *Chem. Soc. Rev.*, **8** (1979) 29; (b) McLauchlan, K. A., *Chem. Brit.*, **21** (1985) 825; (c) McLauchlan, K. A. & Stevens, D. G., *Acc. Chem. Res.*, **21** (1988) 54.
51. (a) Robbins, W. K. & Eastman, R. H., *J. Am. Chem. Soc.*, **92** (1970) 6077; (b) Rizzardo, E. & Solomon, D. H., *Polym. Bull.*, **1** (1979); 529; (c) Solomon, D. H. & Moad, G., *Makromol. Chem., Makromol. Symp.*, **10/11** (1987) 109.
52. Decker, C. & Fizet, M., *Makromol. Chem., Rapid Commun.*, **1** (1980) 637.
53. Carblom, L. H. & Pappas, S. P., *J. Polym. Sci., Polym. Chem. Ed.*, **15** (1977) 1381.
54. (a) Kuhlmann, R. & Schnabel, W., *Polymer,* **18** (1977) 1163; (b) Kuhlmann, R. & Schnabel, W., *Angew. Makromol. Chem.*, **70** (1978) 145.
55. Sato, T., *Plaste Kautsch.*, **29**(6) (1982) 317.
56. (a) Bevington, J. C., Huckerby, T. N. & Hutton, N. W. E., *J. Polym. Sci., Polym. Chem. Ed.*, **20** (1982) 2655; (b) Bevington, J. C., Huckerby, T. N. & Hutton, N. W. E., *Eur. Polym. J.*, **18** (1982) 963; (c) Bevington, J. C., Huckerby, T. N. & Vickerstaff, N., *Makromol. Chem., Rapid Commun.*, **4** (1983) 349.
57. (a) Groenenboom, C. J., Hageman, H. J., Overeem, T. & Weber, A. J. M., *Makromol. Chem.*, **183** (1982) 281; (b) Hageman, H. J., Oosterhoff, P., Overeem, T., Polman, R. J. & Werf, S. v.d., *Makromol. Chem.*, **186** (1985) 2483.
58. Baxter, J. E., Davidson, R. S., Hageman, H. J. & Overeem, T., *Makromol. Chem.*, **189** (1988) 2769.
59. Kornis, G. & de Mayo, P., *Can. J. Chem.*, **42** (1964) 2822.
60. Bradshaw, J. S., Knudsen, R. D. & Parrish, W. R., *J. Chem. Soc., Chem. Commun.*, (1972) 1321.
61. Heine, H. G., *Tetrahedron Lett.*, (1972) 4755.
62. Pappas, S. P. & Chattopadhyay, A., *J. Am. Chem. Soc.*, **95** (1973) 6484.
63. Adam, S., Güsten, H., Steenken, S. & Schulte-Frohlinde, D., *Justus Liebigs Ann. Chem.*, (1974) 1831.
64. Paul, H. & Fischer, H., *Helv. Chim. Acta,* **56** (1973) 1575.
65. Ledwith, A., Russell, P. J. & Sutcliffe, L. H. *J. Chem. Soc., Perkin Trans. II,* (1972) 1925.
66. Hageman, H. J. & Overeem, T., *Makromol. Chem., Rapid Commun.*, **2** (1981) 719.
67. Cocivera, M. & Trozzolo, A. M., *J. Am. Chem. Soc.*, **92** (1970) 1772.
68. Closs, G. L. & Paulsen, D. R., *J. Am. Chem. Soc.*, **92** (1970) 7229.
69. Dominh, T., *Ind. Chim. Belge,* **36** (1971) 1080.
70. Maryama, K., Furuta, H. & Otsuki, T., *Bull. Chem. Soc. Jpn,* **53** (1980) 2421.
71. Baumann, H., Müller, U., Pfeifer, D. & Timpe, H. J., *J. Prakt. Chem.*, **324**(2) (1982) 217.

72. Mochel, W. E., Crandall, J. L. & Peterson, J. H., *J. Am. Chem. Soc.*, **77** (1955) 494.
73. Hutchinson, J. & Ledwith, A., *Polymer*, **14** (1973) 405.
74. Pappas, S. P. & Asmus, R. A., *J. Polym. Sci., Polym. Chem. Ed.*, **20** (1982) 2643.
75. Hageman, H. J., v.d. Maeden, F. P. B. & Janssen, P. C. G. M., *Makromol. Chem.*, **180** (1979) 2351.
76. Lipscomb, N. T. & Tarshiani, Y., *J. Polym. Sci., Polym. Chem. Ed.*, **26** (1988) 529.
77. Osborn, C. L. & Sandner, M. R., *ACS Coat. Plast. Reprints*, **34**(1) (1974) 660.
78. Heine, H. G., *Justus Liebigs Ann. Chem.*, **735** (1970) 56.
79. Heine, H. G. & Rudolph, H., *Justus Liebigs Ann. Chem.*, **754** (1971) 28.
80. Sheehan, J. C. & Wilson, R. M., *J. Am. Chem. Soc.*, **86** (1964) 5277.
81. Sheehan, J. C., Wilson, R. M. & Oxford, A. W., *J. Am. Chem. Soc.*, **93** (1971) 7222.
82. Heine, H. G., Rudolph, H. & Kreuder, H. J., Ger. Offen. 1 919 678 (1970).
83. Hageman, H. J., de Klein, W. J. & Giezen, E. A., European Patent 37 152 (1981).
84. Hageman, H. J. & Jansen, L. G. J., *Makromol. Chem.*, **189** (1988) 2781.
85. Angad Gaur, H., Groenenboom, C. J., Hageman, H. J., Hakvoort, G. T. M., Oosterhoff, P., Overeem, T., Polman, R. J. & v.d. Werf, S., *Makromol. Chem.*, **185** (1984) 1795.
86. Angad Gaur, H., Hageman, H. J., Oosterhoff, P., Overeem, T., Polman, R. J., Verbeek, J. & v.d. Werf, S., *Makromol. Chem.* (submitted).
87. Heine, H. G., Fuhr, K., Rudolph, H. & Schnell, H., S. African Patent 6 905 401 (1970).
88. Askerov, D. B., Tarasov, V. F., Vainer, A. Ya., Stebunova, G. V., Bogdanov, V. A. & Dyumaev, K. M., *Zh. Org. Khim.*, **18** (1982) 1047.
89. Ahn, K. D., Ihn, K. J. & Kwon, I. C., *J. Macromol. Sci., Chem.*, **A23** (1986) 355.
90. Angad Gaur, H., Groenenboom, C. J., Hageman, H. J., Oosterhoff, P., Overeem, T., Polman, R. J., Verbeek, J. & v.d. Werf, S., *Makromol. Chem.* (submitted).
91. Fischer, M., Kuesters, W., Penzel, E., Rehder, W. & Trautmann, W., Ger. Offen. 2 831 263 (1980).
92. Adams, W. R., US Patent 4 037 021 (1977).
93. Brattesani, A. J., US Patent 4 072 694 (1978).
94. Heine, H. G. & Rosenkranz, H. J., Ger. Offen. 2 648 685 (1978).
95. Schönberg, A., Fateen, A. K. & Omran, S. M. A. R., *J. Am. Chem. Soc.*, **78** (1956) 1224.
96. Petropoulos, C. C., *J. Polym. Sci.*, **A2** (1964) 69.
97. Sato, T., Abe, M. & Otsu, T., *Makromol. Chem.*, **180** (1979) 1165.
98. Sandner, M. R. & Osborn, C. L., *Tetrahedron Lett.*, (1974) 415.
99. Osborn, C. L., *J. Radiat. Curing*, **3**(3) (1976) 2.
100. Borer, A., Kirchmayr, R. & Rist, G., *Helv. Chim. Acta*, **61** (1978) 305.

101. Christensen, J. E., Jacobine, A. F. & Scanio, C. J. V., *J. Radiat. Curing*, **8**(3) (1981) 4.
102. Fouassier, J. P. & Lougnot, D. J., *J. Chem. Soc., Faraday Trans. 1*, **83** (1987) 2935.
103. Kirchmayr, R., Berner, G. & Rist, G., *Farbe Lack*, **86** (1980) 224.
104. Kuhn, R. & Trischmann, H., *Chem. Ber.*, **94** (1961) 2258.
105. Bak, C., Praefcke, K., Muszkat, K. A. & Weinstein, M., *Naturforsch. B*, **32** (1977) 674.
106. Fouassier, J. P. & Merlin, A., *J. Photochem.*, **12** (1980) 17.
107. Baxter, J. E., Davidson, R. S., Hageman, H. J., Hakvoort, G. T. M. & Overeem, T., *Polymer*, **29** (1988) 1575.
108. (a) Gehlhaus, J. & Kieser, M., Ger. Offen. 2 722 264 (1978); (b) Herz, C. P. & Eichler, J., *Farbe Lack*, **85** (1979) 933; (c) Herz, C. P., Eichler, J. & Neisius, K. H., *Kontakte* (Merck), **1979**(3), 37; (d) Ohngemach, J., Neisius, K. H., Eichler, J. & Herz, C. P., *Kontakte* (Merck), **1980**(3), 15.
109. Felder, L., Kirchmayr, R. & Hüsler, R., European Patent 3002 (1979).
110. Eichler, J., Herz, C. P., Naito, I. & Schnabel, W., *J. Photochem.*, **12** (1980) 225.
111. Yankelevich, A. Z., Potapov, V. K., Hageman, H. J., Kutznets, V. M., Pershin, A. D. & Buchachenko, A. L., *Izv. Akad. Nauk SSSR, Ser. Khim.*, **1982**(3), 513.
112. Kirchmayr, R., Berner, G., Hüsler, R. & Rist, G., *Farbe Lack*, **88** (1982) 910.
113. Schenk, G. O., Koltzenburg, G. & Grossmann, H., *Angew. Chem.*, **69** (1957) 177.
114. Eichler, J., Herz, C. P. & Schnabel, W., *Angew. Makromol. Chem.*, **91** (1981) 39.
115. (a) Gehlhaus, J. & Ohngemach, J., *Radcure Confer. Lausanne 1983*, SME Techn. Paper FC 83-271; (b) Schornick, G. & Jacobi, M., *Radcure Confer. Lausanne 1983* SME Techn. Paper FC 83-263; (c) LiBassi, G., Cadona, L. & Broggi, F., *Radcure Confer. Munich 1987*, Confer. Proc. Ch. III, p. 15.
116. (a) Baeumer, W., Koehler, M. & Ohngemach, J., *Radcure Confer. Baltimore 1986*, SME Techn. Paper FC 86-815; (b) Koehler, M. & Ohngemach, J., *Radcure Confer. Munich 1987*, Confer. Proc., Ch. III, p. 1.
117. Koehler, M. & Ohngemach, J., *Radtech. Confer. New Orleans 1988*, Confer. Papers, p. 150.
118. LiBassi, G., Broggi, F. & Revelli, A., *Radtech. Confer. New Orleans 1988*, Confer. Papers, p. 160.
119. (a) Rutsch, W., Berner, G., Kirchmayr, R., Hüsler, R., Rist, G. & Buhler, N. In *Organic Coatings: Science and Technology*, Vol. 8, ed. G. D. Parfitt & A. V. Patsis. Marcel Dekker, New York, 1986, p. 175; (b) Meier, K., Rembold, M. W., Rutsch, W. & Sitek, F. Paper presented at 2nd Int. Symp. Radiation Curing of Polymers, University of Lancaster, UK, 1986; (c) Dietliker, K., Rembold, M. W., Rist, G., Rutsch, W. & Sitek, F., *Radcure Confer. Munich 1987*, Confer. Proc., Ch. III, p. 37.

120. Werner, W. & Piquet, A., *Ber. Dtsch. Chem. Ges.*, **37** (1904) 4295.
121. Meisenheimer, J. & Lampaerter, W., *Ber. Dtsch. Chem. Ges.*, **57** (1924) 276.
122. (a) Laridon, U. & Delzenne, G. A., Belgian Patent 717 039 (1969); (b) Delzenne, G. A., Laridon, U. & Peeters, H., *Eur. Polym. J.*, **6** (1970) 933; (c) Green, P. N. & Doherty, J., British Patent 1 537 921 (1979.
123. Hong, S. L., Kurasaki, T. & Okawara, M., *J. Polym. Sci., Polym. Chem. Ed.*, **12** (1974) 2553.
124. McGinniss, V. D., *J. Radiat. Curing* **2**(1) (1975) 3.
125. Baas, P. & Cerfontain, H., *J. Chem. Soc., Perkin Trans. 1* (1979) 156.
126. Danilewicz, J. C., *J. Chem. Soc. (C)*, (1970) 1049.
127. Baas, P. & Cerfontain, H., *J. Chem. Soc., Perkin Trans. 2* (1979) 1653.
128. Amirzadeh, G., Dissertation T. U. Berlin, 1981.
129. Baumann, H., Timpe, H. J., Zubarev, V. E., Fok, N. V. & Mel'nikov, M. Y., *J. Photochem.*, **30** (1985) 487.
130. Groenenboom, C. J., Hageman, H. J., Oosterhoff, P., Polman, R. J., Verbeek, J., Weber, A. J. M. & v.d. Werf, S., forthcoming publication.
131. Buys, Th. S. V., Cerfontain, H., Geenevasen, J. A. J. & Stunnenberg, F., *Recl. Trav. Chim. Pays-Bas*, (submitted).
132. Stunnenberg, F., Dissertation, University of Amsterdam, 1989.
133. (a) Lechtken, P., Buethe, I. & Hesse, A., Ger. Offen. 2 830 927 (1980); (b) Lechtken, P., Buethe, I., Jacobi, M. & Trimborn, W., Ger. Offen. 2 909 994 (1980); (c) Heine, H. G., Rosenkranz, H. J. & Rudolph, H., Ger. Offen. 3 023 486 (1980); (d) Jacobi, M. & Henne, A., *J. Radiat. Curing*, **10**(3) (1983) 16.
134. Sumiyoshi, T., Schnabel, W., Henne, A. & Lechtken, P., *Polymer*, **26** (1985) 141.
135. Baxter, J. E., Davidson, R. S., Hageman, H. J., McLauchlan, K. A. & Stevens, D. G., *J. Chem. Soc., Chem. Commun.*, (1987) 73; cf. also Ref. 143.
136. Kamachi, M., Kuwata, K., Sumiyoshi, T. & Schnabel, W., *J. Chem. Soc., Perkin Trans. 2* (1988) 961.
137. Baxter, J. E., Davidson, R. S., Hageman, H. J. & Overeem, T., *Makromol. Chem., Rapid Commun.*, **8** (1987) 311.
138. Baxter, J. E., Davidson, R. S. & Hageman, H. J., *Polymer*, **29** (1988) 1569.
139. Baxter, J. E., Davidson, R. S. & Hageman, H. J., *Eur. Polym. J.*, **24** (1988) 419.
140. Baxter, J. E., Davidson, R. S. & Hageman, H. J., *Eur. Polym. J.*, **24** (1988) 551.
141. Baxter, J. E., Davidson, R. S., Hageman, H. J. & Overeem, T., *Makromol. Chem.*, **189** (1988) 2769.
142. (a) Sumiyoshi, T. & Schnabel, W., *Makromol. Chem.*, **186** (1985) 1811; (b) Sumiyoshi, T., Schnabel, W. & Henne, A., *J. Photochem.*, **30** (1985) 63.
143. Baxter, J. E., Davidson, R. S., Walker, M. D. & Hageman, H. J., *J. Chem. Res. (S)*, (1988) 164.
144. (a) Ogata, Y. & Tomioka, H., *J. Org. Chem.*, **35** (1970) 596; (b) Terauchi, K. & Sakurai, H., *Bull. Chem. Soc. Jpn*, **43** (1970) 883.

145. Baxter, J. E., Davidson, R. S., Hageman, H. J. & Overeem, T., *Makromol. Chem.* (submitted).
146. (a) Walker, O. J. & Wild, G. L. E., *J. Chem. Soc.*, (1937) 1132; (b) Bevington, J. C. & Lewis, T. D., *Trans. Faraday Soc.*, **54** (1958) 1340; (c) Kochi, J. K. & Krusic, P. J., *J. Am. Chem. Soc.*, **91** (1969) 3940; (d) Sheldon, R. A. & Kochi, J. K., *J. Am. Chem. Soc.*, **92** (1970) 4395; (e) Bax, H. C., Budzinski, E. E. & Freund, H. G., *J. Am. Chem. Soc.*, **92** (1970) 5305.
147. (a) Walling, C. & Gibian, M. J., *J. Am. Chem. Soc.*, **87** (1965) 3413; (b) Kaptein, R., den Hollander, J. A., Antheunius, D. & Oosterhoff, L. J., *Chem. Commun.*, (1970) 1687; (c) Fahrenholz, S. R. & Trozzolo, A. M., *J. Am. Chem. Soc.*, **93** (1971) 251.
148. Smith, W. F., *Tetrahedron*, **25** (1969) 2071.
149. Leffler, J. E. & Miley, J. W., *J. Am. Chem. Soc.*, **93** (1971) 7005.
150. (a) Thijs, L., Gupta, S. N. & Neckers, D. C., *J. Org. Chem.*, **44** (1979) 4123; (b) Gupta, S. N., Gupta, I. & Neckers, D. C., *J. Polym. Sci., Polym. Chem. Ed.*, **19** (1981) 103; (c) Gupta, S. N., Thijs, L. & Neckers, D. C., *J. Polym. Sci., Polym. Chem. Ed.*, **19** (1981) 855; (d) Gupta, I., Gupta, S. N. & Neckers, D. C., *J. Polym. Sci., Polym. Chem. Ed.*, **20** (1982) 147; (e) Abu-Abdoun, I. I., Thijs, L. & Neckers, D. C., *J. Polym. Sci., Polym. Chem. Ed.*, **21** (1983) 3129; (f) Neckers, D. C., *J. Radiat. Curing*, **10**(2) (1983) 19.
151. (a) Martens, J. & Praefcke, K., *Chem. Ber.*, **107** (1974) 2319; (b) Bak, C., Praefcke, K., Muszkat, K. A. & Weinstein, M. *Naturforsch*, **326** (1977) 674.
152. Ogata, Y., Tagaki, K. & Takayanagi, Y., *J. Chem. Soc., Perkin Trans. 2* (1973) 1244.
153. For reviews see: (a) Ref. 37, Ch. 10; (b) Wagner, P. J., *Fortschr. Chem. Forsch.*, **66** (1976) 1.
154. Kuhlmann, R. & Schnabel, W., *Polymer*, **17** (1976) 419.
155. Merlin, A., Lougnot, D. J. & Fouassier, J. P., *Polymer Bull.*, **2** (1980) 847.
156. Cohen, S.G., Parola, A. H. & Parsons, G. H., *Chem. Rev.*, **73** (1973) 141.
157. Davidson, R. S. In *Molecular Association*, vol. 1, ed. R. Foster, Academic Press, London, 1975, p. 216.
158. Davidson, R. S. In *Advances in Physical Organic Chemistry*, Vol. 19, ed. D. Bethel & V. Gold. Academic Press, London, 1983, p. 1.
159. Warmser, C. C., Hammond, G. S., Chang, C. T. & Baylor, C., *J. Am. Chem. Soc.*, **92** (1970) 6362.
160. Davidson, R. S., Lambeth, P. F., McKellar, J. F., Turner, P. H. & Wilson, R., *Chem. Commun.*, (1969) 732.
161. Bartholomew, R. F., Davidson, R. S., Lambeth, P. F., McKellar, J. F. & Turner, P. H., *J. Chem. Soc., Perkin Trans. 2* (1972) 577.
162. Arimitsu, S. & Masuhara, S., *Chem. Phys. Lett.*, **22** (1973) 543.
163. Arimitsu, S., Masuhara, H., Mataga, N. & Tsubomura, H., *J. Phys. Chem.*, **79** (1975) 1255.
164. Merlin, A., Lougnot, D. J. & Fouassier, J. P., *Polym. Bull.*, **3** (1890) 1.
165. Leaver, I. H. & Ramsey, G. C., *Tetrahedron*, **25** (1969) 5669.

166. Roth, H. D. & Mannion, M. L., *J. Am. Chem. Soc.*, **97** (1975) 6886.
167. Thomas, M. J., Wagner, P. J., Mannion-Schilling, M. L. & Roth, H. D., *J. Am. Chem. Soc.*, **99** (1977) 3842.
168. Thomas, M. J. & Wagner, P. J., *J. Am. Chem. Soc.*, **99** (1977) 3845.
169. Inbar, S., Linschitz, H. & Cohen, S. G., *J. Am. Chem. Soc.*, **103** (1981) 1048.
170. Stone, P. G. & Cohen, S. G., *J. Am. Chem. Soc.*, **102** (1980) 5685.
171. Stone, P. G. & Cohen, S. G., *J. Am. Chem. Soc.*, **104** (1982) 3435.
172. Griller, D., Howard, J. A., Marriot, P. R. & Scaiano, J. C., *J. Am. Chem. Soc.*, **103** (1981) 619.
173. Davidson, R. S. & Goodin, J. W., *Eur. Polym. J.*, **18** (1982) 597.
174. Ledwith, A. & Purbrick, M. D., *Polymer,* **14** (1973) 521.
175. Ledwith, A., Bosley, J. A. & Purbrick, M. D., *J. Oil Col. Chem. Assoc.*, **61** (1978) 95.
176. Ghosh, P. & Ghosh, R., *Eur. Polym. J.*, **17** (1981) 545.
177. Ghosh, P. & Ghosh, R., *Eur. Polym. J.*, **17** (1981) 817.
178. Clarke, S. R. & Shanks, R. A., *J. Macromol. Sci., Chem. A,* **14** (1980) 69.
179. Kubota, H. & Ogiwara, Y., *J. Appl. Polym. Sci.*, **27** (1982) 2683.
180. Carlini, C., Ciardelli, F., Donati, D. & Gurzoni, F., *Polymer,* **24** (1983) 599.
181. Carlini, C., *Brit. Polym. J.*, **18** (1986) 236.
182. Carlini, C., Toniolo, L., Rolla, P. A., Barigelletti, F., Bortolus, P. & Flamigni, L., *New Polym. Mat.*, **1** (1987) 63.
183. Block, H., Ledwith, A. & Taylor, A. R., *Polymer,* **12** (1971) 271.
184. Hutchinson, J., Lambert, M. C. & Ledwith, A., *Polymer,* **14** (1973) 250.
185. Droste, W., Scharf, H. D. & Korte, F., *Liebigs Ann. Chem.*, **724** (1969) 71.
186. Amirzadeh, G. & Schnabel, W., *Makromol. Chem.*, **182** (1981) 2821.
187. Göthe, S., Dissertation, Royal Inst. Technol. Stockholm, 1982.
188. Sandner, M. R., Osborn, C. L. & Trecker, D. J., *J. Polym. Sci., Polym. Chem. Ed.*, **10** (1972) 3173.
189. Hageman, H. J., Overeem, T., Polman, R. J. & v.d. Werf, S., *Makromol. Chem.* (submitted).
190. Allen, N. S., Catalina, F., Green, P. M. & Green, W. A., *Eur. Polym. J.*, **22** (1986) 347.
191. Allen, N. S., Catalina, F., Moghaddam, B., Green, P. N. & Green, W. A., *Eur. Polym. J.*, **22** (1986) 691.
192. Allen, N. S., Catalina, F., Green, P. N. & Green, W. A., *Eur. Polym. J.*, **22** (1986) 793.
193. Allen, N. S., Catalina, F., Green, P. N. & Green, W. A., *Eur. Polym. J.*, **22** (1986) 871.
194. Davis, M. J., Gawne, G., Green, P. N. & Green, W. A., *Spec. Chem.*, **6**(4) (1986) 3.
195. Allen, N. S., Catalina, F., Green, P. N. & Green, W. A., *J. Photochem.*, **36** (1987) 99.
196. Allen, N. S., Catalina, F., Peinado, C., Sastre, R., Mateo, J. L. & Green, P. N., *Eur. Polym. J.*, **23** (1987) 985.

197. Davidson, R. S. & Orlon, S. P., *J. Chem. Soc., Chem. Commun.*, (1974) 209.
198. Specht, D. P., Martic, P. A. & Farid, S., *Tetrahedron,* **38** (1982) 1203.
199. Williams, J. R. L., Specht, D. P. & Farid, S., *Polym. Eng. Sci.*, **23**(18) (1983) 1022.
200. Bunbury, D. L. & Chan, T. M., *Can. J. Chem.*, **50** (1972) 2499.
201. Allen, N. S., Catalina, F., Green, P. N. & Green, W. A., *Eur. Polym. J.*, **22** (1986) 49.
202. Sengupta, P. K. & Modak, S. K., *J. Macromol. Sci., Chem. A,* **20** (1983) 789.
203. Sengupta, P. K. & Modak, S. K., *Makromol. Chem.*, **186** (1985) 1593.
204. (a) Muzycko, T. M. & Jones, T. H., US Patent 3 888 671 (1972); (b) Via, F. A., Ger. Offen. 2 639 742 (1976).
205. Leermakers, P. A. & Vesley, G. F., *J. Am. Chem. Soc.*, **85** (1963) 3776.
206. Kendall, D. S. & Leermakers, P. A., *J. Am. Chem. Soc.*, **88** (1966) 2766.
207. Closs, G. L. & Miller, R. J., *J. Am. Chem. Soc.*, **100** (1978) 3483.
208. Schönberg, A., Latif, N., Moubasher, R. & Sina, A., *J. Chem. Soc.*, (1951) 1364.
209. (a) Fujisawa, T., Monroe, B. M. & Hammond, G. S., *J. Am. Chem. Soc.*, **92** (1970) 542; (b) Fujisawa, T. & Hammond, G. S., *J. Am. Chem. Soc.*, **94** (1972) 4175.
210. Leermakers, P. A., Warren, P. C. & Vesley, G. F., *J. Am. Chem. Soc.*, **86** (1964) 1768.
211. (a) Davidson, R. S. & Goodwin, D., *J. Chem. Soc., Perkin Trans. 2* (1982) 993; (b) Encinas, M. V., Lissi, E. A., Zanocco, A., Stewart, L. C. & Scaiano, J. C., *Can. J. Chem.*, **62** (1984) 386.
212. Huyser, E. S. & Neckers, D. C., *J. Org. Chem.*, **29** (1964) 276.
213. Pappas, S. P., Pappas, B. C., Okamoto, Y. & Sakamoto, H., *J. Org. Chem.*, **53** (1988) 4404.
214. Maillard, B., Ingold, K. U. & Scaiano, J. C., *J. Am. Chem. Soc.*, **105** (1983) 5095.
215. McGinniss, V. D., Provder, T., Kuo, C. & Gallopo, A., *Macromolecules,* **11** (1978) 405.
216. Wright, F. R., *J. Polym. Sci., Polym. Lett. Ed.*, **16** (1978) 127.
217. (a) Adam, S., Güsten, H. & Schulte-Frohlinde, D., *Tetrahedron,* **30** (1974) 4249; cf. also (b) Bousquet, J. A., Fouassier, J. P. & Faure, J., *Polym. Bull.*, **1** (1978) 233.
218. Decker, C. & Jenkins, A. D., *Macromolecules,* **18** (1985) 1241.
219. v. Neerbos, A., *J. Oil Col. Chem. Assoc.*, **61** (1978) 241.
220. Rubin, H., *J. Paint Technol.*, **46**(588) (1974) 74.
221. Bolon, D. A. & Webb, K. K., *J. Appl. Polym. Sci.*, **22** (1978) 2543.
222. Hult, A. & Ränby, B., *Am. Chem. Soc. Polym. Prepr.*, **25**(1) (1984) 329.
223. (a) Gruber, G. W., US Patent 4 017 652 (1977); (b) Gruber, G. W., US Patent 4 024 296 (1977).
224. (a) Behrens, B. & Delius, H., US Patent 3 699 022 (1972); (b) Finelli, A. F., US Patent 4 221 646 (1980).
225. Mao, T. J. & Eldred, R. J. *J. Polym. Sci. A-1,* **5** (1967) 1741.

226. Kharasch, M. S., Mosher, R. A. & Bengelsdorf, I. S., *J. Org. Chem.*, **25** (1960) 1000.
227. (a) Gush, D. P. & Ketley, A. D., *Mod. Paint Coat.*, **1978**(11), 58; (b) Morgan, C. R. & Ketley, A. D., *J. Radiat. Curing*, **7**(2) (1980) 10; (c) Morgan, C. R. & Ketley, A. D., *J. Polym. Sci., Polym. Lett. Ed.*, **16** (1978) 75; (d) Morgan, C. R. & Kyle, D. R., *J. Radiat. Curing*, **10**(3) (1983) 4.
228. Baxter, J. E., Davidson, R. S., de Boer, M. A. U., Hageman, H. J. & v. Woerkom, P. C. M., *Eur. Polym. J.*, **24** (1988) 819.
229. (a) Bartholomew, R. F. & Davidson, R. S., *J. Chem. Soc. (C)*, (1971) 2342; (b) Bartholomew, R. F. & Davidson, R. S., *J. Chem. Soc. (C)*, (1971) 2347; (c) Bartholomew, R. F., Davidson, R. S. & Howell, M. J., *J. Chem. Soc. (C)*, (1971) 2804.
230. Hoyle, C. E. & Kim, K. J., *J. Radiat. Curing*, **12**(4) (1985) 9.
231. Hoyle, C. E. & Kim, K. J., *J. Appl. Polym. Sci.*, **33** (1987) 2985.
232. Hoyle, C. E., Keel, M. & Kim, K. J., *Polymer*, **29** (1988) 18.
233. Phan, X. T., *J. Radiat. Curing*, **13**(1) (1986) 23.
234. (a) Decker, C., *Makromol. Chem.*, **180** (1979) 2027; (b) Decker, C., Faure, J., Fizet, M. & Rychla, L., *Photogr. Sci. Eng.*, **23** (1979) 137.
235. Ramamurthy, V., *Tetrahedron*, **42** (1986) 5753.
236. Scaiano, J. C., Johnston, L. J., McGimpsey, W. G. & Weir, D., *Acc. Chem. Res.*, **21** (1988) 22.

Chapter 2

PHOTOINITIATED CATIONIC POLYMERIZATION

S. PETER PAPPAS

Polymers and Coatings Department, North Dakota State University, Fargo, USA

1 INTRODUCTION

Photoinitiated cationic polymerization has not as yet achieved the commercial significance of radical polymerization in important photopolymerization processes, including UV curing and photoimaging. Several reasons for this may be advanced, including: (1) the development of photoinitiators for cationic polymerization was preceded by substantial advances in UV curing technology based on radical polymerization; (2) the early developmental work on UV curing by cationic polymerization utilized aryl diazonium salts as photoinitiators which, while highly photoactive, are thermally unstable, thereby preventing long-term storage of fully formulated compositions; and (3) the discovery of thermally stable onium salts, such as diaryliodonium and triarylsulfonium salts, as effective photoinitiators for cationic polymerization of epoxy-functional resins was made almost simultaneously by several industrial groups, which resulted in an unclear patent situation—a situation which appears, at least in part, to be resolved at present.

Nevertheless, during the past decade, significant advances have been made in the technology of UV curing and photoimaging based on cationic polymerization of epoxy-functional resins with onium salts,[1-3] including the beneficial effects of added polyols,[4,5] and more recently with mixed-ligand arene cyclopentadienyl Fe(II) salts.[6,7]

Although commercially less readily available at present, vinyl-ether functional resins, which tend to exhibit higher reactivity than epoxies, are gaining considerable interest for UV curing by cationic polymerization.[8,9]

2 PHOTOINITIATORS FOR CATIONIC POLYMERIZATION

An important factor which retarded the development of photoinitiated cationic polymerization is the tendency of cationic polymerization to be dominated by termination and chain transfer processes at ambient temperatures.[10] However, this tendency may be retarded by utilizing cationic initiators with highly non-nucleophilic anions, such as PF_6^-, AsF_6^- and SbF_6^-.

The low nucleophilicity of these anions reduces termination processes and allows cationic polymerization to proceed under ambient conditions. Furthermore, reaction of growing polymer cations (P^+) with such complex metal halide anions is expected to yield the corresponding Lewis acid, as shown in eqn (1). To the extent that the Lewis acid becomes involved in initiation of a new chain, eqn (1) represents a chain transfer reaction.

$$P^+ + PF_6^- \rightarrow P\text{—}F + PF_5 \quad (1)$$

Low termination rates in these systems also contribute to continued polymerization following light exposure, a catalytic activity which is not observed in photocrosslinking by radical polymerization. The important absence of air-inhibition further distinguishes cationic from radical polymerization.

Commercially important photoinitiators for crosslinking by cationic polymerization are onium salts[1,11] and mixed-ligand arene cyclopentadienyl metal salts[6] with complex metal halide anions. At present, epoxy-functional monomers and oligomers enjoy major utilization in both UV curing and photoimaging applications. In general for photoinitiated cationic polymerization of epoxides with both onium[1,12] and metal[6] salts, rates, molecular weights and percentage conversions increase in the order $BF_4^- \ll PF_6^- < AsF_6^- < SbF_6^-$. These results may be attributed, at least in part, to (1) increasing stability of the anions in the same order, which is expected to result in decreasing participation in eqn (1); (2) increasing initiating ability of the corresponding Lewis acids in the same order, i.e. $BF_3 < PF_6 < AsF_5 < SbF_5$; and (3) increasing acidity of the corresponding Brönsted acids in the same order.

2.1 Onium Salts

2.1.1 Aryldiazonium Salts

The first commercial system for photoinitiated cationic polymerization in UV curing and imaging applications utilized aryl diazonium salts of complex metal halide anions together with epoxy-functional monomers and oligomers.[2] Photolysis of these salts yields Lewis acids, as shown in eqn (2), which may initiate cationic polymerization of epoxides directly and/or produce protic (Brönsted) acids by reaction with adventitious (or purposefully added) hydroxyl substances, such as water or alcohols, as shown in eqn (3).

$$Ar{-}N_2^+PF_6^- \xrightarrow{h\nu} Ar{-}F + N_2 + PF_5 \qquad (2)$$

$$PF_5 + ROH \longrightarrow H^+PF_5(OR)^- \qquad (3)$$

The spectral response of diazonium salts may be varied throughout the near UV–visible range by appropriate substituents on the aryl ring.[13] Furthermore, both the absorptivity and thermal stability of the diazonium salts are enhanced by electron-donating groups (one of the few exceptions to Murphy's law). Nevertheless, the inherent thermal instability of aryl diazonium salts limits their practical utility as photoinitiators in crosslinking systems. In addition, the evolution of N_2 limits film thickness to *ca.* 15 μm or less due to formation of bubbles and pinholes in thicker films.

2.1.2 Diaryliodonium and Triarylsulfonium Salts

These deficiencies prompted investigations on alternative photoinitiators for cationic polymerization, which led to the development of diaryliodonium (**I**) and triarylsulfonium (**II**) salts with complex metal halide anions.[1,14] In contrast to aryl diazonium salts, the iodonium and sulfonium salts exhibit high thermal stability, as well as hydrolytic stability, and may be considered second-generation photoinitiators for crosslinking by cationic polymerization.

$$\underset{\textbf{I}}{Ar_2I^+X^-} \qquad \underset{\textbf{II}}{Ar_3S^+X^-}$$

$$X^- = PF_6^-, AsF_6^- \text{ and } SbF_6^-$$

Irradiation of these salts results in homolytic cleavage of Ar–I (or Ar–S) bonds to yield reactive radical cations, as shown in eqns (4) and (5), respectively.

$$Ar_2I^+X^- \rightarrow ArI\cdot^+X^- + Ar\cdot \qquad (4)$$

$$Ar_3S^+X^- \rightarrow Ar_2S\cdot^+X^- + Ar\cdot \qquad (5)$$

Direct evidence for this homolytic bond cleavage has been obtained from laser flash photolysis studies.[15] Furthermore, the radical cations have been found to be highly reactive with nucleophiles, including cyclohexene oxide and vinyl ethers.[16,17]

The phenyliodinium radical cation ($PhI\cdot^+$) was found to be >10 times more reactive with cyclohexene oxide relative to other oxygen nucleophiles, including methanol, tetrahydrofuran (THF) and water. These results provide experimental support for direct initiation of cyclohexene oxide polymerization by $PhI\cdot^+$ even in the presence of other oxygen nucleophiles as well as hydrogen-donors (e.g. THF).

A possible reaction mechanism for photoinitiated polymerization of epoxides utilizing a diphenyliodonium salt **I** as photoinitiator, which is consistent with the laser flash photolysis studies as well as with polymer end-group analysis studies (see below), is provided in Scheme 1. (Note that the anion is not included.)

Key features of this scheme are (1) direct reaction of $PhI\cdot^+$ with the epoxide, (2) oxidation of the resulting adduct by a second diphenyliodonium salt, and (3) initiation of polymerization by both the resulting proton (H^+) and activated epoxide. This mechanism predicts both hydroxyl and 4-alkoxyphenyl iodide polymer end groups.

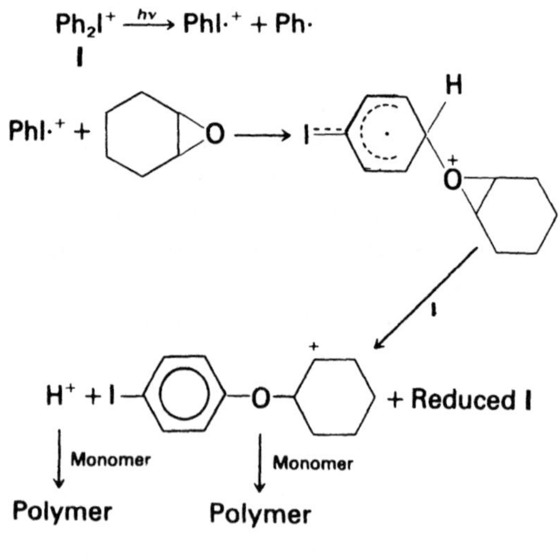

SCHEME 1

Support for photoinitiator fragment incorporation comes from studies with the triarylsulfonium salt **III**.[18] Utilizing **III** as photoinitiator for cationic polymerization of cyclohexene oxide, spectroscopic evidence (absorption and emission) has been obtained for the incorporation of 4-alkoxydiphenyl sulfide end groups in poly(cyclohexene oxide). These results are readily explicable in terms of initiation by the diphenylsulfinium radical cation ($Ph_2S\cdot^+$) in accordance with Scheme 1.

Furthermore, direct evidence for formation of $Ph_2S\cdot^+$ from irradiation of **III**, as shown in eqn (6), has been obtained from laser flash photolysis studies.[15] Thus, the laser flash photolysis and end-group analysis studies are complementary and support direct initiation of cationic polymerization (of cyclohexene oxide) by the reactive phenyliodinium and diphenylsulfinium ions, derived from the corresponding iodonium and sulfonium salt photoinitiators, in accordance with Scheme 1.

$$\underset{\textbf{III}}{Ph_2S^+\text{—}Ar\text{—}SPh} \rightarrow Ph_2S\cdot^+ + Ph\text{—}S\text{—}Ar\cdot \qquad (6)$$

(Ar represents 1,4-disubstituted benzene)

In the absence of reactive monomer, photogenerated $ArI\cdot^+$ and $Ar_2S\cdot^+$ are expected to react with nucleophiles (NuH) and possibly H-donors (RH) to produce protons, as shown in Scheme 2. Alternatively, protons may be generated directly by recombination of photogenerated $Ar_2S\cdot^+$ (or $ArI\cdot^+$) and $Ar\cdot$ at the ring positions of the sulfinium (iodinium) salts, probably within a solvent cage, as shown in Scheme 2 for triarylsulfonium salts. Evidence for the importance of such recombination in the photolysis of triphenylsulfonium salts has been reported recently.[19] Of course, recombination at the heteroatom (S or I) regenerates the starting salt and, thereby, constitutes an energy wastage process.

$$ArI\cdot^+(Ar_2S\cdot^+) + NuH \rightarrow H^+ + \text{unidentified products}$$

$$ArI\cdot^+(Ar_2S\cdot^+) + RH \rightarrow H^+ + ArI(Ar_2S) + R\cdot$$

$$Ar_3S^+ \xrightarrow{h\nu} [Ar_2S\cdot^+ \cdots Ar\cdot] \longrightarrow Ar\text{—}S\text{—}Ar'\text{—}Ar + H^+$$

(Ar′ represents disubstituted benzenes)

SCHEME 2

The reactions in Scheme 2 account for polymerization of monomer introduced after irradiation,[1,11] as well as for photoinduced acid-catalysed reactions utilized in photoimaging applications.[20,21]

Absorptivity of triarylsulfonium salts is enhanced in the 300–400 nm region by the introduction of thiophenoxy substituents, as exemplified by 4-diphenylsulfonio-diphenylsulfide (**III**)[22] and 4,4′-bis-(diphenylsulfonio)-diphenylsulfide (**IV**).[23] Commercially available triarylsulfonium salts are generally mixtures, including **III** and **IV** with $X = PF_6$ or SbF_6.

$$Ph_2\overset{+}{S}\text{—Ar—S—Ar—}\overset{+}{S}Ph_2X^-$$

IV

(Ar represents 1,4-disubstituted benzene)

In addition to the proposals for photogeneration of cationic initiators in Schemes 1 and 2, both of which originate with homolytic bond cleavage, evidence for photoinduced heterolytic bond cleavage of diaryliodonium and triarylsulfonium salts in methanol has been reported,[24] as shown in Scheme 3. However, the reported substantial yields of fluorobenzene appear to be inconsistent with earlier studies on the photochemistry of diaryliodonium[25] and triarylsulfonium salts,[26] carried out under similar conditions. Possibly, fluorobenzene arises from secondary photochemical reactions.

$$Ph_2I^+(Ph_3S^+)PF_6^- \xrightarrow{h\nu} PhI(Ph_2S) + Ph^+PF_6^-$$

$$Ph^+PF_6^- \longrightarrow PhF + PF_5$$

SCHEME 3

2.1.3 Phenacylsulfonium and 4-Hydroxyphenylsulfonium Salts

Dialkyl phenacylsulfonium[27] and 4-hydroxyphenylsulfonium[28] salts, such as **V** and **VI**, respectively, also function as photoinitiators for cationic polymerization. However, in contrast to triarylsulfonium salts, polymerization is reported to cease following light exposure. Furthermore, polymerization does not occur when **V** and **VI** are irradiated in the absence of monomer followed by monomer addition. These results, together with additional evidence (including deuterium exchange studies) have been interpreted in terms of reversible photogeneration of acid, as shown in eqns (7) and (8).

$$Ph\overset{O}{\overset{\|}{C}}CH_2\overset{+}{S}Me_2PF_6^- \xrightarrow{h\nu} Ph\overset{O}{\overset{\|}{C}}CH{=}SMe_2 + H^+PF_6^- \qquad (7)$$

V

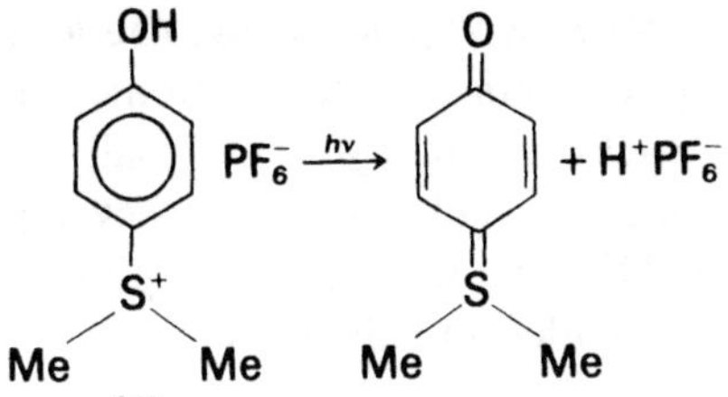

VI (8)

In the case of phenacylsulfonium salts, such as **V**, ylide ($R_2C{=}SR_2$) formation proceeds by intramolecular hydrogen-abstraction by the ketone from the γ-carbon.[27] In both cases, monomer can apparently compete for protons with the highly basic ylide products. However, the lifetime of the protons is reduced in the presence of the basic ylides.

Reversible photogeneration of acid may be advantageous in certain applications where continued polymerization following light exposure is undesirable. A possible example is photoresists where reversible photogeneration of acid may minimize migration into unexposed areas and thereby provide sharper relief images. However, lower lifetime of the protons is expected to reduce the efficiency (sensitivity) of photoinitiated polymerization.

2.1.4 *Sulfoxonium Salts*

Aryloxysulfoxonium salts of general structure **VII** are reported[29] to be efficient photoinitiators for cationic polymerization with activities comparable with or greater than diaryliodonium and triarylsulfonium salts. They apparently undergo irreversible photolysis and acid generation.

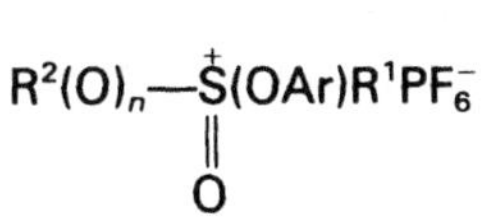

VII ($n = 0$ or 1)

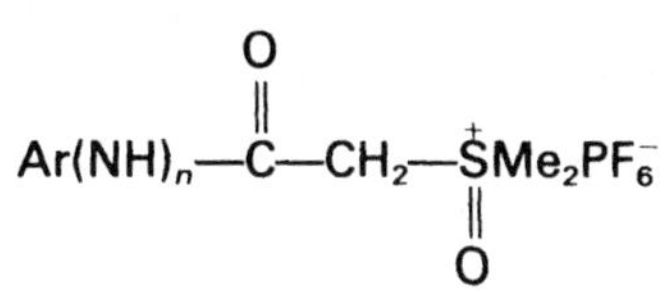

VIII ($n = 0$ or 1)

Sulfoxonium salts of general structure **VIII** are also reported to be efficient photoinitiators for cationic polymerization.[29] With these photoinitiators, cationic polymerization is reported to cease following light exposure, as observed with phenacyl and 4-hydroxyphenylsulfonium salts (see above).

2.2 Mixed-Ligand Arene Cyclopentadienyl Metal Salts

A newer class of photoinitiators for cationic polymerization are mixed-ligand arene cyclopentadienyl metal salts of complex metal halide anions,[30,31] such as (η^6-benzene)(η^5-cyclopentadienyl)Fe(II) hexafluorophosphate (**IX**). A mechanism for photoinitiated polymerization of epoxides is provided in Scheme 4.[6,32]

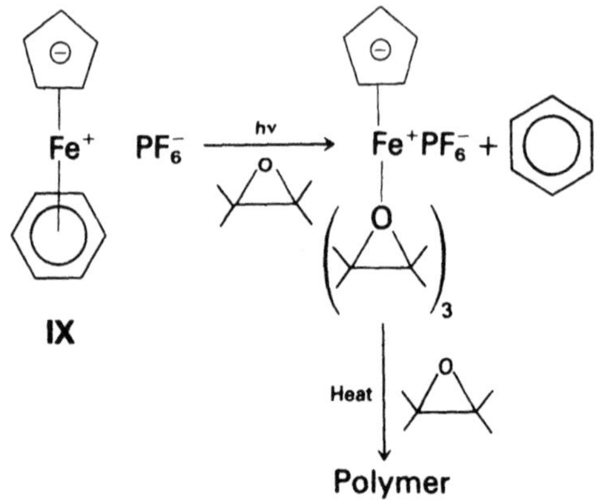

SCHEME 4

The proposed photoinduced ligand exchange of the arene by three epoxide groups is based on earlier ligand exchange studies.[33] With bicyclic epoxides, such as cyclohexene oxide, polymerization is spontaneous under ambient conditions. However, moderate heating (50–150°C) is required with less reactive epoxides, such as glycidyl ethers. The thermal requirement may be reduced substantially in the presence of oxidizing agents, such as peroxides, which apparently oxidize Fe(II) to the more reactive Fe(III) state. It is noteworthy, in this regard, that photoinitiated polymerization of epoxides, utilizing triarylsulfonium salts, is also enhanced by moderate heating, which also reduces the adverse effects of high humidity.[2]

The spectral response of the mixed-ligand arene cyclopentadienyl metal salts may be varied throughout the near-UV–visible region by proper selection of the arene component and also by photosensitization.[34]

A comparative study on UV curing of epoxy-functional resins by triarylsulfonium salts, both **III** and **IV** ($X = PF_6$), and (η^6-cumene) (η^5-cyclopentadienyl) Fe(II) PF_6^- has been reported.[35]

3 PHOTOSENSITIZATION OF CATIONIC POLYMERIZATION

The spectral response of onium salts[1,11] as well as mixed-ligand arene cyclopentadienyl metal salts[34] can be extended to longer wavelengths by photosensitization.

In the broadest sense, a photosensitizer sensitizes or activates a system toward light regardless of the mechanism. More narrowly, a photosensitizer is regarded as an energy transfer agent which absorbs light, transfers the energy to an acceptor and reverts to its original state. In the first reported mechanistic study on photosensitization of onium salts for cationic polymerization, a broadened definition of photosensitization was adopted, including electron transfer as well as exciplex-induced bond cleavage.[36] In common with energy transfer, these processes may also result, ultimately, in regeneration of the photosensitizer or at least its basic chromophoric group. This broadened definition of photosensitization, including all processes which may proceed catalytically, is also adopted herein.

3.1 Onium Salts

Representative photosensitizers for cationic polymerization with diaryliodonium (Ar_2I^+), triarylsulfonium (Ar_3S^+), dialkyl phenacylsulfonium (DPS^+) and dialkyl 4-hydroxyphenylsulfonium (DHS^+) salts are provided in Table 1. The results were obtained from photoinduced polymerization of 3,4-epoxycyclohexylmethyl-3′,4′-epoxycyclohexane carboxylate (**X**), as evaluated by photogelation[11] and photocalorimetry.[1]

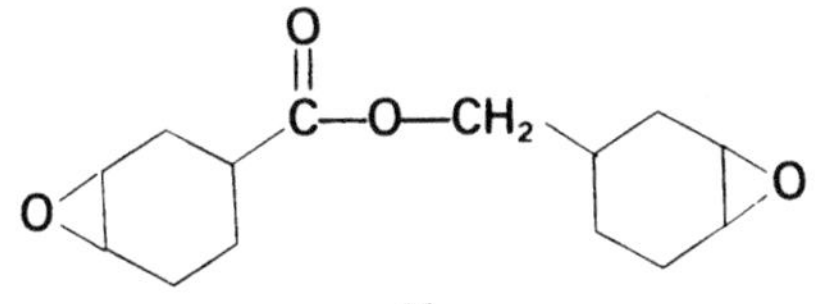

X

3.1.1 Triarylsulfonium Salts

Parallel studies on (1) photosensitized polymerization of di-epoxide **X**, (2) quantum yields of photosensitized photolysis, and (3) luminescence quenching experiments (to identify the photosensitizer excited states) with triphenylsulfonium hexafluoroarsenate, coupled with calculations on the thermodynamic favorability of energy and electron transfer,

TABLE 1
Photosensitizer/Onium Salt Combinations for Cationic Polymerization of Di-epoxide 10[a]

Photosensitizer	Ar_2I^{+} [b]	Ar_3S^{+} [b]	DPS^{+} [c]	DHS^{+} [c]
Anthracene	Yes	Yes	Yes	No
Perylene	Yes	Yes	Yes	No
Phenothiazine	Yes	Yes	Yes	No
Michler's ketone	Yes	Yes	[d]	[d]
Xanthone	Yes	No	[d]	[d]
Thioxanthone	Yes	No	Yes	Yes
Benzophenone	Yes[e]	No	No	Yes
Acetophenone	Yes[e]	No	No	Yes
Fluorenone	[f]	[g]	[h]	Yes
9,10-Anthraquinone	[f]	[g]	[h]	Yes

[a] Yes signifies that polymerization was observed.
[b] Ref. 37.
[c] Ref. 45.
[d] Not reported; predicted to occur.
[e] Requires a hydrogen-donor.
[f] Not reported; predicted to occur with a hydrogen-donor.
[g] Not reported; predicted not to occur.
[h] Not reported; predicted to be borderline with a hydrogen-donor.

have provided strong evidence for the electron transfer mechanism, as shown in Scheme 5.[37]

$$PS \xrightarrow{h\nu} PS^*$$

$$PS^* + Ar_3S^+(Ar_2I^+) \rightarrow \underset{\text{Exciplex}}{(PS \cdots Ar_3S^+)^*} \text{ or } \underset{\text{Exciplex}}{(PS \cdots Ar_2I^+)^*}$$

$$\text{Exciplex} \rightarrow PS\cdot^+ + Ar_2S(ArI) + Ar\cdot \qquad (9)$$

$$PS\cdot^+ + \text{Monomer} \rightarrow \text{Polymer}$$

SCHEME 5

The electron transfer process (via the exciplex), eqn (9), is shown to yield Ar_2S and $Ar\cdot$, directly, in accordance with recent studies which provide evidence for concomitant bond cleavage with one-electron reduction of arylalkylsulfonium salts.[38] The reported irreversible potential for reduction of triphenylsulfonium ion is −1·2 V (vs saturated calomel electrode, SCE).[39] The accompanying photosensitizer radical cation ($PS\cdot^+$) is expected to initiate cationic polymerization directly, as shown, or indirectly by interaction with nucleophiles (or hydrogen-donors) to yield Brönsted acids, which initiate polymerization.

Photosensitization by electron transfer is further supported by spectroscopic evidence for the formation of *N*-ethylphenothiazine radical cation on irradiation of *N*-ethylphenothiazine in the presence of triarylsulfonium (as well as diaryliodonium) salts. Furthermore, phenothiazine and *N*-ethylphenothiazine radical cations are reported to initiate cationic polymerization.[40]

The participating excited states of the photosensitizers were determined by quenching of their emission with triphenylsulfonium hexafluoroarsenate. In the case of anthracene, perylene, phenothiazine and Michler's ketone, the observed quenching is thermodynamically favorable by electron transfer, but not by energy transfer. Thus, photosensitized polymerization of di-epoxide **X** (see Table 1) and sensitized photolysis of the sulfonium salt may be attributed to electron transfer, as shown in Scheme 5.

The triplet excited states $^3(PS)^*$ of xanthone and acetophenone were also quenched by the sulfonium salt, for which electron transfer is thermodynamically unfavorable. In these cases, quenching may be attributed to favorable triplet energy transfer, as shown in eqn (10). However, neither photosensitized polymerization of di-epoxide **X** (see Table 1) nor sensitized photolysis of the sulfonium salt is observed with xanthone or acetophenone, signifying that the resulting excited triplet state of Ph_3S^+ does not participate in the photogeneration of reactive ions, or does so inefficiently.

$$^3(PS)^* + Ph_3S^+ \rightarrow PS + {}^3(Ph_3S^+)^* \quad (10)$$

These results also signify that excited singlet states of Ph_3S^+ are primarily responsible for photoinitiator activity in direct irradiation, as well, and that intersystem crossing to triplet states represents an energy wastage process.

These conclusions reasonably follow from the results with xanthone and acetophenone. However, quenching of xanthone and acetophenone phosphorescence by the sulfonium salt is relatively inefficient, probably resulting from similar triplet excitation energies of the ketones and sulfonium salt. Consequently, these results do not preclude the possibility that more efficient photosensitized photolysis of triphenylsulfonium salts may be observed with higher-energy (or longer-lived) triplet photosensitizers.

An alternative mechanism for photosensitization, which may be considered, is exciplex-induced bond cleavage, shown in Scheme 6.

$$PS \xrightarrow{h\nu} PS^*$$

$$PS^* + Ph_3S^+ \longrightarrow \underset{\text{Exciplex}}{(PS \cdots Ph_3S^+)^*}$$

$$\text{Exciplex} \longrightarrow PS + Ph\cdot + Ph_2S\cdot^+ \qquad (11)$$

$$Ph_2S\cdot^+ + \text{Monomer} \longrightarrow \text{Polymer}$$

SCHEME 6

In this case, the shared excitation energy within the exciplex may be utilized to cleave the Ph–S bond of the sulfonium salt, as shown in eqn (11), to yield the reactive $Ph_2S\cdot^+$ initiating ion (which is also formed by direct irradiation of the sulfonium salt). This mechanism requires sufficient energy in the exciplex for homolytic Ph–S bond cleavage. However, the apparent non-reactivity of triplet-state sulfonium salt with an excitation energy $>290\ \text{kJ mol}^{-1}$ (from emission studies) casts doubt on the likelihood of this mechanism for photosensitizers with lower excitation energies, including perylene and phenothiazine.[11]

A feature which clearly distinguishes the electron transfer mechanism (Scheme 5) from exciplex-induced bond cleavage (Scheme 6) is the contrasting role of the photosensitizer in generation of the active initiator. In the electron transfer mechanism, the reactive cationic species ($PS\cdot^+$) is derived from the photosensitizer; by contrast, in the bond-cleavage mechanism, the reactive cationic species ($Ar_2S\cdot^+$) is derived from the onium salt.

3.1.2 Diaryliodonium Salts

Photosensitization studies analogous to those discussed above with triphenylsulfonium hexafluoroarsenate were also carried out with diphenyliodonium hexafluoroarsenate.[37] The results support the electron transfer mechanism (Scheme 5) for all effective photosensitizers in Table 1, except benzophenone and acetophenone. In these cases, a hydrogen-donor is required and the results may be explained by an indirect electron transfer process, as outlined in Scheme 7 for benzophenone and 2-propanol.

$$Ph_2C{=}O \xrightarrow{h\nu} (Ph_2C{=}O)^*$$

$$(Ph_2C{=}O)^* + Me_2CH{-}OH \longrightarrow Ph_2\dot{C}{-}OH + Me_2\dot{C}{-}OH$$

$$Ph_2\dot{C}{-}OH + Ph_2I^+ \longrightarrow Ph_2\overset{+}{C}{-}OH + PhI + Ph\cdot \qquad (12)$$

$$Me_2\dot{C}{-}OH + Ph_2I^+ \longrightarrow Me_2\overset{+}{C}{-}OH + PhI + Ph\cdot \qquad (13)$$

$$Ph\cdot + Me_2CH{-}OH \longrightarrow Me_2\dot{C}{-}OH + Ph{-}H \qquad (14)$$

$$Ph_2\overset{+}{C}{-}OH \longrightarrow Ph_2C{=}O + H^+$$

$$Me_2\overset{+}{C}{-}OH \longrightarrow Me_2C{=}O + H^+$$

$$H^+ + \text{Monomer} \longrightarrow \text{Polymer}$$

SCHEME 7

Key features of this proposal are (1) photogeneration of ketyl radicals by hydrogen-abstraction, and (2) oxidation of the ketyl radicals by the iodonium salt to produce carbocations (protonated ketones), which dissociate to yield protons for initiation of cationic polymerization.

Reduction of the iodonium salt (by the ketyl radicals) is shown to occur with concomitant dissociation into PhI and Ph·, in accordance with electrochemical reduction of diphenyliodonium salts for which an irreversible potential of −0·2 V (vs SCE) is reported.[41]

Hydrogen-abstraction by phenyl radicals from 2-propanol, eqn (14), regenerates the corresponding ketyl radical. Thus, eqns (13) and (14) constitute a radical chain reaction leading ultimately to protons.

The proposal in Scheme 7 predicts that quantum yields of formation of protons (as well as PhI) may exceed one. That is, more protons may be produced than photons absorbed by benzophenone. This predicted amplification of photons in the formation of protons has been observed and evidence for a radical chain reaction has been obtained.[37]

Radicals produced photochemically with free-radical photoinitiators (as well as thermally) have also been shown to participate in the generation of reactive cations by the reduction of diaryliodonium[42] (as well as aryldiazonium)[42,43] salts, as illustrated in Scheme 8.

Triarylsulfonium salts do not participate efficiently in these processes (Schemes 7 and 8),[18,43] which undoubtedly reflects their less favorable reduction potentials. The free energy change (ΔG) for the redox reactions in Scheme 7, eqns (12) and (13), may be estimated from eqn (15), where multiplying by 97 converts the redox potentials

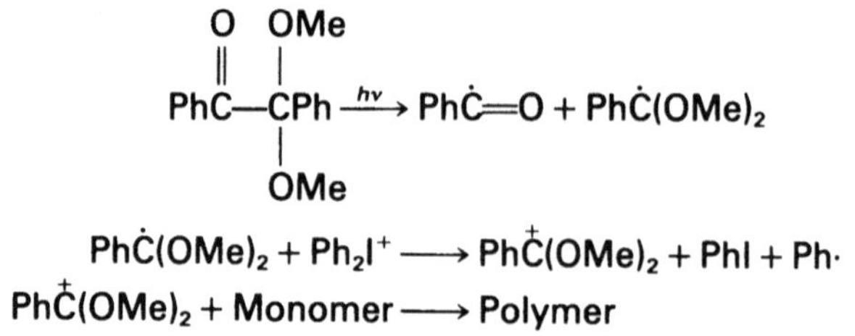

SCHEME 8

from volts into kJ mol^{-1}. Utilizing oxidation potentials ($E^{ox}_{1/2}$) of −0·8 and −1·0 V for acetone and benzophenone ketyl radicals, respectively,[44] together with reduction potentials ($E^{red}_{1/2}$) of −0·2 and −1·2 V for diphenyliodonium and triphenylsulfonium salts, respectively, ΔG values are seen to be highly favorable (exergonic) for reduction of the iodonium salt (−58 to −78 kJ mol^{-1}) and unfavorable for reduction of the sulfonium salt (+19 to 39 kJ mol^{-1}).

$$\Delta G = 97[E^{ox}_{1/2} - E^{red}_{1/2}] \qquad (15)$$

3.1.3 Phenacylsulfonium Salts

Photosensitization of dialkyl phenacylsulfonium salts by anthracene, perylene, phenothiazine and thioxanthone (see Table 1) is reported to occur by the electron transfer mechanism, in accordance with Scheme 5, and spectroscopic evidence for the perylene radical cation has been obtained.[45] Irreversible reduction potentials of phenacylsulfonium salts are *ca.* −0·8 V (vs SCE),[38] intermediate between diaryliodonium and triarylsulfonium salts. The more favorable reduction potential for phenacylsulfonium salts accounts for photosensitization by thioxanthone, which is not observed with triarylsulfonium salts.[1]

Whereas direct irradiation of phenacylsulfonium salts results in reversible acid formation, as discussed in Section 2.1.3, photosensitization produces acid irreversibly. The photosensitization results are in accord with electron transfer, which is expected to occur irreversibly, as shown in eqn (16).

(16)

3.1.4 4-Hydroxyphenylsulfonium Salts

Photosensitization of dialkyl 4-hydroxyphenylsulfonium salts by thioxanthone, benzophenone, fluorenone and anthraquinone (see Table 1) has been attributed to hydrogen-abstraction from the phenol group followed by electron transfer between the resulting ketyl radical and phenol radical cation.[45] This mechanism is outlined in Scheme 9, in which the excited-state photosensitizer is represented by $(Ar_2C{=}O)^*$.

Photosensitization by aromatic ketones results in reversible acid formation, as observed in direct irradiation of hydroxyphenylsulfonium salts (see Section 2.1.3). Reversible acid generation is consistent with Scheme 9, since formation of the highly basic ylide is proposed, which is expected to limit the lifetime of protons; see eqn (17) in Scheme 9.

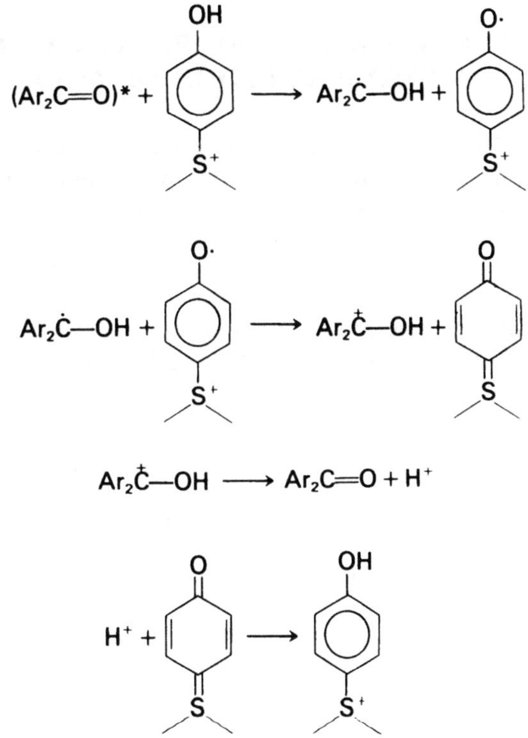

(17)

SCHEME 9

Polynuclear aromatic hydrocarbons are reported to be inactive as photosensitizers for 4-hydroxyphenylsulfonium salts,[45] in contrast to their effectiveness with phenacylsulfonium as well as with triarylsulfonium salts (see Table 1). The lack of photosensitization by polynuclear aromatics, such as anthracene and perylene, which should be capable of undergoing electron transfer to 4-hydroxyphenylsulfonium salts in accordance with Scheme 5, is an interesting observation, which does not appear to have an obvious explanation.

3.2 Mixed-Ligand Arene Cyclopentadienyl Metal Salts

Photosensitizers, including anthracene and anthrone,[46] have been claimed for photoinitiated cationic polymerization with mixed-ligand arene cyclopentadienyl metal salts. Studies on the mechanism of photosensitization have, apparently, not been reported.

4 REVERSIBLE AND IRREVERSIBLE PHOTOGENERATION OF ACID

Direct and sensitized photolyses of onium salts provide versatile systems for reversible and irreversible photogeneration of cationic initiators. The systems are summarized in Table 2.

TABLE 2
Reversible and Irreversible Photogeneration of Cationic Initiators by Direct and Sensitized Photolysis of Onium Salts

Onium salt	*Direct irradiation*	*Photosensitization*
Diaryliodonium	Irreversible	Irreversible
Triarylsulfonium	Irreversible	Irreversible
Phenacylsulfonium	Reversible	Irreversible
4-Hydroxyphenylsulfonium	Reversible	Reversible

5 CONCURRENT RADICAL–CATIONIC POLYMERIZATION: HYBRID SYSTEMS

Direct and sensitized photolyses of diaryliodonium and triarylsulfonium salts, as well as sensitized photolysis of phenacylsulfonium salts, yield both cationic and radical initiators. Accordingly, triarylsulfonium salt **III** has been utilized for photoinitiated polymerization of mixed acrylate- and epoxy-functional resins, as well as mixed acrylate- and vinyl ether-functional resins.[8] Suppression of air inhibition relative to radical polymerization of acrylated resins was observed.

Enhanced cure rates and cure depths[47] as well as desirable application and film properties[48] have been reported for photocrosslinking mixtures of acrylate- and epoxy-functional resins utilizing mixtures of photoinitiators for radical and cationic polymerization. In general, properly selected photoinitiator combinations for concurrent radical–cationic polymerization are expected to provide enhanced light absorption and utilization. Furthermore, hybrid systems are expected to provide greater flexibility in monomer and oligomer utilization for achieving desirable application and film properties, such as viscosity control, hardness, flexibility, gloss and adhesion.

A related marriage of radical and cationic systems is probably involved in the cationic polymerization of vinyl ether-functional resins by electron beams (EB curing) in the presence of onium salts.[49] The results could be attributed to the conversion of EB-generated radicals into cations by reduction of onium salts along the lines of Scheme 8.

UV and EB curing represent important complementary technologies of radiation curing.[50] In contrast to light, electron beams may penetrate highly opaque media and may, therefore, be utilized to cure highly pigmented compositions, such as magnetic disks and tapes.[51]

REFERENCES

1. Crivello, J. V., *Adv. Polym. Sci.*, **62** (1984) 1.
2. Watt, W. R. In *UV Curing: Science and Technology*, Vol. 2, ed. S. P. Pappas. Technology Marketing Corp., Norwalk, CT, 1985, pp. 247–82.
3. Zopf, R. F., *Radiat. Curing*, **9**(4) (1982) 10.
4. Smith, G. H. & Olofson, P. M. (3M Co.), US Patent 4 173 476 (1979).
5. Crivello, J. V., Conlin, D. A., Olson, D. R. & Webb, K. K., *J. Radiat. Curing* **13**(4) (1986) 3.
6. Lohse, F. & Zweifel, H., *Adv. Polym. Sci.*, **78** (1986) 61.
7. Meier, K. & Zweifel, H., *J. Radiat. Curing*, **13**(4) (1986) 26.
8. Crivello, J. V., Lee, J. L. & Conlin, D. A., *J. Radiat. Curing*, **10**(1) (1983) 6.
9. Dougherty, J. A., Vara, F. J. & Anderson, L. R., *Radcure '86, Conf. Proc.*, Soc. Manuf. Eng., Dearborn, MI, 1986, pp. 15/1–15/13.
10. Ledwith, A. & Sherrington, D. C. In *Reactivity, Mechanism and Structure in Polymer Chemistry*, ed. A. D. Jenckins & A. Ledwith. Wiley–Interscience, New York, 1974, pp. 244–309.

11. Pappas, S. P., *Progr. Org. Coat.*, **13** (1985) 35.
12. Bal, T. S., Cox, A., Kemp, T. J. & Pinot de Moire, P., *Polymer,* **21** (1980) 423.
13. Schlesinger, S. I., *Photogr. Sci. Eng.*, **18** (1974) 387.
14. Crivello, J. V. In *UV Curing: Science and Technology,* ed. S. P. Pappas. Technology Marketing Corp., Norwalk, CT, 1978, pp. 23–77.
15. Pappas, S. P., Pappas, B. C., Gatechair, L. R., Jilek, J. H. & Schnabel, W., *Polym. Photochem.* **5** (1984) 1.
16. Pappas, S. P., Pappas, B. C., Gatechair, L. R. & Schnabel, W., *J. Polym. Sci., Polym. Chem. Ed.*, **22** (1984) 69.
17. Yagci, Y. & Schnabel, W., *Macromol. Chem., Macromol. Symp.*, **13/14** (1988) 161.
18. Ledwith, A., *22nd Fall Symp. Prepr., Photogr. Sci. Eng.*, Arlington, VA, Nov. 1982, pp. 44–5.
19. Dektar, J. L. & Hacker, N. P., *J. Chem. Soc., Chem. Commun.*, (1987) 1591.
20. Crivello, J. V. In *Polymers in Electronics,* ed. T. Davidson ACS Symp. Ser. **242,** American Chemical Society, Washington, DC, 1984, pp. 3–10.
21. Ito, H. & Willson, C. G. In *Polymers in Electronics,* ed. T. Davidson, ACS Symp. Ser. **242,** American Chemical Society, Washington, DC, 1984, pp. 11–23.
22. Crivello, J. V. & Lam, J. H. W., *J. Polym. Sci., Polym. Chem. Ed.*, **18** (1980) 2677, 2697.
23. Watt, W. R., Hoffman, H. T., Pobiner, H., Schkolnick, L. J. & Yang, L. S., *J. Polym. Sci., Polym. Chem. Ed.*, **22** (1984) 1789.
24. Davidson, R. S. & Goodin, J. W., *Eur. Polym. J.*, **18** (1982) 589.
25. Knapczyk, J. W., Lubinkowski, J. J. & McEwen, W. E., *Tetrahedron Lett.*, (1972) 3739.
26. Knapczyk, J. W. & McEwen, W. E., *J. Org. Chem.*, **35** (1970) 2539.
27. Crivello, J. V. & Lee, J. L., *Macromolecules,* **16** (1983) 864.
28. Crivello, J. V. & Lam, J. H. W., *J. Polym. Sci., Polym. Chem. Ed.*, **18** (1980) 1021.
29. Stark, B. P. & Irving, E., *Ind. Chem. Bull.*, **1** (1982) 164.
30. Meier, K., Buehler, N., Zweifel, H., Berner, G. & Lohse, F., (Ciba-Geigy AG), Eur. Pat. Appl. 94 915 (1983); *Chem. Abstr.* **100** (1984) 105 179.
31. Palazzotto, M. C. & Hendrickson, W. A., (3M Co.), Eur. Pat. Appl. 109 851 (1984); *Chem. Abstr.*, **101** (1985) 153 584.
32. Roloff, A., Meier, K. & Riediker, M., *Pure & Applied Chem.*, **58** (1986) 1267.
33. Gill, T. P. & Mann, K. R., *J. Organometal. Chem.*, **216** (1981) 65.
34. Meier, K. & Zweifel, H., *J. Imaging Sci.*, **30** (1986) 174.
35. Gaube, H. G., *Radcure '86,* Conf. Proc., Soc. Manuf. Eng., Dearborn, MI, 1986, pp. 15/27–15/42.
36. Pappas, S. P. & Jilek, J. H., *Photogr. Sci. Eng.*, **23** (1979) 140.
37. Pappas, S. P., Gatechair, L. R. & Jilek, J. H., *J. Polym. Sci., Polym. Chem. Ed.*, **22** (1984) 77.
38. Saeva, F. D. & Morgan, B. P., *J. Am. Chem. Soc.*, **106** (1984) 4121.

39. McKinney, P. S. & Rosenthal, S., *J. Electroanal. Chem.*, **16** (1968) 261.
40. Ledwith, A., *Am. Chem. Soc., Polym. Prepr.*, **23**(1) (1982) 323.
41. Bachofner, H. E., Beringer, F. M. and Meites, L., *J. Am. Chem. Soc.*, **80** (1958) 4269.
42. Ledwith, A., *Makromol. Chem. Suppl.*, **3** (1979) 348.
43. Baumann, H. and Timpe, H.-J., *J. Prakt. Chem.*, **326** (1984) 529.
44. Rao, P. S. & Hayon, E., *J. Am. Chem. Soc.*, **96** (1974) 1287.
45. Crivello, J. V. and Lee, J. L., *Macromolecules*, **14** (1981) 1141.
46. Meier, K. & Zweifel, H. (Ciba-Geigy AG), Eur. Pat. Appl. 152 377 (1985); *Chem. Abstr.*, **104** (1986) 59 420.
47. Tsao, J.-H. & Ketley, A. D. (W. R. Grace), US Patent 4 156 035 (1979).
48. Perkins, W. C., *J. Radiat. Curing*, **8**(1) (1981) 16.
49. Lapin, S. C., *Radcure '86*, Conf. Proc., Soc. Manuf. Eng., Dearborn, MI, 1986, pp. 15/15–15/26.
50. Senich, G. A. & Florin, R. E., *J. Macromol. Sci., Rev. Macromol. Chem. Phys.*, **C24** (1984) 239.
51. Rand, W. M., Jr, *J. Radiat. Curing* **11**(3) (1984) 18.

Chapter 3

THE CHEMISTRY OF PHOTORESISTS

S. RICHARD TURNER

Corporate Research Laboratories, Eastman Kodak Company, Rochester, New York, USA

&

ROBERT C. DALY

Electronics Research Laboratories, Eastman Kodak Company, Rochester, New York, USA

1 INTRODUCTION

Polymers and polymeric systems that can undergo imagewise light-induced reactions are of high research interest and technological importance. The process of forming images in photopolymers and then transferring these images into the substrate or to other materials is called photolithography. Traditionally, photolithography has had great commercial impact on the printing industry, photomachining of fine parts, and so on. More recently, the electronics revolution has been made possible by the photopolymers used to delineate the tiny features that make up modern integrated circuits. It is our objective in this chapter to discuss the chemistries of the photoresist polymers and polymeric systems used in photolithography. We will focus this chapter on microresists and microlithography for integrated circuit manufacture because of the high technological and scientific interest in this area and because of our known experiences and biases.

Photoresists are polymers or polymeric systems (polymer binders with dispersed or dissolved photoactive compounds present) which undergo photochemical reactions in the solid film that affect a chemical or physical change that can be developed, i.e. selective removal of either the exposed or unexposed area of the film. The images generated in such polymer films generally range in width from *c.* 0·5 μm (in some microresist work) up to *c.* 25 μm in photoprinting systems. Since these images are raised above the substrate on which the polymer has been coated, they are called relief images. In order for the image to be transferred to the substrate, the polymer must resist the chemical activity of the particular etchant which can be either a wet (solvent) or dry (plasma) process. Thus, these polymers are called 'photoresists'.

Most of the photoresists that will be discussed in this chapter are developed by wet processes. The exposed areas of the polymer film are rendered more or less soluble in a particular solvent called a developer. When the exposed areas of the film are more soluble than the unexposed areas in the developer, a positive image results (Fig. 1).

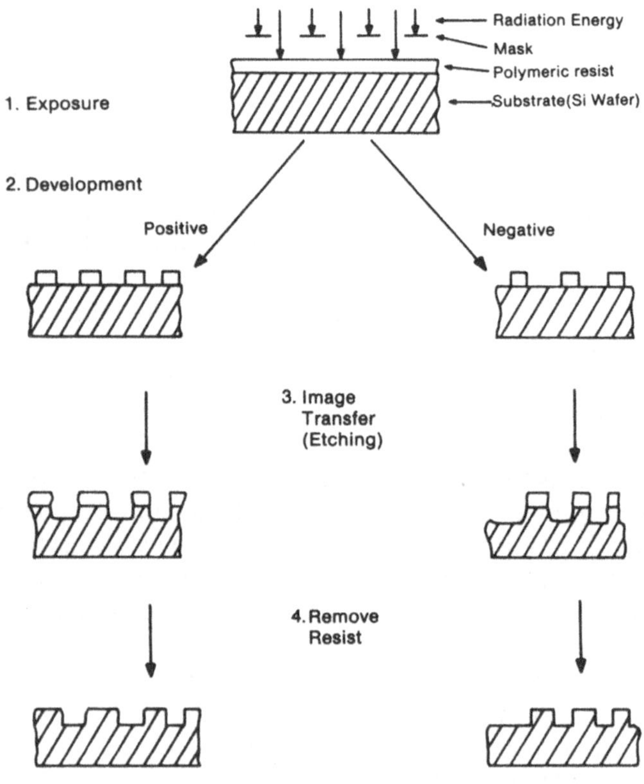

FIG. 1. Positive and negative modes for a photoresist film.

There are two classes of chemical reactions that can lead to enhanced dissolution of the exposed areas. The first is a backbone cleavage of the resist polymer such that the exposed area of the polymer film is degraded to a molecular weight range (or all the way back to monomer) that has a dissolution rate significantly faster than the rate for the unexposed higher-molecular-weight area in an appropriate developer. The second chemical reaction that can occur is a photofunctional change that converts the exposed area to a different polarity to the unexposed areas. Proper choice of the developer can then lead to the selective dissolution of the exposed areas.

When exposure produces an image which is less soluble in the developer, a negative image results. Negative images are generally produced by photochemical reactions that lead to crosslinking and network formation in the polymer film. Such network polymers are totally insoluble in the developer solvent. A second means of generating a negative image is by producing a photofunctional change which makes the exposed areas less soluble in a selective developer.

There are two important aspects of the performance of a photoresist: lithography (imaging) and image transfer (resistance). An image must be made on the substrate and then the image must be transferred permanently to the substrate by etching, deposition, etc. Whilst the requirements for resistance are varied and difficult to quantify, the lithographic parameters are easily measured and quantified. The most important lithographic parameters for a resist are its wavelength sensitivity, speed and contrast. Wavelength sensitivity is rarely measured by itself for microlithography because there is such a limited number of available exposure wavelengths and exposure sources. Speed or sensitivity is signified by the exposure dose required to yield a particular response in the resist coating after a development process. The speed point or sensitivity for a positive resist is the dose necessary to clean out the exposed areas completely during development, D_p^0. For a negative resist the most commonly defined speed point or sensitivity is the dose required to leave 50% of the original coating thickness in place after development and is represented by $D_n^{0.5}$. These points are shown in Fig. 2. The data for Fig. 2 were obtained by measuring the film thickness of the resist remaining after a series of exposures followed by a common development process. The film thickness remaining is then plotted against the log of the dose. The contrast, γ, of the photoresist is taken from this plot as the slope of the curve. In almost all applications, a high-contrast, high-speed photo-

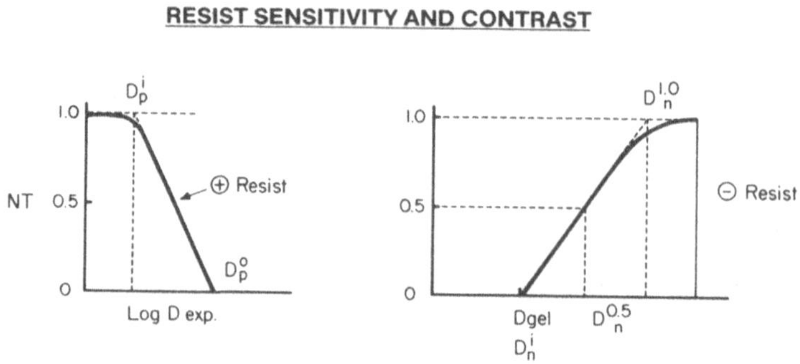

FIG. 2. Schematic plot of normalized thickness (NT) versus log exposure dose (D) for positive and negative photoresists.

resist is desired. Taylor has discussed these definitions and the need to characterize uniformly various photoresist systems.[1]

The microlithographic applications of photopolymers have been the driving force behind the majority of the research in photopolymers over the last few years. An examination of a plot of the decrease in device feature size (the smallest feature that must be resolved in the photoresist to produce the device) versus the year that the device was first manufactured indicates how fast this technology has progressed (Fig. 3). It is fair to say that photoresists have been an enabling component of this technology. This chapter will survey the variety of resist systems which have been reported for microelectronic applications. However, it should honestly be said that almost all of the devices which have ever been manufactured have used either of two resist systems. Negative working resists based on cyclized rubbers and bisazides were the original mainstays of this industry. Although these resists are still used today, for reasons that will be discussed in the text, novolac/naphthalene diazoquinone based positive photoresists are used for most features less than 3 μm in size. The major portion of microlithography is done at this size level today.

The drive to smaller image dimensions has forced two changes in the way that microresists are used. The first trend has been toward shorter wavelengths for the exposure radiation. To a very rough approximation, the resolution limit of an optical exposure source is given by 2λ, where λ is the wavelength. As the wavelength decreases, the potential resolution becomes smaller (better). This, of course, requires resists sensitive at these shorter wavelengths and light sources with sufficient

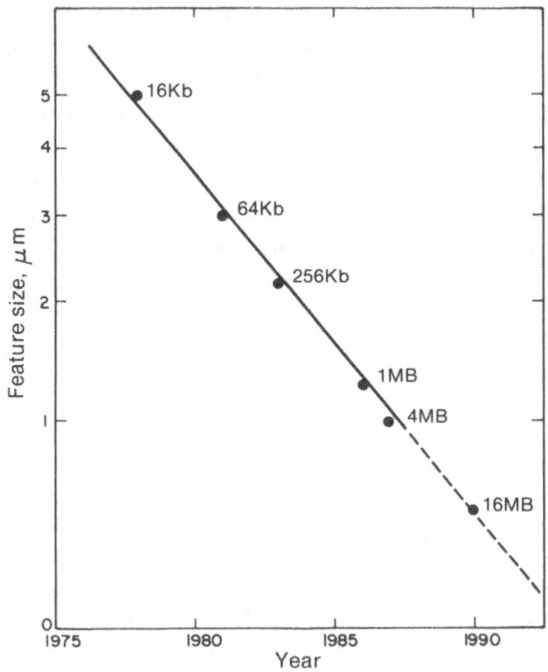

FIG. 3. Plot of device feature size versus approximate year of initial production.

output at these wavelengths. The other change in resist processing involves the use of multilayer schemes in place of a simple photoresist coating. These multilayer schemes serve to separate the lithographic function of the photoresist from the problems of topography, reflectance and chemical resistance. This chapter will discuss short wavelength resists specifically. Information concerning multilayer materials and technology can be found in several recent reviews.[2,3]

Only a very limited number of exposure wavelengths have been investigated in any detail because of the limited number of point-source, high-intensity lamps available. Almost every commercial exposure source uses a super-high-intensity Hg or Hg/Xe lamp. The high-intensity outputs are 436, 405, 365, 330 and 313 nm. The first three are normally grouped and called the near-UV (NUV). The second two wavelengths are called the mid-UV (MUV). Light from these lamps and other sources with wavelengths below 280 nm comprises the deep-UV (DUV). Only with the coming of excimer lasers have high-output DUV sources become readily available.

There are alternative high-energy exposure sources that have the demonstrated potential for high resolution. These include electron-beam (e-beam), X-ray, and ion-beam lithography. Each of these technologies has specific resist requirements and various polymeric resist systems have been developed and are under development for each of these lithographic technologies. Again, however, this chapter will be limited to polymers and polymeric resists that are sensitive to ultraviolet and visible light sources. Up-to-date reviews can be readily found that focus on polymeric resists sensitive to these other high-energy sources.[4–6]

2 POSITIVE RESISTS

2.1 Near-UV–Visible Chemistries (436, 405, 365 nm)

The most commonly used photoresists for delineating features of 5 μm and below are based on the photochemistry of diazonaphthoquinone sulfonate esters (NDS) as the photoactive compound (PAC) in a cresol novolac (cresol–formaldehyde condensation polymer) binder resin. Photolysis of the PAC affects the solubility of the binder in the aqueous base solution used for development. For proper function the three components, PAC, novolac, and developer, must be finely tuned. The PAC is designed to be sensitive at near-UV–visible wavelengths. Upon exposure it undergoes a Wolf–Sus rearrangement,[7,8] via a ketene intermediate which reacts with water in the film, to form an indene carboxylic acid. Nitrogen is given off in this reaction. In the absence of water the ketene intermediate has been shown to react with the phenolic —OH groups on the novolac to form pendant ester linkages.[9] These reactions are shown in eqn (1). Under normal ambient working conditions enough water is present in the novolac binder to ensure the formation of the carboxylic acid. The rate of development in an aqueous base is much higher for the exposed areas. Good developer discrimination between the exposed and unexposed areas results in high-contrast, high-resolution images. An example of the high-resolution images that can be developed in these resists is shown in Fig. 4. The following discussion concerns the primary components of this important class of photoresists and their mechanistic role in generating images.

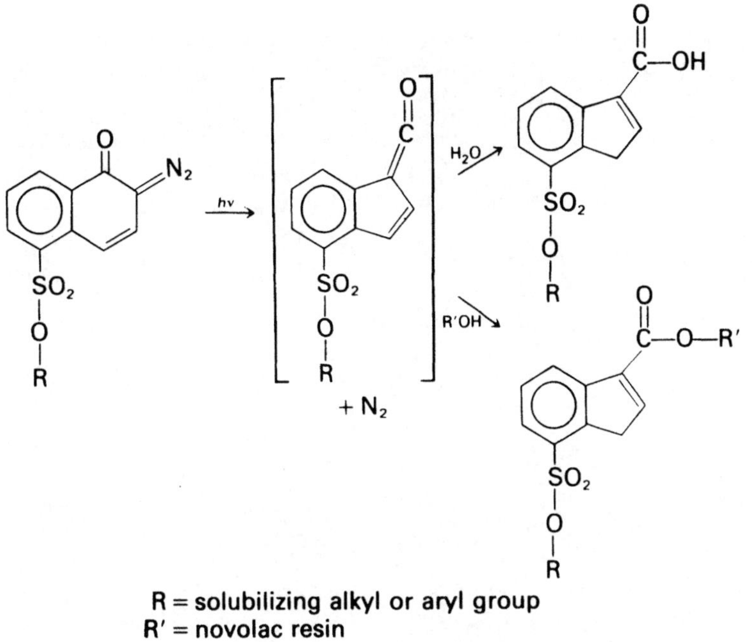

(1)

The detailed understanding of the mechanism of image discrimination in these resists continues to be an area of active research. This discrimination results from the large differences in dissolution rate between the exposed and unexposed areas. The dissolution rates of thin polymer films can readily be studied by laser interferometry[10] or by a quartz crystal microbalance technique.[11] The use of either technique has established that the NDS-PAC is a dissolution inhibitor for the novolac binder, presumably due to its hydrophobic nature. An earlier proposal that the unexposed NDS in the masked areas of the resist readily coupled with the novolac resin in the presence of the aqueous base to form diazo dyes and that these actually inhibited the dissolution of the resin has been shown to be incorrect.[12] The indene carboxylic acid that is formed from the photolysis reaction is a dissolution accelerator for the novolac. Therefore, this hydrophobic-to-hydrophilic switch in the resist has been accepted for several years as the mechanism that enables the high discrimination and high contrast obtainable in this type of resists.

It has been proposed that this simplistic picture may not represent all of the factors which lead to the image discrimination.[13] A

FIG. 4. Micrograph of high-resolution image developed in NDS–novolac positive photoresist.

monofunctional PAC, **Ia**, and its indene carboxylic acid photoproduct, **IIb**, were formulated separately into a polymeric binder as a model resist system. The dissolution rates of a film of the original novolac binder, the resist film before and after exposure and the novolac film with the added acid were then monitored with the quartz crystal

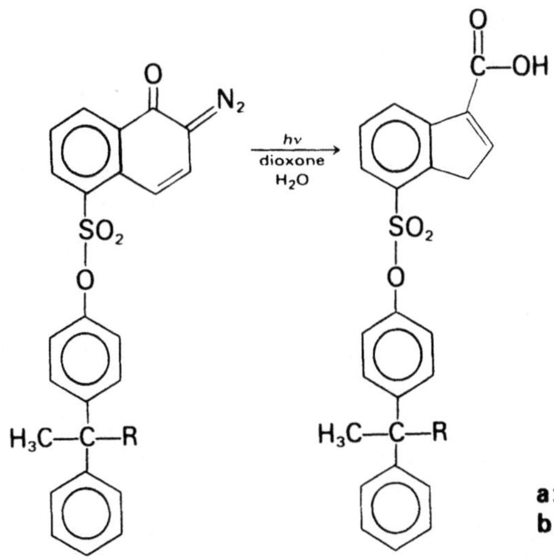

a: R = H
b: R = CH_3

Ia, b **IIa, b**

microbalance technique. These results are shown in Table 1. Comparison of the dissolution rate of the unexposed resist film with the novolac film shows the enormous dissolution inhibition effect of the dispersed NDS in the film. Surprisingly, however, the novolac film containing the externally synthesized and 'back-doped' indene carboxylic did not show a dissolution increase over the novolac resin itself. The exposed resist, however, showed approximately 30% increase in dissolution rate over the back-doped mixture and the pure novolac film. It was proposed that this enhancement in dissolution rate is a result of the gaseous nitrogen which is formed during the NDS photolysis [see eqn (1)]. It is hypothesized that the nitrogen generates microvoids and stresses in the novolac film and that this enhances the

TABLE 1
Dissolution Rates of Model Photoresist Coatings[13]

Coating	*Dissolution rate* (Å/*s*)[a]
Novolac resin	1100
Novolac resin + **I** (unexposed)	130
Novolac resin + **I** (exposed)	1370
Novolac resin + **II**	1000

[a] Quartz crystal microbalance technique.

diffusion of the developer into the polymer film and leads to an increase in the dissolution rate. Such a mechanism has been previously proposed for the development characteristics of poly(methyl methacrylate).[14]

The role of the evolved nitrogen gas and the precise mechanism appear to remain open questions. A recent publication[15] indicates that, indeed, the formation of the acid is the important step leading to dissolution rate enhancement in the exposed areas of the resist. Again, back-doping experiments with presynthesized indene carboxylic acids were used. In these experiments the novolac resin dissolved about five times more slowly than that used in the previous experiment, i.e. at 185 Å/s vs 1000 Å/s (Table 2). The unexposed resist with NDS-PAC gave a rate of 11 Å/s whereas the dissolution rate of the fully exposed resist was 640 Å/s. This *c.* 60-fold increase between exposed and unexposed is in good agreement with others' work.[16] Back-doping the resin with the acid **IIb** gave a dissolution rate of 2650 Å/s. The authors draw two conclusions from these results. First, they feel that the evolution of nitrogen is insignificant; secondly, apparently only about half of the PAC is photolyzed to the carboxylic acid in the fully exposed resist because the dissolution rate of this resist is about 4 times slower than the novolac back-doped with an acid concentration equivalent to 100% conversion of the NDS-PAC to acid. It is proposed that the other 50% of the PAC couples with the polymeric phenol through the ketene to form an ester pendant to the novolac.

This latter proposal appears to contradict the results of another recent report[16] where FTIR was used to follow directly the formation of ketene from NDS-PAC photolysis in a photoresist film. Under ambient conditions the ketene appeared to react quickly with water in the film to convert to the indene carboxylic acid. Under dry condi-

TABLE 2
Dissolution Rates of Model Photoresist Coatings[15]

Coating	*Dissolution rate* (Å/*s*)[a]
Novolac resin	185
Novolac resin + **I** (unexposed)	11
Novolac resin + **I** (exposed)	640
Novolac resin + **II**	2650

[a] Laser interferometry technique.

tions, with water excluded, the ketene appeared to be quite stable. Although these experiments were qualitative, they indicate that the ketene reaction with the polymeric phenol group is quite slow in comparison to the reaction with water. Thus, the exact details of the image discrimination appear yet to be resolved despite the intense research interest in these technologically important resists.

The structure of the NDS molecule has an important role in the performance of these photoresists. For several years, many of the commercial positive photoresists have contained multifunctional, principally trifunctional, NDS derivatives. Compounds such as **III** and **IV** are examples of such structures.[17,18] Higher-contrast, faster-speed resists have resulted from use of such sensitizers. A recent publication[19] has proposed a quantitative model based on 'polyphotolysis' for the efficacy of this type of compound. A practical consideration for use of such multifunctional PACs is their solubility. These multifunctional PACs are notoriously insoluble and at the heart of almost every one of the proprietary formulations utilizing these compounds is a specific compound or combination of compounds and isomers that possess the requisite solubility in the limited number of environmentally allowable spin-coating solvents that are presently available.

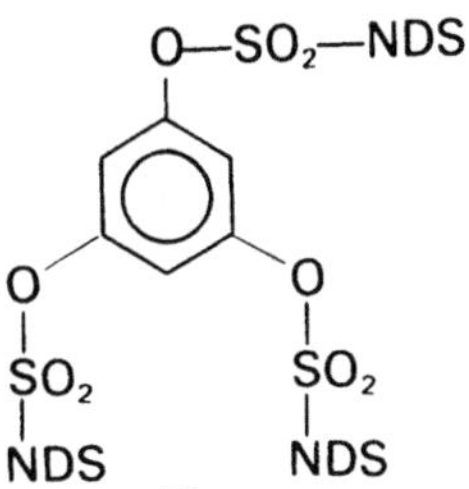

III

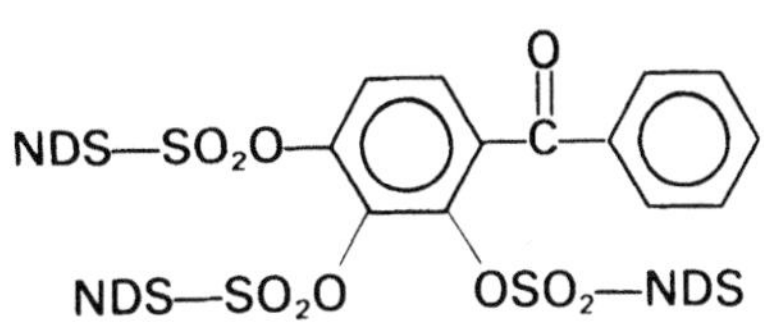

IV

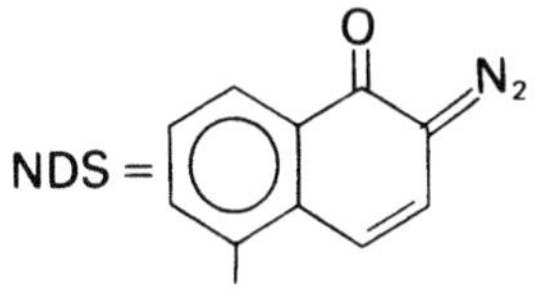

In addition to the enhanced image formation and development properties of the multifunctional PACs, they also contribute to enhanced image stability during the image transfer step in the microlithographic process. In normal resist processing, after the development step, the resist image is hard-baked at *c.* 130°C for

30 min. This 'hardens' the novolac image. At this temperature, the NDS decomposes to the ketene in the polymer film and a large portion of it reacts with the phenolic groups on the novolac to form ester linkages [eqn (2)]. The multifunctional NDS-PACs actually form a crosslinked network in the image area and consequently lead to considerable improvement of the thermal flow. This improves the performance of these images to subsequent image transfer steps such as ion implantation and plasma etching by keeping the image from flowing or distorting at high temperature.

The binder resin has an intricate and demanding function in the performance of the photoresist coating. It must first be soluble in the proper spin-coating solvents. The spin-coated films must next adhere well to the underlying substrate and must have physical integrity to form pinhole-free coatings. In order for the NDS-PAC to produce the discrimination necessary for high-resolution images, the binder polymer–PAC solid solution must have the proper base solubility.

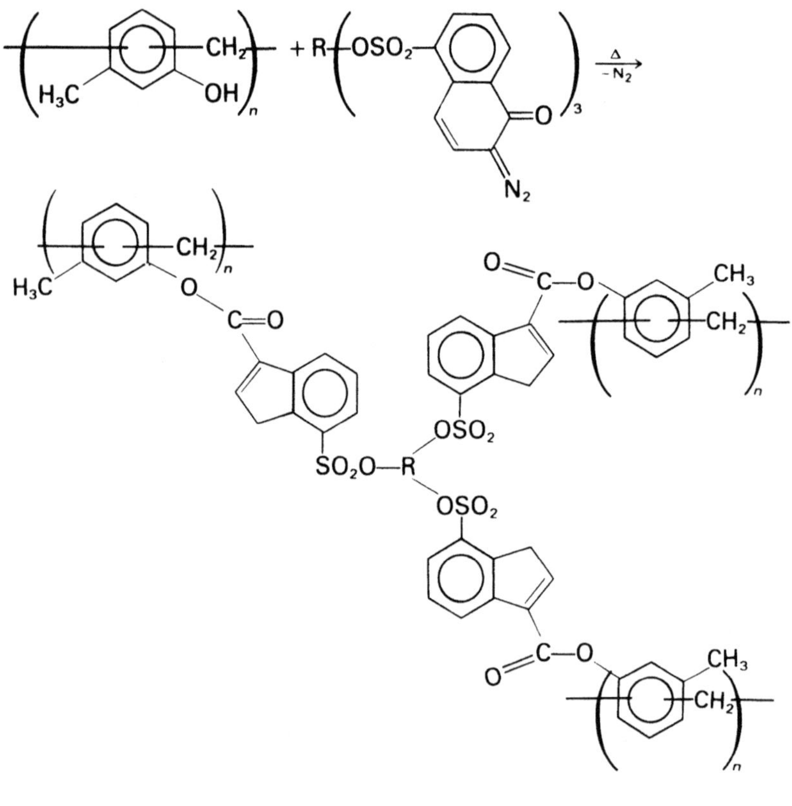

(2)

After the image has been formed, the polymer must provide the physical integrity to resist the image transfer process. Finally, the polymer must be strippable when the image and image transfer steps are finished.

Cresol–formaldehyde condensation polymers (novolacs) have been the polymers of choice for the NDS photochemistry.[20,21] These novolac resins are typically synthesized from commercial cresol mixtures which contain about 60% *m*-cresol and 30% *p*-cresol with a 10% of various other aromatic phenols present. The exact composition of this mixture can vary from lot to lot. The condensation polymerization of the cresol with formaldehyde is difficult to control and results in a relatively low-molecular-weight polymer with M_n around 1000 and with a broad dispersity (*c*. 20–40). These polymers are soluble in most spin-coating solvents, provide excellent adhesion to most substrates and exhibit good film-forming and coating properties in general. They have low glass transition temperatures (T_g), generally in the range 20–120°C depending on molecular weight (Russell, G., unpublished).

These novolac resins have exhibited enough balance in the needed binder properties to be used almost exclusively in all commercial NDS positive photoresist formulations. However, as the demands on these formulations have increased due to smaller imagery and new image transfer steps such as ion implantation and plasma etching, new improved binder resins are needed. Novolacs fall short in two areas. First, a specific chemical composition and molecular size is difficult to reproduce because of inconsistent composition of the cresol starting materials and because the condensation polymerization is difficult to control. These variations can cause irreproducible lithographic performance of the high-resolution resists. Second, the low T_g values of these polymers often lead to unacceptable image distortions when the image is subjected to plasma etching or ion implantation process steps.

Poly(4-hydroxystyrene) (PHS), a compositional isomer of novolac, would appear to be an excellent choice for an improved binder resin. First, it can be prepared from pure monomer via a conventional, controllable, free-radical solution polymerization.[22] Thus, better molecular weight control and greater resin purity are easily achievable. Second, PHS has a much higher T_g (*c*. 180°C) and would be expected to have better physical characteristics than the low-molecular-weight novolacs. Despite these advantages, PHS has been found to perform poorly as a photoresist in NDS-PAC formulations due to excessive dissolution of the unexposed areas during the development step with

TABLE 3
Development Rates for Novolac and Poly(*p*-hydroxystyrene) Model Resists[a], [24]

	Novolac ($\mu m/min$)	*PHS* ($\mu m/min$)
Polymer film	0·15	6·15
Polymer film + 15 wt% PAC[b] unexposed	0·02	2·50
Polymer film + 15 wt% PAC[b] exposed	1·70	10·10
Discrimination (exposed/unexposed)	85	4

[a] Developer 1·5 wt% $(CH_3)_4NOH$ at 20°C.
[b] A model PAC, prepared by the condensation of m-cresol with 2-diazo-1-naphthol-5-sulfonyl chloride.

aqueous base.[20,23] The development rates of photoresist films containing 15 wt% of the NDS-PAC (**III**) in novolac or PHS with 1·5 wt% tetramethylammonium hydroxide developer are shown in Table 3. Both the high base solubility of the PHS and reduced dissolution inhibition by the NDS appear to contribute to the 20-fold difference in discrimination between the two systems.

Other base-soluble binder polymers designed to correct some of the deficiencies of the novolac resins have been reported. A family of alternating high-T_g copolymers based on *N*-(*p*-hydroxyphenyl)maleimide and various olefins has been shown to be useful with the NDS photochemistry to yield images that have thermal stabilities to temperatures of up to 250°C.[24] The dissolution rates of these copolymers in aqueous base were substantially faster than those of novolac resins. This made formulation of functional resists based on these copolymers as binders difficult. However, the α-methylstyrene or stilbene copolymers with this maleimide were found to be compatible with novolac and the solubility characteristics of the blends were found to be similar to novolac over a wide range of blend compositions. An example of this is plotted in Fig. 5 for the blend of poly[α-methylstyrene-co-*N*-(*p*-hydroxyphenyl)maleimide] and novolac.[25] The development of resist images was found to be similar to the conventional novolac resist with acceptable thinning in the unexposed areas. Photoresist images produced with these blends were shown to exhibit resistance to thermal flow at 250°C. These images were demonstrated to have remarkable resistance to various image transfer processes such as plasma etching and ion implantation.[25] High T_g

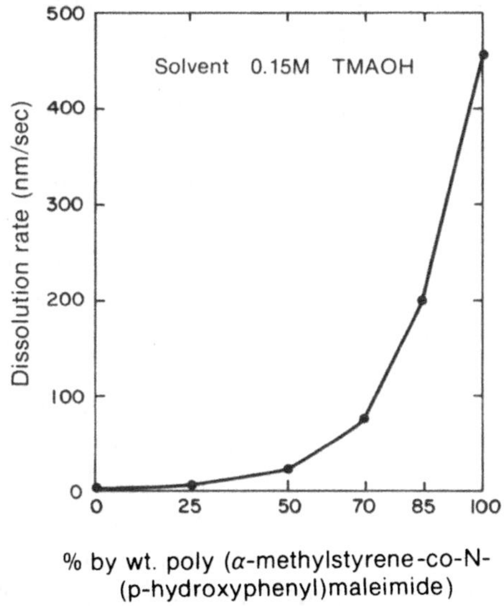

FIG. 5. Plot of dissolution rate versus blend composition for novolac and poly[α-methylstyrene-co-*N*-(*p*-hydroxyphenyl)maleimide].

resins based on alternating copolymers or maleimide and styrene have been reported as novolac replacements.[26] These resins were also reported to have an extremely high rate of aqueous base dissolution.

The developer used to dissolve or develop out the exposed regions of the photoresist coating is an integral component of the overall chemistry of the photoresist process. As stated earlier for the NDS-PAC based positive photoresists, it is an aqueous base. There are two general types of these aqueous base developers. The first class, the metal-ion containing developers, is based on metallic hydroxides. The elimination of metal ion contaminants in semiconductor processing has led to the development of a second class of aqueous base developers, the so-called metal-ion free developers. Aqueous solutions of quaternary alkylammonium hydroxides, such as tetramethylammonium hydroxide, tetraethylammonium hydroxide, etc., are the most important members of this class.

Both types of developers have been shown to exhibit some similar development trends. Cation size has been shown to affect photoresist dissolution with the dissolution rate decreasing the larger the cation for both the metal ion and ammonium ion bases.[27,28] However, resolution and contrast are thought to be generally poorer with the quaternary ammonium developers. The addition of various tetra-alkylammonium salts has been found to enhance contrast of tetra-methylammonium hydroxide developers.[29]

The positive imaging mode of these NDS/novolac-based resists can be reversed to give negative images by a post-exposure treatment of the undeveloped photoresist film with a reactive amine.[30,31] This amine-treated film is next heated and then flood-exposed. Development with a conventional aqueous base at this stage produces a high-resolution negative image of the original positive image. This process is shown in Fig. 6. An amine (monazoline, imidazole, triethanolamine, *inter alia*) reacts with the indene carboxylic acid in the originally exposed areas to form an ammonium salt which is decomposed by heat treatment to indene. Subsequent flood exposure gives the indene carboxylic acid in the previously unexposed areas which are then developed out in the aqueous base developer. The hydrocarbon indene from the original exposure and subsequent process protects the novolac just as did the original NDS.

2.2 Mid- and Deep-UV Chemistries (334, 313 and 254 nm)

The previously mentioned push to higher resolution has caused equipment manufacturers to produce exposure tools with selective output at 405 and 365 nm. Conventional positive resists have performed well at these wavelengths. However, the potential resolution gain has been small and the push toward still shorter wavelengths continues.

There are two principle problems in trying to use a conventional positive photoresist in the deep- and mid-UV regions. First, the quinone diazide photoproduct absorbs at these wavelengths. Second, the novolac resin absorbs strongly at wavelengths below 295 nm. For near-UV exposures, neither problem exists since the novolac and the photoproduct are essentially transparent in the near-UV. However, since the photochemistry and the lithographic processes for this type of resist are so well developed, there have been many attempts to make a short-wavelength quinone diazide resist.

The problems of conventional positive resists with mid-UV exposure tools ($\lambda = 330$ or 313 nm) have been well documented.[32,33] However, since scanning exposure tools with a mid-UV capability are readily available, new resists have been investigated. Naphthoquinone PACs which have the sulfonyl group at the 4-position rather than at the 5-position have a more appropriate absorption in the mid-UV and

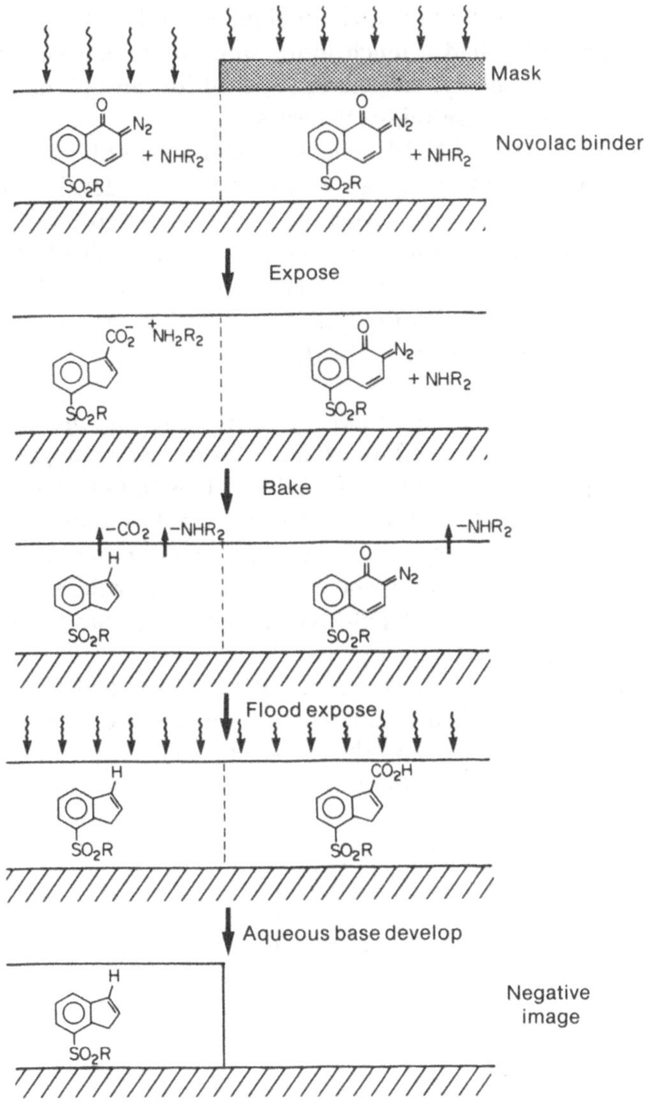

FIG. 6. Schematic of image reversal process of NDS–novolac positive photoresist.

yield a photoproduct with less absorption in the mid-UV.[34,35] A more exotic PAC used both the 4- and 5-sulfonyl-substituted naphthoquinone diazide attached through an aliphatic diol.[36] It was found that the alkyl sulfonates had a much more appropriate exposure response than did the normal aryl sulfonates. Special low-absorbance novolacs were used in all of these mid-UV resists.

When a conventional NDS/novolac resist is exposed in the deep-UV ($\lambda \approx 250$ nm) a negative image is formed upon development with organic solvents.[37] Novolac resins themselves undergo photocrosslinking when exposed at these wavelengths.[20] Almost all of the light is absorbed in the top of the resist. Because of this, the developed image may be wider at the top of the image than at the base. This overhung image may be advantageous. On the other hand, the developer may undercut the image completely.

A positive image using NDS chemistry can be made with the deep-UV if the resin itself does not absorb. Surprisingly novolacs made from the condensation of formaldehyde with pure *p*-cresol had a very transparent window at about 250 nm. NDS/*p*-cresol novolac gave an overhung positive image after deep-UV exposure and development with aqueous base.[38] a copolymer of styrene and maleimide was also used as a binder for an NDS-PAC to give a positive image after deep-UV exposure and development.[26]

The types of diazo chemistry used for near- and mid-UV can work well in the deep-UV if there is no interference caused by the absorbance of the photoproduct. Several 1,3-diacyl-2-diazo compounds, including 5-diazo Meldrum's acid (**V**), have been synthesized

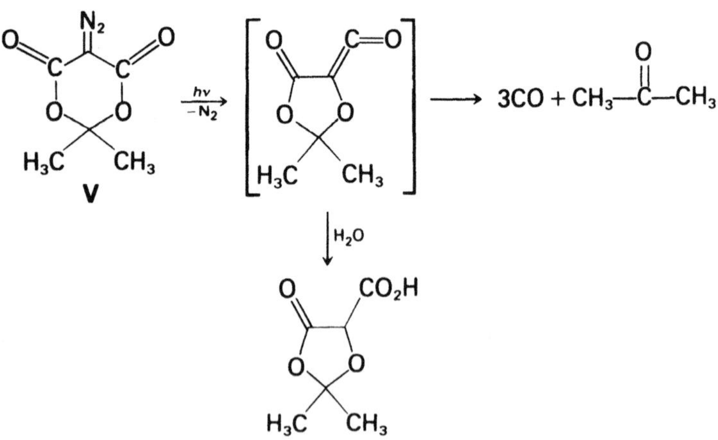

(3)

and formulated into novolac-based resists.[32,39] Just as with the NDS-PACs, the starting material can inhibit the base solubility of a novolac, whilst either set of products should accelerate the dissolution. This chemistry is shown in eqn (3).

o-Nitrobenzyl photochemistry is the basis for another dissolution inhibitor approach to making deep-UV resists[40,41] [eqn (4)]. In this case the polymer binders are base-soluble copolymers containing methacrylic acid and the dissolution inhibitor is an *o*-nitrobenzyl ester (**VI**) of a large organic molecule such as cholic acid. Upon exposure, the ester cleaves to provide products which will not interfere with the resin dissolution.

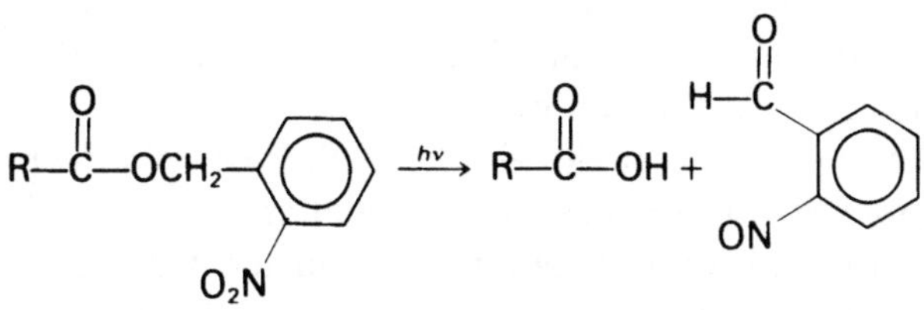

VI

(4)

The most commonly studied and best understood resist for the deep-UV is poly(methyl methacrylate), PMMA (**VII**).[42] Exposure around 230 nm causes chain scission to occur. Absorption occurs at the

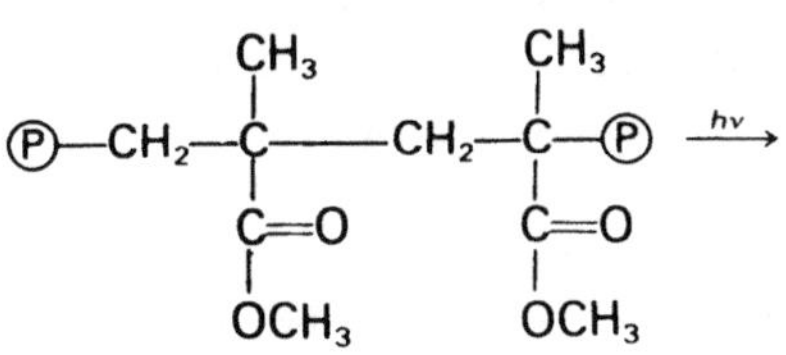

VII

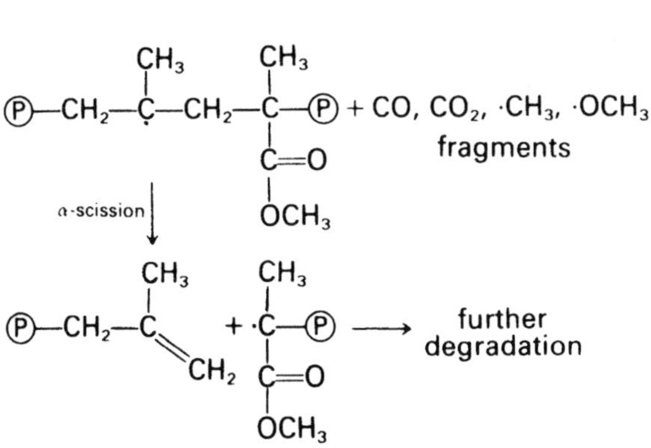

(5)

ester carbonyl and the ester group fragments off until there is a tertiary backbone radical [eqn (5)]. Cleavage alpha to the radical then causes the chain to break. The resulting exposed areas are lower in molecular weight than the unexposed areas and a positive relief image results after development with an appropriate organic solvent such as chlorobenzene. Studies of the dissolution rate of PMMA as a function of molecular weight indicate that more has happened than simple scission. The dissolution rate of an unexposed film of PMMA is much slower than the rate for a PMMA film of high molecular weight which has been irradiated and scissioned to the lower molecular weight of the first coating.[14] Possible explanations for this include plasticization of the film by fragments from the photochemistry, voids in the film caused by exposure and limited depolymerization of the PMMA to give plasticization of the film by monomer.

There are several major drawbacks of PMMA as a deep-UV resist. First, it has a very low extinction coefficient and the transition between no absorbance and total absorbance within the film occurs in a very narrow waveband. Also, the quantum efficiency for scission, G_s, is low (1·3). Another consideration is that the PMMA resist images do not stand up well to the plasma etching treatments which are necessary for producing high-resolution images.

However, PMMA is capable of very high resolution and many groups have worked on copolymers of methyl methacrylate and analogues of PMMA in hopes of improving the speed and physical performance of PMMA resists which retain its resolution.

Efforts to improve upon the deep-UV sensitivity of PMMA by altering the ester group have been largely unsuccessful. Bulky alkyl ester groups did not alter the absorption or the photospeed of methacrylate resists.[43] However, using a perfluorobutyl group gave a seven-fold speed increase over PMMA.[44] This increase may simply represent a more highly discriminating development process for the fluorinated resist.

A related polymer, poly(methyl isopropenyl ketone), PMIPK (**VIII**), has been found to have improved spectral response and speed in the deep-UV. This would be predicted on the basis of the more efficient photochemistries of ketones. PMIPK was five times faster than PMMA under the same conditions. In addition, the PMIPK could be chemically sensitized with substituted benzoic acids to give speeds which were more than 25 times faster than PMMA depending upon the wavelength of exposure.[44,45]

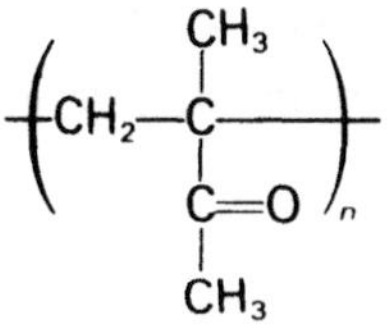

VIII

PMIPK does not, however, have improved plasma resistance over PMMA. One way to get a positive deep-UV resist with improved etch resistance is to blend in a more resistant polymer.[46] The addition of 10% of an aromatic polymer to PMMA caused a four-fold decrease in the etch rate of the PMMA in a fluorine plasma and there was no loss in photographic speed. PMIPK had a 25% speed increase and a five-fold decrease in etch rate in a chloride plasma using 10% of an added aromatic. However, in general, the addition of aromatic comonomers tends to reduce the speed of scissioning resists.

Copolymers (**IX**) of glycidyl methacrylate and methyl methacrylate have been found to be fast, positive deep-UV resists and have the additional advantage that the developed image may be thermally crosslinked.[47] Even COP, a fast, negative working e-beam resist containing glycidyl methacrylate and ethyl acrylate, behaves as a positive resist in the deep-UV.[48]

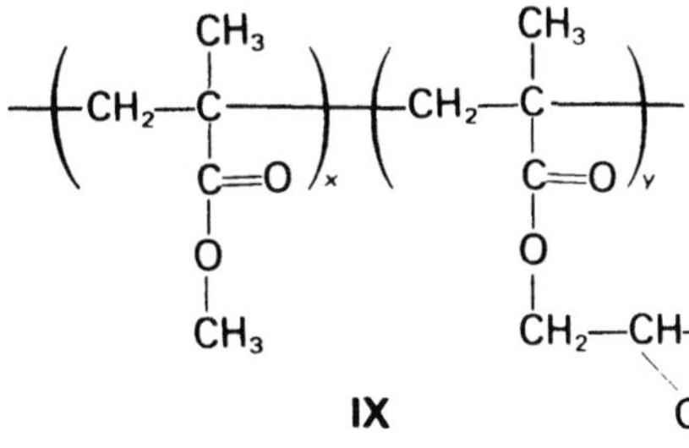

IX

An extensively studied copolymer (**X**) of PMMA contains a 3-oximino-2-butanone methacrylate comonomer.[49] The α-keto-oxime group provides a broad absorption band in the deep-UV, is photolabile and is relatively thermally stable. The addition of 37% of the oximino monomer gave a 50-fold speed increase over PMMA itself.[50] Further improvement in speed with no loss in resolution was obtained with a terpolymer containing the monomers of **X** along with acrylonitrile. Chemical sensitization of these copolymers was seen using *p*-*t*-butylbenzoic acid.[51] Indenone has also been used as a comonomer with MMA to produce a polymer (**XI**) with improved absorption and in hopes that the steric strain in the five-membered ketone ring would give a more effective scission.[52]

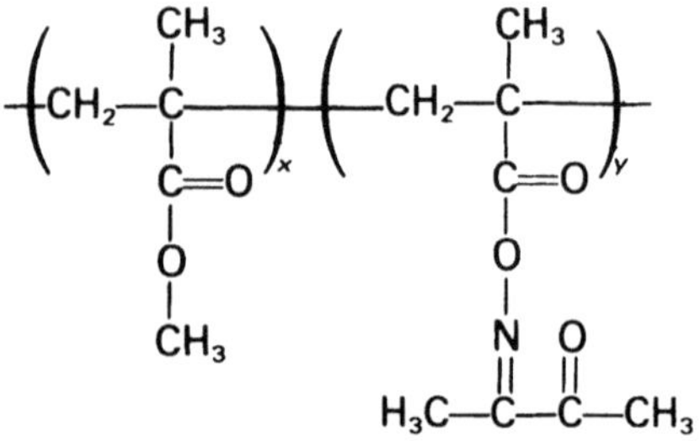

X

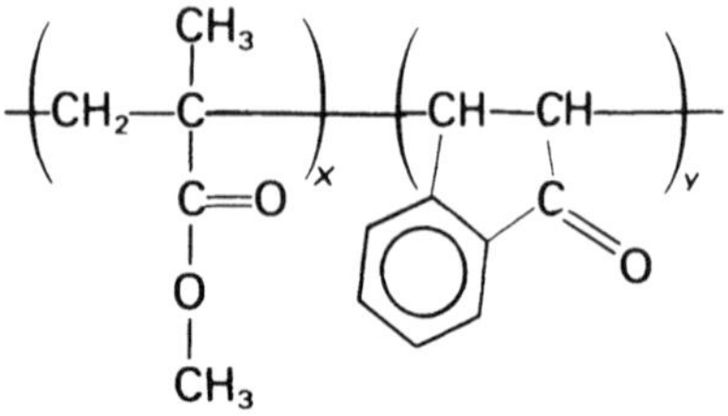

XI

Copolymers (**XII**) of methacrylonitrile with methacrylic acid have also been used as a deep-UV resist, but only after baking.[53] Prior to the pre-exposure bake the copolymer has very little absorbance above 225 nm. Upon baking, the coating has a marked increase in absorbance with a λ_{max} about 240 nm. A cyclized structure is postulated. The resist gives positive images in the deep-UV by normal backbone scission but produces negative images with mid-UV exposure by a hydrogen transfer mechanism. A hard post-bake of the developed images gives greatly improved resistance to fluorine plasmas.

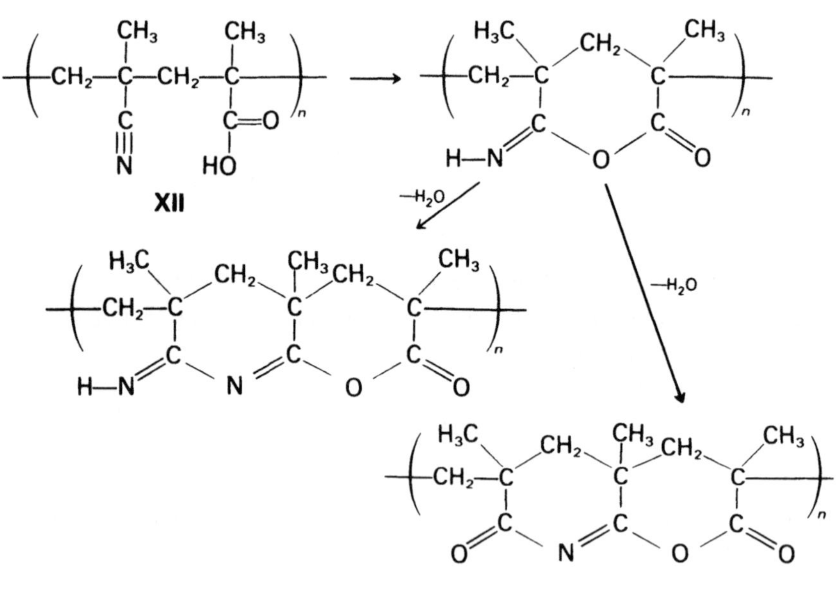

XII

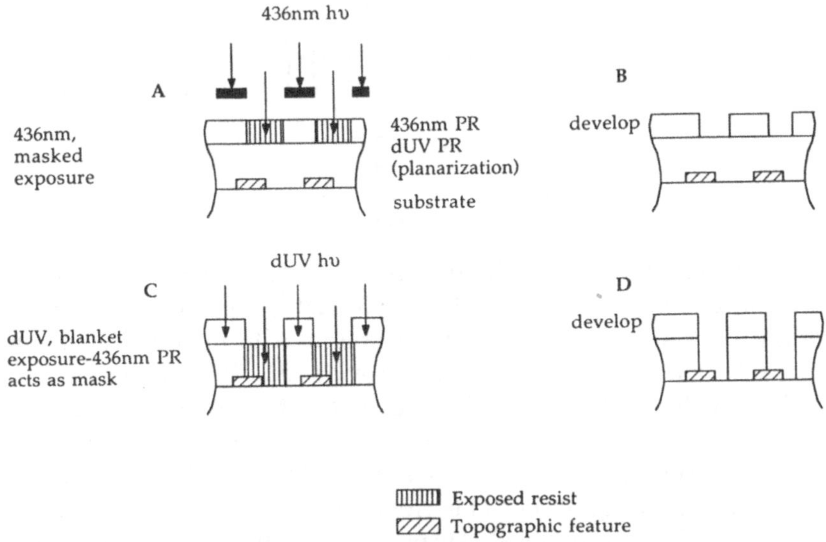

FIG. 7. Schematic of the portable conformable mask (PCM) bilayer process. (PR = photoresist)

Deep-UV resists are used in two principal configurations: as normal single-layer resists, or as the thick planarizing layer in portable conformable mask (PCM) bilayer imaging systems. Improved resolution is the driving force for the use of DUV resists in the single-layer mode. In the PCM mode, the combination of a DUV resist with a conventional resist makes use of the wavelength selectivity of the two types of resists.[54] The PCM process (Fig. 7) first exposes and develops the conventional thin top resist at 436 nm. A second, full wafer, exposure in the deep-UV, 220–250 nm, is masked by the absorption of the novolac-based resist image. The unmasked, positive working deep-UV resist is developed to give a thick, high-resolution image for subsequent processing.

Most of the work on PCM systems has used PMMA as the planarizing layer. Because of the blanket exposure, speed is not a problem. Because the thin novolac cap may remain in place, the poor fabrication resistance of PMMA is less of a problem. A terpolymer (**XIII**) of methyl methacrylate, methacrylic acid and methacrylonitrile provides an improved planarizing deep-UV resist with a two-fold speed increase and a 40°C increase in T_g.[55]

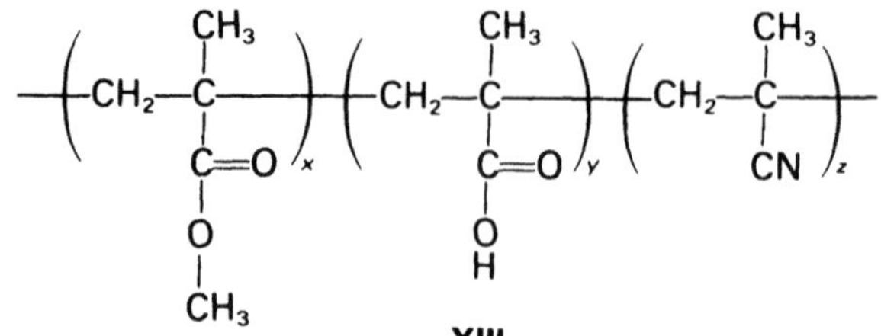

XIII

A further improvement in the physical performance of the planarizing layer was realized with poly(dimethyl glutarimide), PMGI (**XIV**).[56,57] Actually, a copolymer is present with the relative amounts of —H and —alkyl chosen to provide the best compromise of solubility for coating and development. The PMGI behaves as a scissioning resist, very similar to PMMA. However, it absorbs at longer wavelengths, 250 nm rather than 220 nm. It has an additional advantage of being developable from the same aqueous base which is used for the conventional 436 nm resist rather than requiring an organic developer.

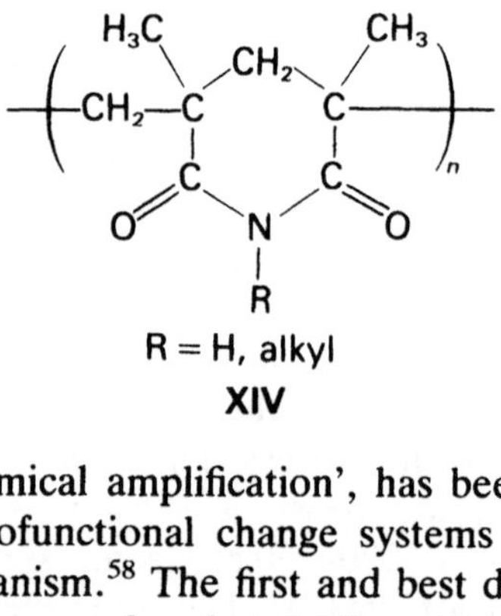

R = H, alkyl

XIV

A new term, 'chemical amplification', has been coined to describe any of several photofunctional change systems which proceed by a chain reaction mechanism.[58] The first and best documented system of this type involves the use of a photolabile acid generator to deprotect imagewise a functional group or depolymerize a polymer.[59,60] Many acid generators are possible, but the most widely used have been iodonium and sulfonium salts.[61] The several-step mechanism for acid production may vary according to the matrix but, generally, is as follows:

$$AR_2I^+X^- \xrightarrow{h\nu} Ar\cdot + ARI\cdot^+X^-$$

$$\xrightarrow{RH} ArI^+HX^- + R\cdot + Ar\cdot \longrightarrow ArI + HX + AR{-}R$$

These iodonium and sulfonium salts have their natural sensitivity between 200 and 300 nm but can be substituted and sensitized out into the visible. Polymers with acid-labile protecting groups such as *t*-butyl carbonate and *t*-butyl esters have been formulated with these salts. For example, the hydrophobic polymer poly(4-*t*-butoxycarbonyloxy)-styrene (**XV**) becomes base-soluble poly(4-hydroxystyrene) when treated with acid and heated. About 40% of the mass of the polymer is volatilized. Imagewise exposure to generate acid and then heating produces a relief image. Either positive or negative images may be developed depending on the solvent chosen. Several other copolymers (**XVI**, **XVII**) with *t*-Boc (*t*-butyloxycarbonyl) protecting groups have been shown to produce workable resists with good physical properties.[26,62]

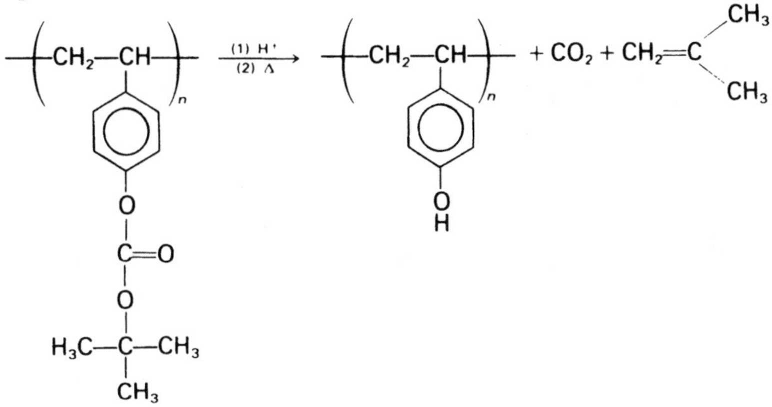

XV

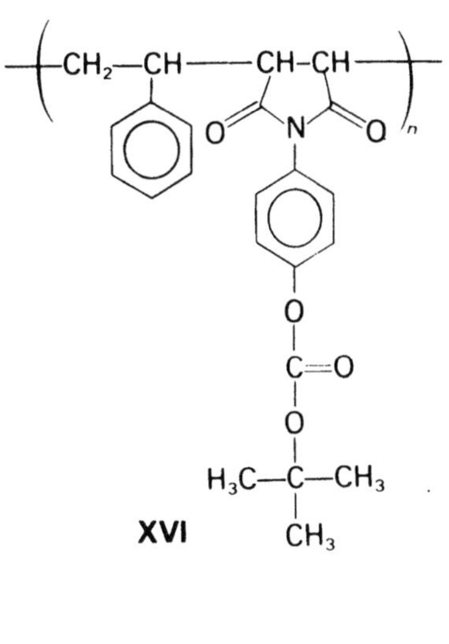

XVI

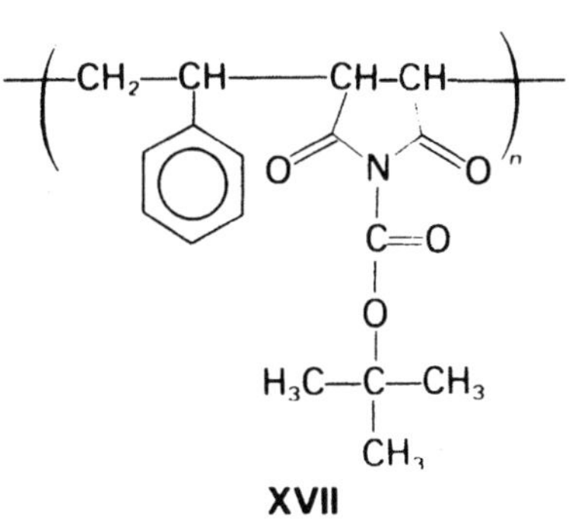

XVII

There have been several attempts to cause scission in condensation polymer backbones, with a resulting reduction in molecular weight and the possibility of forming new, solubilizing groups. The design and synthesis of polymers with carbonate units in the backbone that allow a facile degradation of the polymer by the chemical amplification scheme have recently been reported.[63] An example of such a polymer is a copolycarbonate (**XVIII**) based on bisphenol A and 2,5-dimethyl-2,5-hexanediol. Because of the steric hindrance of the aliphatic tertiary diol in the structure it was necessary to use highly activated bis-*p*-nitrophenyl carbonate derivatives of the aliphatic diol under phase transfer conditions to obtain high polymer [eqn (6)]. These polymers showed thermal degradation at around 160°C and acid-catalyzed thermal degradation by an E_i mechanism at less than 100°C. When

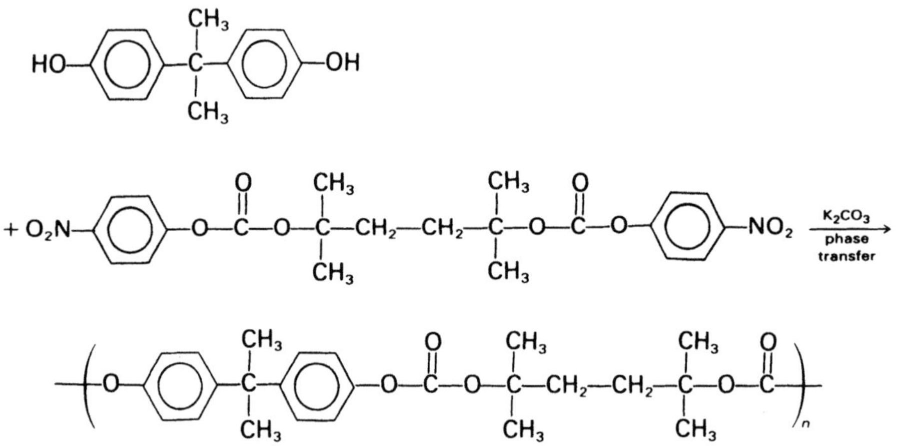

XVIII

(6)

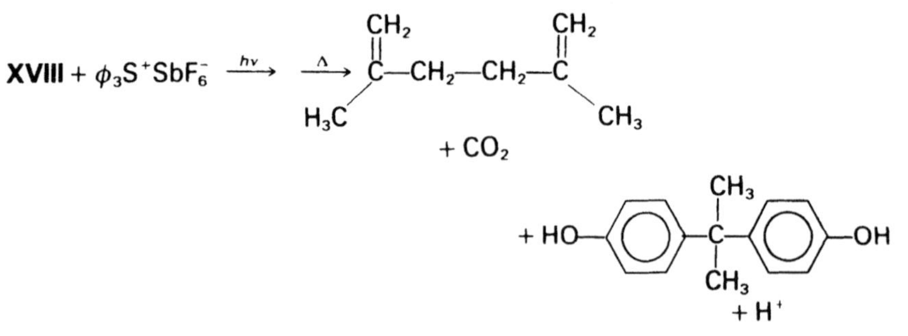

films of this polymer were doped with 5–15% triphenylsulfonium hexafluoroantimonate and exposed at 254 nm with a 5 mJ/cm^2 dose, thermal treatment at 60°C for 2 min followed by solvent development gave high-resolution positive images.

Polymers containing precursors to more volatile fragments such as copolycarbonate (**XIX**) were found to be fully dry-developable. After exposure in the presence of a photoacid and thermal treatment as above, complete image cleanout was observed.[64]

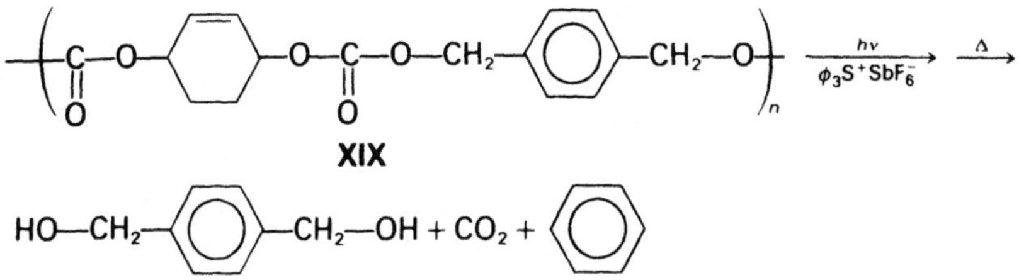

The *o*-nitrobenzyl chemistry can be placed in a polymer backbone and used as a site for depolymerization.[58] When *o*-nitrobenzyl groups are used in a polyformal (**XX**) or as endcaps on a polyformal, exposure will cleave the polymer backbone or uncap the end. The unprotected polyformal can further depolymerize with heat or acid. It is not clear, however, that these depolymerizations are facile in a polymer coating.

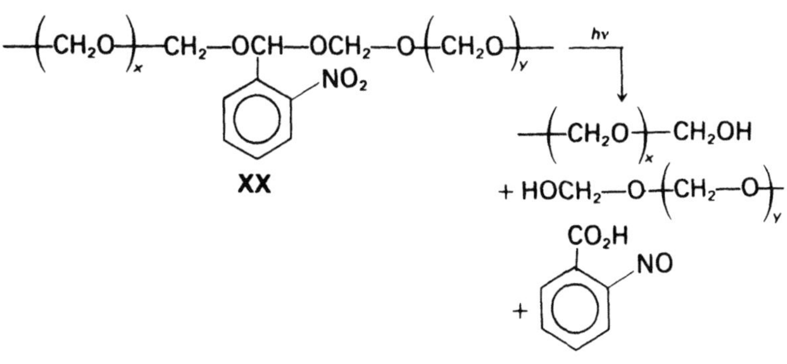

One polymer where depolymerization was successful is polyphthaldehyde (**XXI**). Endcapped polyphthaldehyde is stable to a temperature of 150°C. However, its ceiling temperature is −40°C. When polyphthaldehyde is irradiated with a high-powered excimer laser, it is 'self-developing', i.e. it is blown away as monomer during the exposure.[65] In the presence of an onium salt, the depolymerization can be catalyzed imagewise.[57]

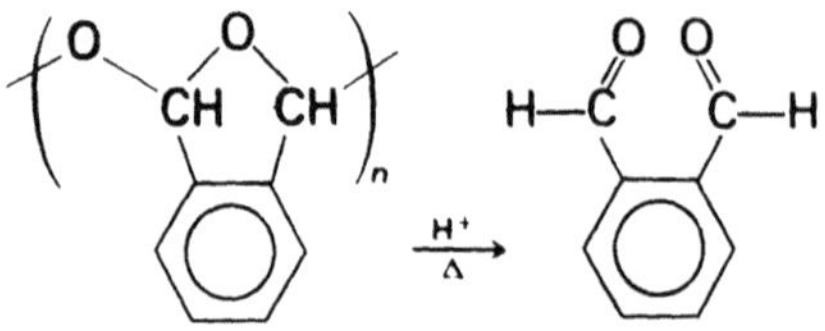

XXI

One of the long-standing drawbacks of deep-UV lithography has been the lack of a workable exposure source. The excimer laser mentioned above may turn out to fill this need.[66,67] As simply a light source, the excimer laser can deliver large amounts of light at several deep-UV wavelengths, depending upon the mixture in the laser gas chamber. Several of the possible outputs are listed in Table 4 according to the excited complex which emits light. Resists which operate with other deep-UV light sources are readily exposed with this tunable tool.

The extremely high-energy output of the laser also allows the potential for the kind of ablative imaging mentioned above.[67–69] When PMMA is used in this manner with an ArF (193 nm) laser, the exposure occurs in the top surface and produces a large amount of scissioning. The kinetic energy of the photopulse explodes the fragments out of the exposed hole. Continued pulses will 'drill' through the resist. Generally, the power is so great that almost any polymer can be ablated.[70,71] Even condensation polymers, polyimides and polyesters, which show very little inherent photospeed, can be ablated at reasonable rates.

TABLE 4
Excimer Laser Outputs

Complex	*Wavelength (nm)*
ArF	193
KrCl	222
KrF	248
XeBr	282
XeCl	308
XeF	351

Resistance to an oxygen plasma is a necessity for many kinds of multilayer schemes. The easiest way to improve the oxygen plasma resistance of an organic is to incorporate silicon in the polymer. For example, trimethylsilyl groups can be attached to the common kinds of PMMA resists. Copolymer **XXII** requires about 10% of the exposure of PMMA and etches at about 3% of the rate of PMMA in an oxygen reactive ion enhanced (RIE) etch.[72] The silicon is simply an inert diluent; many other methods of putting silicon into a resist are possible.

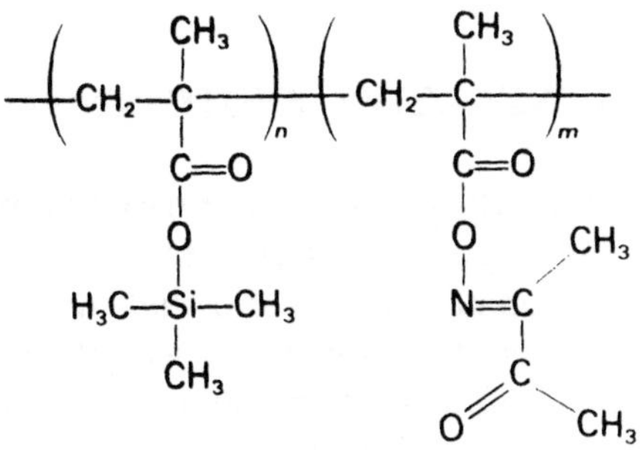

XXII

Of more interest are polymers which contain silicon as part of the backbone or as the active component of the resist. Polysiloxanes in general crosslink upon exposure to the deep-UV, but improved speed can be obtained if vinyl (**XXIII**) or chloromethyl groups (**XXIV**) are appended to the chain.[73,74] Aromatic groups are required in order to obtain absorbance in the deep-UV region.

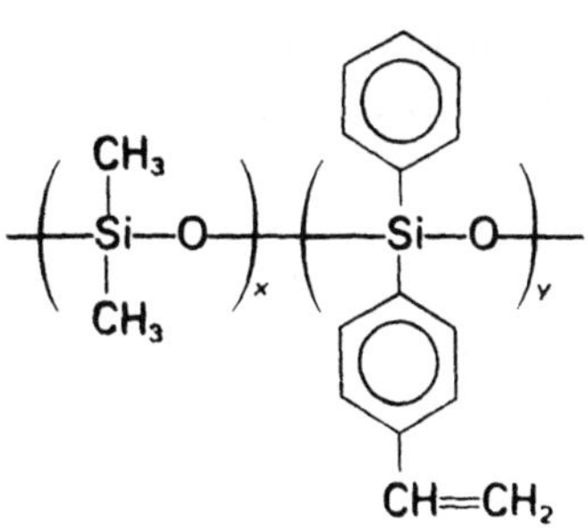

XXIII

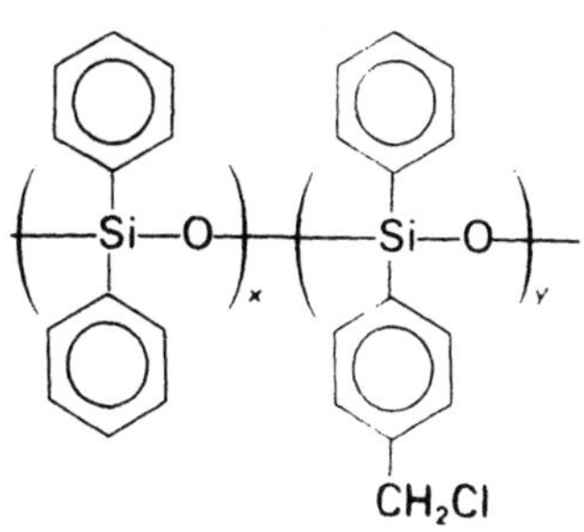

XXIV

Polysilanes (**XXV**) undergo scission in the backbone during UV exposure. The absorption curve is determined by molecular weight

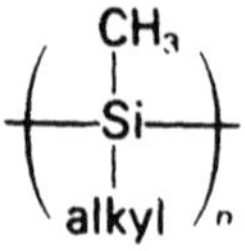

XXV

and substituents but generally the λ_{max} is at the high edge of the deep-UV. As exposure continues, the λ_{max} shifts to lower wavelengths.[75]

3 NEGATIVE PHOTORESISTS

Most negative working resist systems are based upon the fact that the dissolution rate for a polymer decreases as the molecular weight increases. In the ultimate case, a crosslinked polymer network is insoluble in any solvent. High-molecular-weight linear polymers may be crosslinked, or multifunctional monomers may be polymerized, to give the network. These technologies are fairly old and have been extensively reviewed.[76–78] However, new problems and opportunities have caused a resurgence of interest in negative imaging systems.

Early polymer imaging systems were based upon Judean asphalt, bichromated colloids or diazo resin salts.[79,80] Probably the first photographic images were recorded by Niepce in Judean asphalt. Light caused the unsaturated bitumen coating to become insoluble in mineral spirit. A more widely used system took advantage of the photoinduced reduction of dichromate to chromic acid and chromate. In these dichromate resists, the chromic acid may act to crosslink the gelatin or poly(vinyl alcohol) binder or to assist in oxidative crosslinking of the binder resin. The diazo resist systems are based not on crosslinking but on photofunctional change. A diazonium salt and a binder resin are coated from an aqueous solution. Upon exposure, the diazonium salt is destroyed and a neutral product is formed. The salt enhances the binder solubility in an aqueous developer while the neutral compound retards the solubility. All of these imaging systems suffer from limited stability and limited exposure wavelength sensitivity, and are very slow photographically.

These processes were all displaced for resist applications by better characterized resists using photodimerization (cinnamate chemistries), photocrosslinking (azide chemistries) and photoinitiated polymerization (free-radical or cationic) to yield insoluble polymer networks.

The simplest photodimerizable resist is poly(vinyl cinnamate) (**XXVI**), made by the esterification of poly(vinyl alcohol) with cinnamoyl chloride. Upon irradiation, the cinnamate groups can dimerize to yield a truxillate or truxinate. If cinnamates from two chains are involved, the cyclic product represents a crosslink.

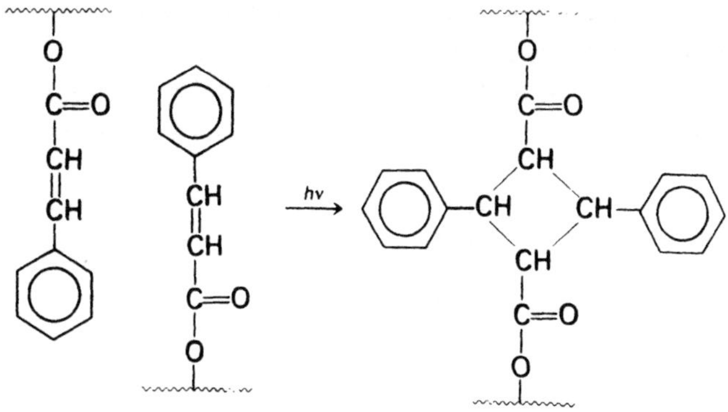

XXVI

This crosslinking reaction can take place through either the singlet or the triplet excited state. The physical chemistry of the singlet process has been studied in some detail.[81] The gel dose, D_g, for the crosslinking is expressed by eqn (7),

$$D_g = \frac{rd}{A\bar{M}_w\Phi} \tag{7}$$

where r = resist density, d = film thickness, A = absorbance, $\bar{M}_w$ = weight-average molecular weight and Φ = quantum yield for crosslinking. Experimentally the gel dose can be found by exposing a coating through the back of a transparent substrate. Subsequent development leaves a coating whose thickness is related to the exposure. A plot of thickness remaining versus log dose produces information about both the gel dose and the resist contrast, γ. This is shown in Fig. 2.

Very few of the cinnamate-type resists have been used in micro applications. One recent example used a bisstyrylpyridinium as the photocrosslinkable group in a polyamide backbone. The polymer had extremely good thermal stability and high speed.[82]

Only one negative working resist chemistry has seen extensive use for microelectronics applications. Resists which use a cyclized polyisoprene (CPI) polymer and a bisazide photoactive compound were the basis for all of the early microlithography and continue to be used where high resolution is not required.[83] The cyclized rubber has excellent adhesion to electronic substrates and plenty of reaction sites for the azide. The cyclized units increase the T_g of the rubber while reducing the solution viscosity so that high-solids coatings are possible.[84] The well-known photoreaction of azide to nitrene provides a highly reactive chemical fragment which can react with the CPI. Since a bifunctional azide is used, two polymer chains can be linked and eventually a crosslinked network is formed.[85,86] These chemistries are shown in eqns (8) and (9). Because of solubility limitations, it has been difficult to make a bis(azide) resist which is sensitive beyond 436 nm.

$$\underset{\text{azide}}{R{-}N_3} \xrightarrow{h\nu} \underset{\text{nitrene}}{R{-}\ddot{\bar{N}}\cdot} \qquad (8)$$

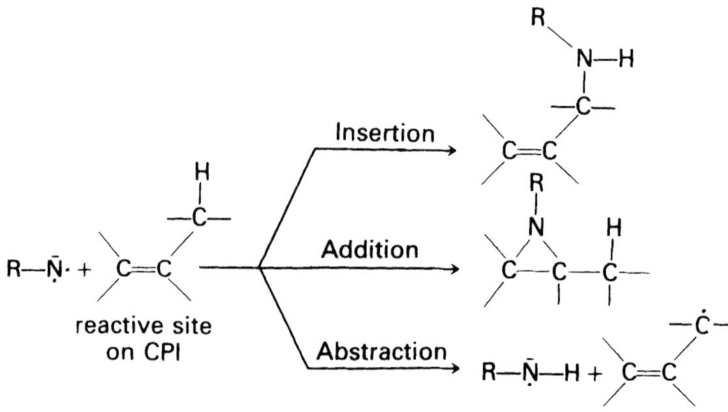

(9)

The drawback of this imaging chemistry comes during development. A good solvent must be used to remove the unexposed high polymer. This solvent penetrates and swells the images which are to be left behind. With large (>10 μm) spaces between images, the image returns to its original position, but has been distorted at the partially exposed, partially crosslinked edges. As the spaces become small (<3 μm), two neighboring images may swell, touch and coalesce to leave a polymeric bridge between the two images. This bridge will, of course, translate into a short in the electronic circuit. The problem is inherent in this type of image system but is particularly bad for the CPI resist and its hydrocarbon developers, since there is a very large gel layer present during development.

Resists which use bis(azides) along with aqueous base-soluble polymeric binders have been shown to produce superior resolution properties.[87] Novolac resins and acyclic acid copolymers were used as binders and showed little swelling during development. The bisazides were tuned for deep-UV absorbance in order to get the highest possible resolution. Poly(4-hydroxy styrene) can be crosslinked in the deep-UV using 3,3′-diazidodiphenylsulfone as the photoactive compound (**XXVII**).[88]

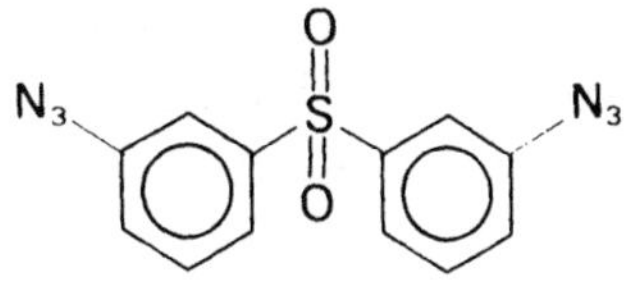

XXVII

One unusual bis(azide)-based negative resist uses a copolymer containing glycidyl methacrylate and a quaternary ammonium salt function.[89] The image containing the quaternary ammonium group can then be reacted with a dye to form a colored image. These images are used on top of charged coupled device (CCD) arrays to allow color images to be recorded.

A newer approach for general lithography uses a monoazide with an aqueous base-soluble polymer, poly (4-hydroxystyrene).[90] The monoazide (**XXVIII**) can be tuned readily for absorbance and the image network does not swell strongly during development. The wavelength sensitivities of this type of resist are determined solely by the azide. It has been shown that novolac resins can be sensitized to all common exposure sources, including the deep-UV, simply by altering the azide structure.[91]

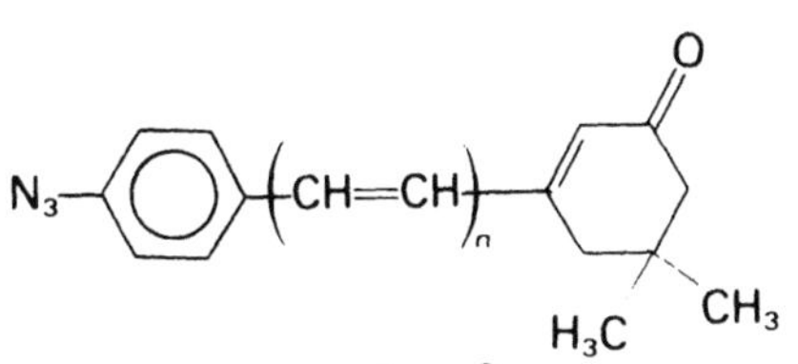

$n = 1$ or 2

XXVIII

Several simple negative working deep-UV resists have been reported but have not been widely used because they have not been capable of high resolution. Resists based on bisazides with cyclized polyisoprene can be made sensitive in the deep-UV but cannot resolve at the necessary level due to swelling.[92] A resist with less swell was formulated from a short-wavelength absorbing bisazide and poly(4-hydroxy styrene).[93]

Many negative working imaging systems based upon photoinitiated polymerization have been investigated.[94] Most of these systems have an inert binder and a multifunctional acrylate or methacrylate monomer along with a photolabile polymerization initiator. Because these systems are difficult to spin-coat and are sensitive to the amount of oxygen in the coatings, they have been little used in microlithography.

One microlithographic application for photoinitiated polymerization involves locking inorganic atoms into the exposed areas.[95] Silicon-containing monomers are generally used because of availability and because silicon forms a non-volatile oxide (SiO_x) in the presence of an oxygen plasma. A heating step after exposure removes the monomer from the unexposed areas and densifies the polymer network in the exposed areas. Treatment with an oxygen RIE then gives the desired relief image without any solvent development step. This process is represented in Fig. 8.

The oxygen sensitivity of photoinitiated polymerization may be avoided by using cationic rather than free-radical polymer systems. There has been considerable activity to prepare sulfonium and iodonium salts which decompose to form a strong acid during

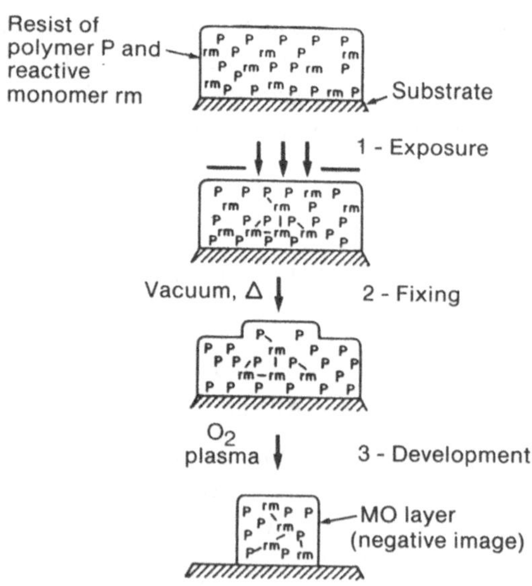

FIG. 8. Schematic of the 'photolocking' process used to bind etch-resistant molecules imagewise in a polymer film.

exposure.[96] Subsequent heating of the coating can cause several types of acid-catalyzed reactions, including polymerization.[65]

Images which are stable to high temperature have been produced by the photoinitiated polymerization of methacrylate groups pendant from a polyamic acid/ester backbone.[97] Polymerization of the methacrylates crosslinks the chain, allowing a developable image to be obtained. Subsequent heating decomposes the methacrylate esters and a polyimide image remains.[98]

4 SUMMARY

Optical microresists continue to be the patterning system of choice for the fabrication of VLSI devices. Improvements and extensions of diazonaphthoquinone-based resists provide the primary thrust for submicron lithography. The other most prominent new lithography concepts are the use of silicon in resists, and multilayer technologies. The design of new materials to be used in these three ways will drive manufacturing for the next few years. Further out on the manufacturing timeline, DUV resists (possibly using excimer lasers) and chemically amplified resist systems will have an impact.

REFERENCES

1. Taylor, G. N. *Solid State Technol.*, (June 1984) 105.
2. Lin, B. J. In *Introduction to Microlithography,* ACS Symposium Series No. 219, ed. L. F. Thompson, C. G. Willson & M. J. Bowden. American Chemical Society, Washington, DC, 1983, 287.
3. McDonnell Bushness, L. P., Gregor, L. V. & Lyons, C. F., *Solid State Technol.*, (June 1986) 133.
4. Schnabel, H. & Sotobayashi, H., *Prog. Polym. Sci.*, **9** (1983) 297.
5. Bowden, M. J., *Crit. Rev. Solid State Sci.*, (1979) 223.
6. Willson, C. G. & Bowden, M. J., In *Electronic and Photonic Applications of Polymers,* Adv. Chem. Series No. 218, ed. M. J. Bowden & S. R. Turner. American Chemical Society, Washington, DC, 1988, Ch. 2, p. 75.
7. Wölff, W., *Annalen,* **394** (1912) 25.
8. Süs, O., *Annalen,* **556** (1944) 65.
9. Pacansky, J. & Lyerla, J. R., *IBM J. Res. Develop.*, **23** (1979) 42.
10. Willson, C. G., In *Introduction to Microlithography*, ACS Symposium Series No. 219, ed. L. F. Thompson, C. G. Wills & M. J. Bowden. American Chemical Society, Washington, DC, 1983, Ch. 3, p. 87.

11. Hinsberg, W. D., Willson, C. G. & Kanazawa, K. K., *J. Electrochem. Soc.*, **133** (1986) 1448.
12. DeForest, W. S., *Photoresist Materials and Processes*, McGraw-Hill, New York, 1975.
13. Hinsberg, W. D., Willson, C. G. & Kanazawa, K. K., *Proc. SPIE, Int. Soc. Opt. Eng.*, **539** (1985) 6.
14. Ouano, A. C. In *Polymers in Electronics*, ACS Symposium Series No. 242, ed. T. Davidson. American Chemical Society, Washington, DC, 1984, Ch. 7.
15. Blum, L., Perkins, M. E. & McCullough, A. W., *Proc. SPIE, Int. Soc. Opt. Eng.*, **771** (1987) 148.
16. Daly, R. C., DoMihn, T., Arcus, R. A. & Hanrahan, M. J., In *Polymers for High Technology*, ACS Symposium Series No. 346, ed. M. J. Bowden & S. R. Turner. American Chemical Society, Washington, DC, 1987, Ch. 20, p. 237.
17. Lussier, B. (to Eastman Kodak Company), US Patent 4 684 597 (1987).
18. Pacansky, J., *IBM J. Res. Develop.*, **23** (1979) 42.
19. Trefonas, P., III & Daniels, B. K., *Proc. SPIE, Int. Soc. Opt. Eng.*, **771** (1987) 194.
20. Pampalone, T. R., *Solid State Technol.*, (June 1984) 115.
21. Kalle Corp., DBP 879 205 (14 Sept. 1949). (German Patent No.)
22. Frechet, J. M. J., Eichler, E., Ito, H. & Willson, C. G., *Polymer*, **24** (1983) 995.
23. Hanrahan, M. J. & Hollis, K. S., *Proc. SPIE, Int. Soc. Opt. Eng.*, **771** (1987) 128.
24. Turner, S. R., Arcus, R. A., Houle C. G. & Schleigh, W. R., *Polym. Eng. Sci.*, **26** (1986) 1096.
25. Turner, S. R., Schleigh, W. R., Arcus, R. A. & Houle, C. G., *Proc. Interface '86, Microelectronics Seminar*, Eastman Kodak Company, Rochester, NY, 1986, p. 158.
26. Osuch, C. E., Brahim, K., Hopf, F. R., McFarland, M. J., Mooring, A. M. & Wu, C. S., *Proc. SPIE, Int. Soc. Opt. Eng.*, **631** (1986) 68.
27. Hinsberg, W. D. & Gutierrez, M. L., *Proc. Interface '83, Microelectronics Seminar*, Eastman Kodak Company, Rochester, NY, 1983, p. 52.
28. Arcus, R. A., *Proc. SPIE, Int. Soc. Opt. Eng.*, **631** (1986) 124.
29. Guild, J. (to Eastman Kodak Company), US Patent 4 423 138 (1984).
30. MacDonald, S. A., Miller, R. D., Willson, C. G., Feinberg, G. M., Gleason, R. T., Halverson, R. M., MacIntyre, M. W. & Motsiff, W. T., *Proc. Interface '82, Microelectronics Seminar*, Eastman Kodak Company, Rochester, NY, 1982, p. 114.
31. Moritz, H. & Pool, G. (to IBM Corporation), US Patent 4 104 070 (1978).
32. Grant, B. D., Clecak, N. J., Twieg, R. G. & Willson, C. G., *IEEE Trans. Electron Devices*, **ED28** (1981) 1300.
33. Hofer, D. C., Willson, C. G., Newreuther, A. R. and Hakey, M., *Proc. SPIE, Int. Soc. Opt. Eng.*, **334** (1982) 196.
34. McCullough, A. W., *J. Electrochem. Soc.*, **128** (1981) 225.
35. Hakey, M. and Straub, W., *J. Electron. Mater.*, **11** (1982) 813.

36. Willson, C. G., Miller, R., McKean, D., Clecak, N., Thompkins, T., Hofer, D., Michl, J. & Downing, J., *Polym. Eng. Sci.*, **23** (1983) 1004.
37. Itoh, T., Yomashita, Y., Kawaza, R., Kawamura, K., Ohno, S., Asano, T., Kobayashi, K. & Nagamatsa, G. In *Photopolymers principles—processes and materials*, Proc. SPE Mid-Hudson Sec., (1985) 65.
38. Gipstein, E., Ouano, A. C. & Thompkins, T., *J. Electrochem. Soc.*, **129** (1981) 201.
39. Willson, C. G., Miller, R. D., McKeon, D. R. & Pederson, A., *Proc. SPIE, Int. Soc. Opt. Eng.*, **771** (1987) 2.
40. Reichmanis, E., Wilkens, C. W., Jr & Chandross, E. A., *J. Vac. Sci. Technol.*, **19** (1981) 1338.
41. Reichmanis, E., Gooden, R., Wilkens, C. W., Jr & Schonhorn, H., *J. Polym. Sci., Polym. Chem. Ed.*, **21** (1983) 1075.
42. Lin, B. J., *J. Vac. Sci. Technol.*, **12** (1975) 1317.
43. Nakane, Y., Tsumori, T. & Mifume, T., *Proc. Interface '83, Microelectronics Seminar*, Eastman Kodak Company, Rochester, NY, 1983.
44. Mimura, Y., Ohkubo, T., Takeuchi, T. & Sekikawa, K., *Jap. J. Appl. Phys.*, **17** (1978) 541.
45. Tsuda, M., Oikawa, S., Nakamura, Y., Nagata, H., Yokota, A., Nakane, H., Tsumori, T., Nakane, Y. & Mifune, T., *SPSE*, **23** (1979) 290.
46. van Pelt, P., *Proc. SPIE, Int. Soc. Opt. Eng.*, **275** (1981) 150.
47. Yomashita, Y., Ogura, K., Kanishi, M., Kawaza, R., Ohno, S. & Mizokami, Y., *J. Vac. Sci. Technol.*, **16** (1979) 2026.
48. Nakane, Y., Tsumori, T. & Mifumi, T., *Proc. Interface '78, Microelectronics Seminar*, Eastman Kodak Company, Rochester, NY, 1978.
49. Wilkins, C. W., Jr, Reichmanis, E. & Chandross, E. A., *J. Electrochem. Soc.*, **127** (1980) 2510.
50. Chandross, E. A., Reichmanis, E., Wilkins, C. W., Jr & Hartless, R. L., *Solid State Technol.*, (Aug. 1981) 81.
51. Reichmanis, E., Wilkins, C. W., Jr & Chandross, E. A., *J. Electrochem. Soc.*, **127** (1980) 2514.
52. Hartless, R. L. & Chandross, E. A., *J. Vac. Sci. Technol.*, **19** (1981) 1333.
53. Hiracka, H., Welsh, W. L., Jr & Bargon, J., *J. Vac. Sci. Technol.*, **21** (1983) 1062.
54. Lin, B. J., Bassons, E., Chao, V. W. & Petrillo, K. E., *J. Vac. Sci. Technol.*, **19** (1981) 1313.
55. Lin, B. J., Chao, V. W., Petrillo, K. E. & Yang, B. J. L. In *Photopolymers principles—processes and materials*, Proc. SPE Mid-Hudson Sec., (1985) 75.
56. deGrandpre, M. P., Vidusek, D. A. & Legenza, M. W., *Proc. SPIE, Int. Soc. Opt. Eng.*, **539** (1985) 103.
57. McCullough, A. W., Vidusek, D. A., Legenza, M. W., deGrandpre, M. & Imhof, J., *Proc. SPIE, Int. Soc. Opt. Eng.*, **631** (1986) 316.
58. Ito, H. & Willson, C. G. In *Polymers in Electronics*, ACS Symposium Series No. 242, ed. T. Davidson. American Chemical Society, Washington, DC, 1984, pp. 11–23.
59. Willson, C. G., Ito, H., Frechet, J. M. J., Tessier, T. G. & Houlihan, F. M., *J. Electrochem. Soc.*, **133** (1986) 181.

60. Ito, H., Willson, C. G. & Frechet, J. M. J. (to IBM). US Patent 4 491 628 (1985).
61. Crivello, J. V. In *Polymers in Electronics,* ACS Symposium Series No. 242, ed. T. Davidson. American Chemical Society, Washington, DC, 1984, pp. 3–11.
62. Turner, S. R., Ahn, K. D. & Willson, C. G. In *Polymers for High Technology,* ACS Symposium Series No. 346, ed. M. J. Bowden & S. R. Turner. American Chemical Society, Washington, DC, 1987, Ch. 17, p. 200.
63. Frechet, J. M. J., Bouchard, F., Houlihan, F. M., Kizka, B., Eichler, E., Clecak, N. & Willson, C. G., *J. Imaging Sci.,* **30** (1986) 59.
64. Frechet, J. M. J., Eichler, E., Stancialesca, M., Iizada, T., Bouchard, F., Houlihan, F. & Willson, C. G. In *Polymers for High Technology,* ACS Symposium Series No. 346, ed. M. J. Bowden & S. R. Turner. American Chemical Society, Washington, DC, Ch. 12, p. 138.
65. Jain, K., Willson, C. G. & Lin, B. J., *IEEE Electron Device Lett.,* **1–3** (1982) 53.
66. Lawes, R. A., *Semiconductor Int.,* (July 1986) 76.
67. Znotim, T., *Solid State Technol.,* (Sept 1986) 99.
68. Srinivasan, R., *J. Vac. Sci. Technol.,* **B-1** (1983) 932.
69. Yeh, J. T. C., *J. Vac. Sci. Technol.,* **A-4** (1986) 653.
70. Brarenoud, B. & Srinivasan, R., *J. Vac. Sci. Technol.,* **B-3** (1985) 913.
71. Srinivasan, R., *Science* (Oct. 1986) 559.
72. Reichmanis, E. & Smolinsky, G., *J. Electrochem. Soc.,* **132** (1985) 1178.
73. Shaw, J. M., Hatzakis, M., Paraszczak, J., Liatkus, J. & Babich, E. In *Photopolymers principles—processes and materials,* Proc. SPE Mid-Hudson Sec., (1982) 285.
74. Morita, M., Tanaka, A., Imamara, S., Tamamura, T. & Kogure, O., *Jap. J. Appl. Phys.,* **22** (1983) L659.
75. Hofer, D. C., Miller, R. D. & Willson, C. G., *Proc. SPIE, Int. Soc. Opt. Eng.,* **469** (1984) 16.
76. Williams, J. L. R., *Fortschritte der chemischen Forschung.* Springer-Verlag, Berlin, Bd 13(2) (1969) pp. 227–50.
77. Delzene, G. A., *Eur. Polym. J., Suppl.,* (1969) 55–91.
78. Delzene, G. A., *Makromol. Chem.,* Suppl. 2 (1979) 169–88.
79. Hephor, M., *J. Photogr. Sci.,* **12** (1964) 181.
80. Kosar, J., *Light-Sensitive Systems,* John Wiley, New York, 1965.
81. Reiser, A., *J. Chim. Phys. Phys.-Chim. Bull.,* **77** (1980) 469.
82. Li, M. Y., Pearce, E. M. & Narang, S., *Proc. SPIE, Int. Soc. Opt. Eng.,* **771** (1987) 46.
83. Deforest, W. S., *Photoresist Materials and Processes,* McGraw-Hill, New York, 1975, ch. 2.
84. Aquihatri, R. K., Falcon, D. L., Hood, F. P., Lesoine, L. G., Needham, C. D. & Offenback, J. A., *Photopolymers: Principles Processes and Materials,* Proceedings of the 1970 SPE Conference, 1970, p. 33.
85. Sagura, J. J. & van Allen, J. A. (to Eastman Kodak Company), US Patent 2 732 301 (1960).
86. Delzene, G. A., *Photopolymers: Principles, Processes and Materials,* Proceedings of the 1967 SPE Conference, 1967, p. 5.

87. Yang, J.-M., Chiong, K., Yan, H.-J. & Chow, M.-F., *Proc. SPIE, Int. Soc. Opt. Eng.*, **469** (1984) 117.
88. Tomioka, H., *Proc. SPIE, Int. Soc. Opt. Eng.*, **539** (1985) 151.
89. Sanada, S. I. & Miyamura, M., *Proc. SPIE, Int. Soc. Opt. Eng.*, **631** (1986) 187.
90. Nonagaki, S., Hashimoto, M., Iwayanagi, T. & Shiraishi, H., *Proc. SPIE, Int. Soc. Opt. Eng.*, **539** (1985) 189.
91. Koibuchi, S., Isobe, A., Makimo, D., Iwayanagi, T., Hashimoto, M. & Nonagaki, S., *Proc. SPIE, Int. Soc. Opt. Eng.*, **539** (1985) 182.
92. Iwayanagi, T., Kohashi, T. & Nonagaki, S., *J. Electrochem. Soc.*, **127** (1980) 2759.
93. Iwayanagi, T., Kohashi, T., Nonagaki, S., Matsuzawa, T., Douta, K. & Yonazawa, H., *IEEE Trans. Electrochem. Devices,* **28** (1981) 1306.
94. Labana, S. S., *J. Macromol. Sci., Rev. Macromol. Chem.*, **C11** (1974) 299.
95. Taylor, G. N., Wolf, T. M. & Stillwagon, L. E., *Solid State Technol.*, (Feb. 1984) 145.
96. Crivello, J. V. In *UV Curing: Science and Technology*, ed. S. P. Pappas. Technology Marketing Corp., Stanford, CT, 1978, pp. 23–77.
97. Rubner, R., Ahnu, H., Kuhn, E. & Kolodzig, G., *J. Photogr. Sci. Eng.*, **23** (1979) 305.
98. Rohde, O., Riediker, M., Schaffner, A. & Bateman, J., *Proc. SPIE, Int. Soc. Opt. Eng.*, **539** (1985) 175.

Chapter 4

PHOTOGRAFTING PROCESSES ONTO POLYMERS

IGNAZIO RENATO BELLOBONO & ELENA SELLI
Università degli Studi di Milano, Italy

1 INTRODUCTION

Chemical or physical modification of the surface properties of polymeric materials is undoubtedly a fascinating field for research as well as a practical way to improve their value. Surface and not bulk properties determine the most relevant aspects of the behaviour in many technologically important fields, such as dyeing, printing, adhesion, membrane separation processes, and so on. Their influence, which is already so decisive in films and moulded materials, becomes of the utmost importance in fibres.

Among the physical methods of modification, polymerization and grafting by high-energy radiation have been intensively studied during the last three decades. Classical research work by Dainton and others has first shown that the mechanism was free radical in nature;[1] but also an ionic polymerization could be demonstrated in the liquid state using high-energy radiation[2] and fundamental processes in radiation chemistry could be rapidly established.[3–6]

Observations that chemical changes were induced onto a natural polymer, such as cellulose, by exposure to low-energy electromagnetic radiation in the visible and ultraviolet region date back to 1883.[7] But it was not until the late 1950s (the pioneering work of Oster *et al.*[8,9] and of Stannett *et al.*[10,11] has to be cited in this context) that a substantial amount of research was carried out on photoinitiated grafting and graft polymerization onto polymer surfaces, particular attention being focused on fibrous materials.

With the wide variety of photochemically reactive monomers available today, photografting promises to become established as a method for modifying efficiently the surface properties of fibres and polymers in general: from antistatic behaviour to thermal stability, from adhesion to bactericidal properties, from water repellency to hygroscopicity, from soil resistance to flame-retardant properties, from optical imaging to dyeability, from separation ability towards microfiltration and ultrafiltration to selectivity towards gases. Photochemistry may consequently play a very important role in this field, in promoting basic knowledge relating to photografting processes, in studying applications, as well as in optimizing conditions of industrial processes potentially capable of being developed.

An extensive review concerned primarily with photoinitiated reactions of cellulose, particularly with graft polymerization reactions, and covering most of the existing literature up to the late 1970s appeared in 1980.[12] This same argument has been dealt with, in a more general frame of chemistry and technology of cellulosic copolymers, together with characterization and properties of graft copolymers (such as proof of grafting, molecular structure, molecular masses and morphology of grafted branches).[13] Furthermore, systematic reporting of the current literature related to photografting processes is easily available.[14–16]

Consequently, the present review will be devoted mainly to some more recent literature aspects of photografting processes, as well as to a critical examination of some relevant variables.

A high level of research interest and activity in ultraviolet radiation-curable treatments for textiles and polymer fibres was achieved in the late 1960s and in the 1970s.[17,18] However, only in the late 1970s and in the 1980s was distinct and methodical attention drawn towards photophysical parameters which characterize grafting.

2 INFLUENCE OF PHYSICOCHEMICAL PARAMETERS ON PHOTOGRAFTING

Three main different methods of grafting onto natural and synthetic polymers using high-energy radiation have been developed:

(1) *the mutual method*, by which the polymer is irradiated directly in

the presence of the monomer, the latter being in the liquid phase (alone or in solution) or in the vapour phase;

(2) *the preirradiation method,* by which the polymer is irradiated and successively treated with the monomer or monomer solution;

(3) *the peroxide method,* by which the polymer is irradiated in air, leading to the production of peroxides and hydroperoxides, the latter being successively decomposed in the presence of monomer or monomer solution.

Only method (2) is able to eliminate almost completely the undesired production of homopolymers.

These same three methods can, in principle, be used also for photografting, produced by low-energy radiation. For photografting, however, method (1) is by far the most commonly employed, even if method (3) deserves attention.

One physical factor differentiates high- and low-energy radiation. Penetration of the latter into polymer samples is considerably restricted because of its lower energy compared with that of gamma rays or electron beams. This restriction results in a localized reaction limited to the polymer surface. But this is not necessarily a drawback. For many applications, including modification of surfaces, membrane properties and so on, it may result in a selective advantage. Rationalization of physicochemical parameters able to influence photografting processes undoubtedly leads not only and obviously to a better understanding of underlying phenomena, but also to a spread of useful and interesting applications.

2.1 Solvent Effects

Solvent effects in the liquid phase have been previously observed in photoinduced systems,[19–21] as well as in gamma-ray-induced graft copolymerization onto cellulose.[22] With the high-energy radiation and using styrene as a typical monomer, grafting efficiency onto cellulose was found[22] to decrease rapidly as the molecular weight of the solvent alcohol increased. Methanol and ethanol were the most suitable alcohols; little copolymerization was achieved with butanol or higher-molecular-weight alkanols. In a parallel photosensitized study,[23] a Trommsdorff effect was observed, similar to that of a high-energy radiation study, but the peak in ultraviolet grafting was shifted from 30% concentration of monomer in methanol to about 80–90%. This difference in position of the peaks was attributed[23] to diffusional limitation, of the sensitizer and/or solvent radicals into the polymer to abstract hydrogen atoms, in the photografted system. Another characteristic feature of photografting[23] was that grafting was enhanced when a poor grafting solvent (isobutanol or *n*-octanol) was added to a solvent where copolymerization readily occurred, such as methanol. The explanation could be a compromise between the role of methanol, which is capable of swelling the trunk polymer, and of the higher-molecular-weight alcohol, which is a better solvent for the homopolymer and decreases the turbidity effect of the latter which inhibits grafting. A further plausible explanation based on monomer-solvent complexes with long-chain alcohols was suggested.[23] The role of solvent should not be separated from its interaction with the substrate, as has previously been underlined for UV grafting onto polypropylene.[24] In any grafting system, a model has been proposed[25] which considers an equilibrium concentration of monomer adsorbed within the grafting region of the backbone polymer. This grafting region may be continually changing, as grafting proceeds, as a consequence of the surface modification by grafting itself. Monomer sorption and partitioning depend on the polarities of monomer, substrate and solvent, as well as on concentration of ionic and non-ionic additives.

Some aspects of the photografting process have been pointed out by the study[26–28] of self-sensitized grafting of acrylated diazenes **I–IV** as model molecules onto polypropylene, polycaprolactam and poly(ethylene terephthalate) films or fibres, in dilute *N,N*-dimethylformamide (DMF) or toluene solutions. One feature which has been found substantially similar in all cases investigated was the general shape of the graft yield curve (Fig. 1). Two consecutive kinetic processes may be recognized; the first has been shown to be grafting of monomeric or oligomeric species, up to a more or less pronounced induction period (plateau of graft yield curve vs time of Fig. 1). Then a step-by-step and slower graft polymerization follows. When operating in solution, the dependence of quantum yields Φ_1 and Φ_2 for the two processes (see Fig. 1) on acrylated dye concentration closely followed Stern–Volmer plots, by which lifetimes of intermediate excited states for the two kinetic processes (τ_1 and τ_2 respectively) could also be obtained. The quantum yields extrapolated to zero concentration by Stern–Volmer plots, $(\Phi_0)_1$ and $(\Phi_0)_2$ respectively, as a function of irradiation wavelength are illustrated graphically in Fig. 2. In this

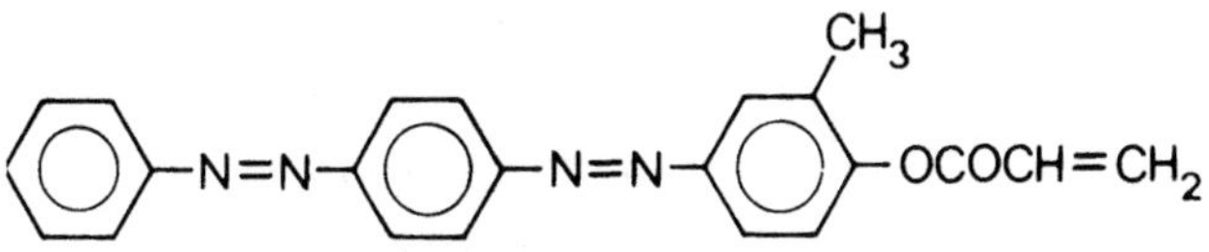

I

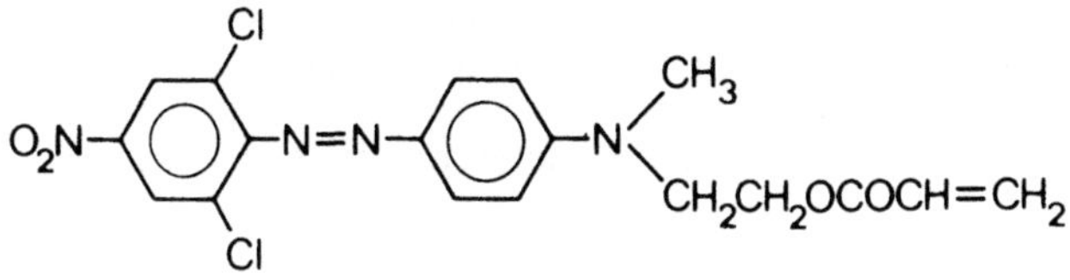

II

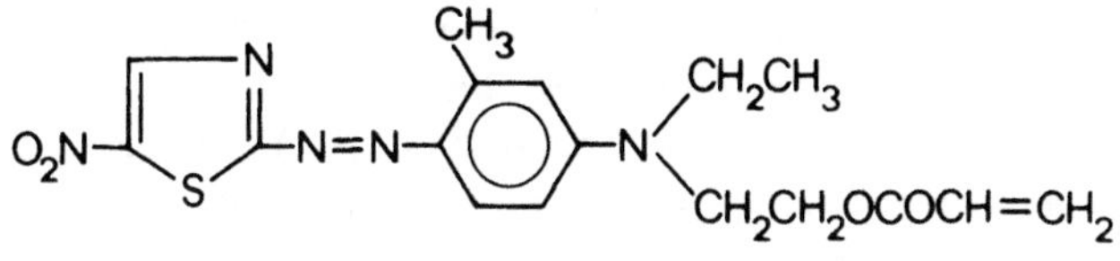

III

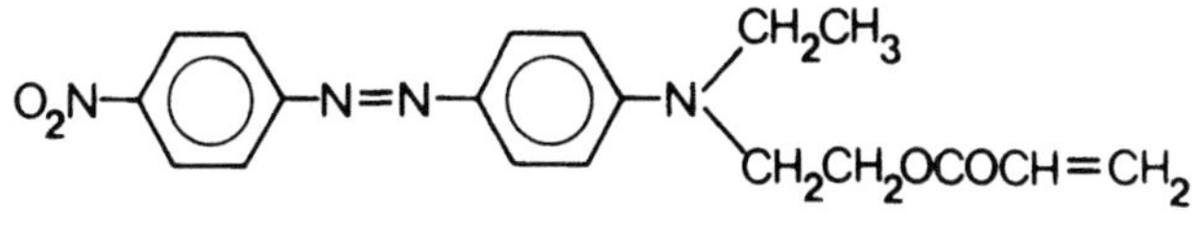

IV

figure, the molar absorptivities of the investigated dyes in DMF are also plotted as a function of wavelength for comparison. A parallel study carried out in toluene solution with acrylated dye **IV**[27] enabled the effect of solvent on quantum yields to be studied.

The effect of concentration on quantum yields Φ_1 and Φ_2 and the relatively long lifetimes of the intermediate excited states (of the order of microseconds) are only consistent with triplet intermediates. This was already observed in some cases of reactions of compounds which

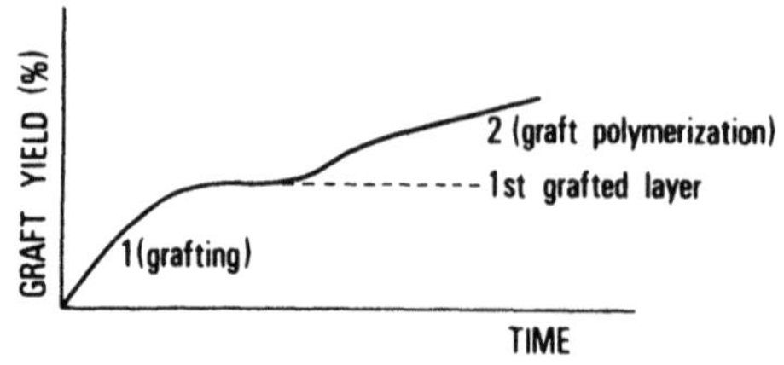

FIG. 1. Typical shape of kinetic curve for grafting (1) and successive step-by-step graft polymerization (2) of monomers onto the first grafted layer, for acrylated dyes **I–IV** in dilute solutions. Rate equation: $(1/S)\,dn/dt = \Phi_{1,2}I$; S = apparent surface of polymer; n = moles of dye at time t; Φ = mean quantum yield; I = radiation intensity effectively absorbed.

have a carbonyl group and a conjugated double bond capable of twisting in an excited state to a non-planar geometry.[29] Furthermore, the constancy of lifetimes τ_1 and τ_2, even as a function of wavelength, for these two processes supports this view. A plausible explanation of the unusual concentration quenching lies in the assumption that this intermediate triplet is an upper excited state T_2 deriving from the lowest excited singlet (S_1) populated by light absorption and leading to hydrogen abstraction from the trunk polymer:

$$S_0 \rightarrow S_1 \rightarrow T_2 \rightarrow \text{hydrogen abstraction } (\Phi_1, \Phi_2) \qquad (1)$$

The long lifetime of the reactive triplet should imply that internal conversion from $T_2\,^3(n, \pi^*)$ to the lowest excited triplet $T_1\,^3(\pi, \pi^*)$ is apparently inefficient. The spin-allowed triplet energy transfer between similar molecules in different states, such as:

$$S_0 + T_2 \rightarrow T_1 + S_0 \qquad (2)$$

contributes significantly to the process. The fate of T_1, formed by eqn (2), would have no effect on the quenching behaviour, provided T_1 fails to undergo the grafting reaction. The occurrence of hydrogen abstraction from only one of the two triplets may then be related to the strong effect of wavelength, which has been observed.[26,27] Grafting reactions in the investigated system thus appear to be initiated by long wavelengths (self-sensitizing action of acrylated dyes) but proceed exclusively through upper excited T_2 states. Low but finite quantum yields were in fact observed (see Fig. 2) around 350 nm, whereas higher efficiencies occur both at shorter and longer wavelengths, which do not reflect (at high frequency) the absorption spectrum. This

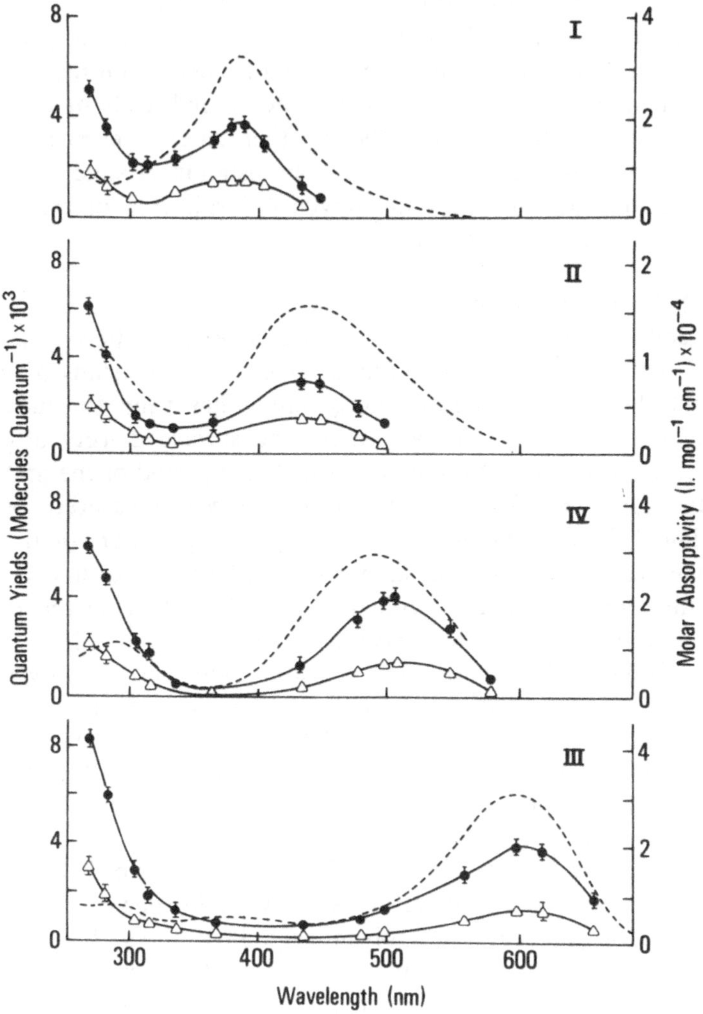

FIG. 2. Ordinate to the left: quantum yields (molecules quantum^{-1}), during the first, $(\Phi_0)_1$●, and second, $(\Phi_0)_2$△, constant-rate periods for the photochemical grafting and step-by-step graft polymerization of acrylated dyes **I–IV** from DMF solutions onto polypropylene fibres, at 25°C, extrapolated to zero concentration of dye by Stern–Volmer plots, as a function of wavelength (nm). Ordinate to the right: molar absorptivities of **I–IV** in DMF solutions (broken lines), as a function of wavelength (nm).[26]

behaviour is again consistent with the explanation that hydrogen abstraction occurs entirely from the triplet state T_2.

Experimental data[26–28] do not provide direct evidence of geometrical isomerization, but it is quite probable that isomerization reactions may proceed either directly from S_1 or S_2 states or indirectly by crossing of S_2 to T_1 states. Unfortunately the fast thermal *cis*→*trans* reaction in this system, as well as the shift of the photostationary state towards the *trans* isomer, do not allow unequivocal clarification of the possible roles of these states in this way.

2.2 Influence of Substrate

The connection between solvent and substrate effects has been mentioned in the preceding section. A convenient means of making evident these effects during photografting has been to study, with model molecules **I–IV** already cited, the surface density of grafted molecules at the end of the first constant-rate period of the graft yield kinetic curve in Fig. 1. This quantity was measured both in solution,[26–28] with different solvents and trunk polymers (and also by varying the physical form of the polymer surface, film or fibre), and by operating without solvent, with dyes **I–IV** adsorbed on the polymer surface itself.[30,31] The dye concentration is the determining influence on this value. Of course, when operating with adsorbed dyes (dyes deposited on the polymer surface), on the contrary, the parameter which may be varied is the mass of dye adsorbed per unit apparent surface. In order to compare the two cases, consequently, concentration has to be expressed for both as n/S, i.e. mass (in moles, n) per unit apparent surface (S).

Schematically, as represented in Fig. 3, two types of behaviour can be distinguished. When operating with dyes adsorbed on the polymer surface in the absence of any solvent [Fig. 3(a)], the surface density of grafted molecules is dependent on the kind of polymer as well as on n/S. This latter dependence has been found fairly linear, in the range of conditions employed, but the slope is independent of the kind of polymer. This clearly expresses the contribution of grafting of an oligomolecular layer during the first constant-rate period of the graft yield curve.

When operating in solution, on the contrary, the concentration of dye has apparently no effect on the surface density of the grafted molecules in the first grafted layer [Fig. 3(b)], and this supports the idea of a nearly monomolecular layer for this stage, when grafting is

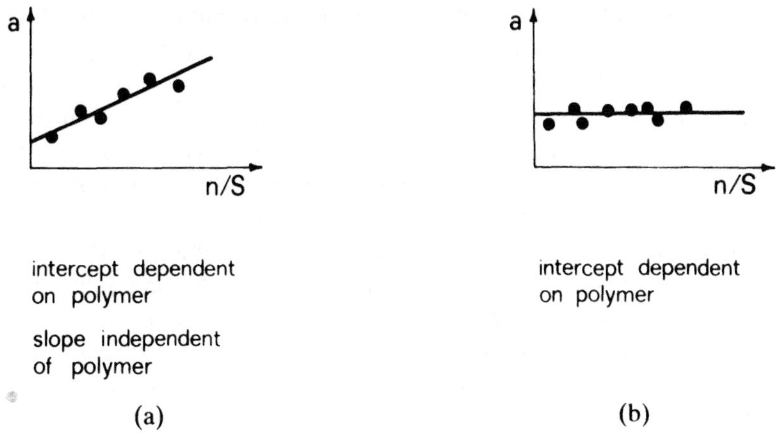

FIG. 3. Typical shapes of the plot of surface density of grafted molecules at the end of the first constant-rate period of photografting, a, as a function of n/S.[31] Dye adsorbed (a) in the absence of solvent; (b) in solution.

carried out from dilute dye solution where the homopolymer has a negligible possibility of being grafted.

Finally, as this value is based on apparent surface, it is also a rough estimate of the ratio between 'true' surface area and apparent geometrical surface area, in the case when films and fibres of a woven fabric of the same polymer are compared. For acrylated dye **IV**, as an example, the surface density from DMF solutions[28] at the end of the first constant-rate period onto polypropylene and polycaprolactam films is $1{\cdot}2 \pm 0{\cdot}3$ and $1{\cdot}8 \pm 0{\cdot}3$ molecules/100 Å^2 respectively, compared with 2·7 and 3·8 molecules/Å^2 for woven fabrics of these same materials,[30] the latter values being extrapolated at $n/S \rightarrow 0$ from plots such as that to the left of Fig. 3. In this way the chemical as well as the physicochemical nature of the polymer surface may be made manifest.

The influence of the chemical nature of the polymer has been confirmed[28,30,31] to decrease in the order: polypropylene > polycaprolactam > poly(ethylene terephthalate) for quantum yields Φ_1. Quantum yields Φ_2, on the contrary, have been found to be independent of the nature of the polymer.

2.3 Influence of Photochemical Parameters

The determination and the use of quantum yields of grafting, which have not been so common and widespread in the literature, is of primary importance to obtain mechanistic information as the discussion in Section 2.2 has already underlined. Self-sensitized grafting of acrylated dyes **I–IV**, for example, occurs, in the absence of any added photoinitiator, with very low quantum yields, as may be seen in Fig. 2. The effect of 1,2-diphenyl-2,2-dimethoxyethanone as photoinitiator on grafting of diazenes **I-IV** onto polypropylene and polycaprolactam fibres has been investigated by polychromatic irradiation.[31] Mean quantum efficiencies have been evaluated as a function of the ratio R between n/S of the photoinitiator and the corresponding value of the dye. While Φ_2 values were practically unaffected by the presence of the photoinitiator, Φ_1 values, on the contrary, were markedly influenced with an acceleration factor of about 30, when R was of the order of 10. That is, in eqn (3)

$$\Phi_1 = \Phi_1^0 + mR \tag{3}$$

where Φ_1^0 denotes the quantum yield of grafting of block copolymer in the absence of the photoinitiator, the ratio between the slope m and Φ_1^0 was about 3,[31] and was independent both of the kind of polymer and of the molecular structure of acrylated dyes **I–IV**.

In a successive study,[32] the validity of eqn (3) was ascertained even during the photoinitiated grafting of several diacrylates in the presence of dyes **I–IV**, with an acceleration factor of the order of magnitude of 10^4, as expressed by the m/Φ_1^0 ratio. As an example, for dye **IV** and 1,6-hexanediol diacrylate as comonomer, with $R = 0{\cdot}005$, a grafting yield of 72 ± 7 molecules of comonomer per quantum may be attained onto poly(ethylene terephthalate) film, corresponding to about this mean number of monomer units in the block copolymer obtained at the end of the first constant-rate period. By means of quantum yield measurements, the fraction resulting from dye sensitization could also be evaluated.[32]

Very high quantum yields may equally be evaluated from a study of photografting of epoxy–diacrylate oligomer onto poly(vinyl chloride),[33] as could be obtained kinetically in laser-induced polymerization of multifunctional acrylate systems.[34]

Information deriving from photochemical parameters enabled formulation of some mechanistic details for the photografting process.[32]

The primary photochemical event presumably consists in a dye-sensitized [(4a)] and/or photoinitiator radical-induced hydrogen abstraction [(4b), (4c)] from the reactive sites of the polymer surface [cf.

eqn (1)].

D = dye skeleton —(CH_2CH_2OCO)—
R_1R_2 = photoinitiator
A = monomer skeleton —(OCO)—
MH = photochemically reactive sites of the polymer surface

$$\mathrm{MH} + \underset{\mathrm{D}}{\underset{|}{\mathrm{CH}}}{=}\mathrm{CH_2} \xrightarrow{h\nu} \mathrm{M}\cdot + \cdot\underset{\mathrm{D}}{\underset{|}{\mathrm{CH}}}{-}\mathrm{CH_3} \tag{4a}$$

$$\mathrm{R_1R_2} \xrightarrow{h\nu} \mathrm{R_1}\cdot + \mathrm{R_2}\cdot \tag{4b}$$

$$\mathrm{MH} + \mathrm{R_1}\cdot(\mathrm{R_2}\cdot) \longrightarrow \mathrm{M}\cdot + \mathrm{R_1H}(\mathrm{R_2H}) \tag{4c}$$

Photochemically produced radical fragments from initiator R_1R_2, moreover, initiate polymerization in the homogeneous phase formed by comonomers alone (or their solution, if any):

$$\underset{\mathrm{D(A)}}{\underset{|}{\mathrm{CH}}}{=}\mathrm{CH_2} + \mathrm{R_1}\cdot(\mathrm{R_2}\cdot) \rightarrow \mathrm{R_1(R_2)}{-}\mathrm{CH_2}{-}\underset{\mathrm{D(A)}}{\underset{|}{\dot{\mathrm{C}}\mathrm{H}}}$$

$$\rightarrow \mathrm{R_1(R_2)}{-}(\mathrm{CH_2}{-}\underset{\mathrm{D(A)}}{\underset{|}{\mathrm{CH}}})_x{-}\mathrm{CH_2}{-}\underset{\mathrm{D(A)}}{\underset{|}{\dot{\mathrm{C}}\mathrm{H}}} \tag{4d}$$

$(x = 1, 2, 3 \cdots)$

The grafting mechanism (first constant-rate period of Fig. 1), may proceed substantially following two parallel reaction pathways: (1) grafting of a 'monomolecular layer' of acrylated monomer [reaction (5) followed by (6)]; (2) grafting of an oligomolecular layer of block copolymer formed either by photoinitiator [reaction (4d) followed by (8b)] or by dye-sensitized copolymerization [reaction (7) followed by (8a)].

$$\cdot\underset{\mathrm{D}}{\underset{|}{\mathrm{CH}}}{-}\mathrm{CH_3} + \underset{\mathrm{A}}{\underset{|}{\mathrm{CH}}}{=}\mathrm{CH_2} \rightarrow \cdot\underset{\mathrm{A}}{\underset{|}{\mathrm{CH}}}{-}\mathrm{CH_3} + \underset{\mathrm{D}}{\underset{|}{\mathrm{CH}}}{=}\mathrm{CH_2} \tag{5}$$

$$\mathrm{M}\cdot + \cdot\underset{\mathrm{D(A)}}{\underset{|}{\mathrm{CH}}}{-}\mathrm{CH_3} \rightarrow \mathrm{M}{-}\underset{\mathrm{D(A)}}{\underset{|}{\mathrm{CH}}}{-}\mathrm{CH_3} \tag{6}$$

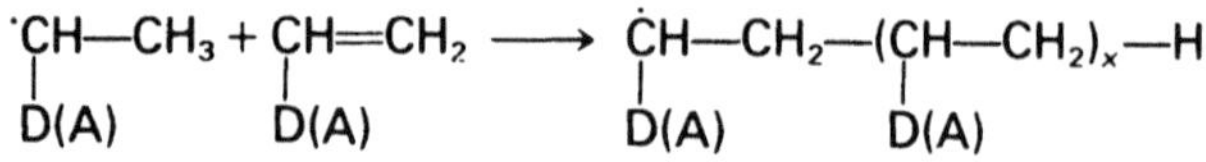

(7)

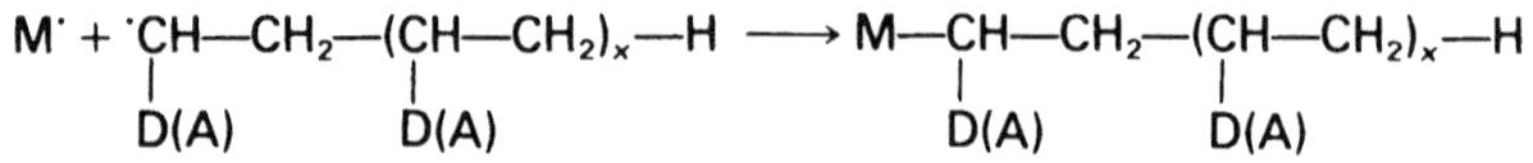

(8a)

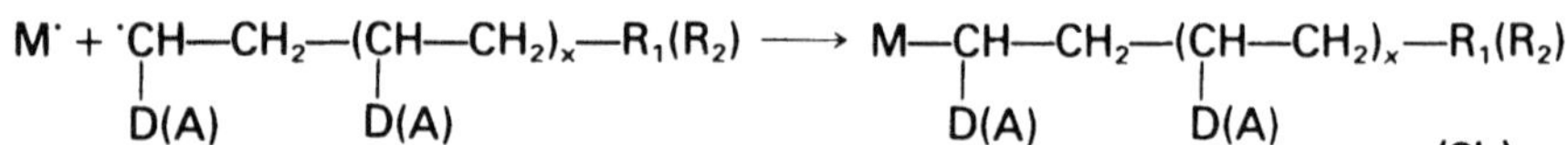

(8b)

$(x = 1, 2, 3 \cdots)$

When the yield Φ_1 is very high and strongly dependent on R, the prevailing reaction should be grafting of growing oligomeric chains onto the substrate [reactions (8a) and (8b), whether dye-sensitized or not]. The mechanism above requires that the composition of the first grafted layer simply reflects the ratio between comonomers. This has always been found in the absence of any solvent[32,35] with acrylated dyes **I–IV** as 'tracers', and also, within certain ranges of experimental conditions, for the photografting of donor and acceptor monomer couples, such as styrene–acrylonitrile, onto cellulose, in the presence of 20 wt% water.[36]

During the second constant-rate period (see Fig. 1), the most probable mechanism is a step-by-step polymerization on the first grafted layer. A reaction scheme such as the following one may be outlined:

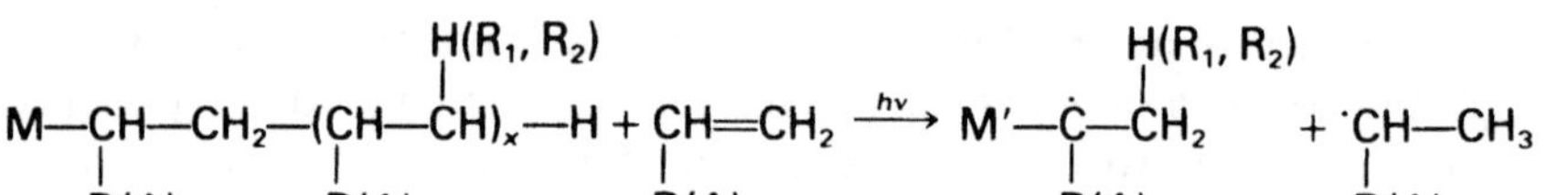

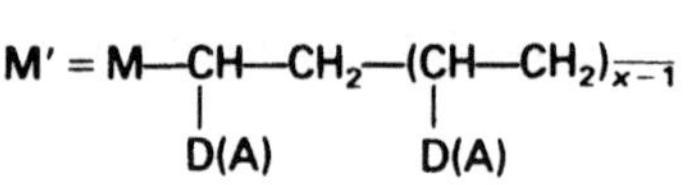

and $x = 1, 2, 3 \ldots$ (9)

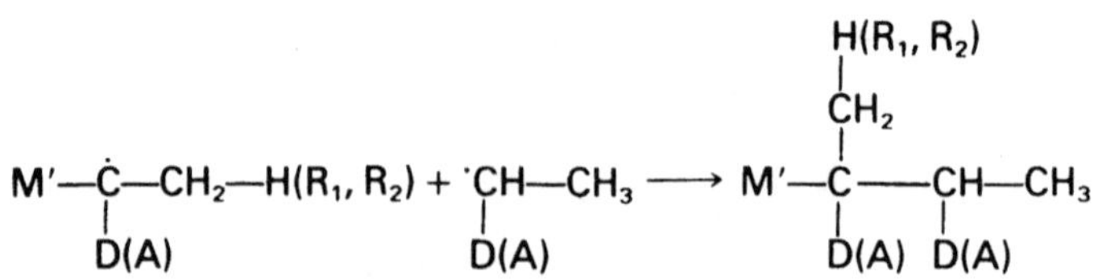

(10)

and so on, for other secondary carbon atoms of the first grafted polymeric chain.

The limiting unit value[32] of Φ_2, when $n/S \rightarrow 0$ and when grafting is performed in a condensed viscous medium of monomers or comonomers, suggests that the mechanism of reaction (9) should be that of a monomolecular process leading finally to chain termination. In other words, chain propagation by radical transfer, such as reaction (11) requiring $\Phi_2 > 1$, would not be favoured.

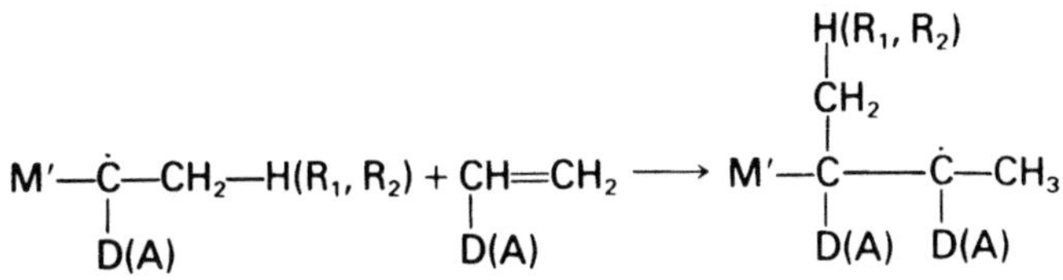

(11)

The value of Φ_2, much lower than Φ_1 in the above conditions,[32,37] allowed the experimental detection of a small quenching effect, of the Stern–Volmer type, consistent with that observed in dilute solutions of dyes **I–IV**, [28] as well as with lifetimes of intermediate excited states measured in solution, if account is taken of the medium effect, qualitatively at least.

The experimental finding (cf. Fig. 1)[30,37] of an induction time after the first grafting period, no matter how brief, indicates that the reactions which predominate in the second stage are different from those involved in the first stage. Otherwise a smooth curvature would be observed in the transition between the two processes, rather than a sharp variation in slope preceded by an induction phenomenon. Moreover, Φ_2 is independent of the photoinitiator concentration. This rules out the possibility that the photoinitiator acts as a radical producer in the second stage.

Among possible hypotheses, two appear worthy of consideration: (1) either radical formation from residual monomer molecules, possibly via energy transfer by photoinitiated triplets;[38] or (2) a similar process involving the pendant acrylate groups of the grafted chains, when using, for example, di- or tri-functional acrylates. Both these reactions should lead to a step-by-step graft polymerization onto the first grafted layer.

Kinetic analysis of unsaturation[39] (see Fig. 4) revealed that the quantum yields Φ_1 and Φ_2 obtained from gravimetric data on grafting were identical with those obtained from the residual acrylate functions

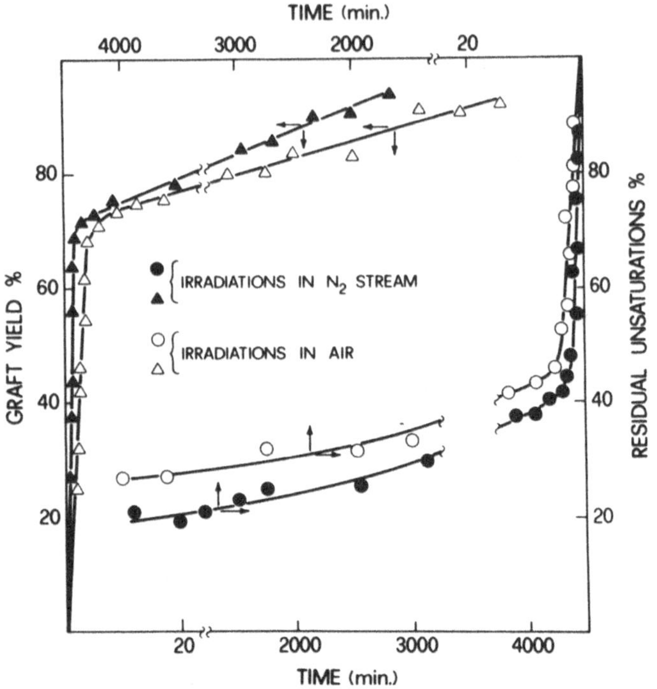

FIG. 4. Percentage graft yield (gravimetric analysis) and percentage residual unsaturation (acrylate functions, by multiple internal reflection IR spectroscopic analysis) of 35·0 mg cm^{-2} of prepolymer (containing 75 wt% of oligomer derived from glycidyl ether of bisphenol A and 25 wt% of trimethylolpropane triacrylate) onto cellulose, as a function of irradiation time, in the presence of 7·0 wt% 1,2-diphenyl-2,2-dimethoxyethanone as the photoinitiator (irradiation flux $(20{\cdot}5 \pm 0{\cdot}7) \times 10^{-8}$ einstein s^{-1} cm^{-2}). Irradiations were carried out either under a stream of nitrogen or in air.[39]

determined by multiple internal reflection IR spectroscopy, provided that one out of two acrylate functions of the epoxydiacrylate compound was considered as grafted, and two out of three acrylate functions of the trimethylolpropane triacrylate were similarly considered as grafted. A structure with one pendant acrylate group per monomer unit may then be assigned to the grafted polymer at the end of the first grafted period.

Further irradiation of the grafted polymer, for which a practically quantitative graft yield, as determined gravimetrically, had already been reached, reduced the residual unsaturation as illustrated in Fig. 4. In this way an even more highly crosslinked polymer could be obtained, even after the end of the second grafting period. The experimental rates r_c of photocrosslinking, calculated at various incident radiation fluxes, are shown in Fig. 5. Their dependence on the square root of the radiation intensity (classic bimolecular termination mechanism) and their values lower than those relating to Φ_2 which are linearly dependent on I, suggest that hypothesis (1) above should be

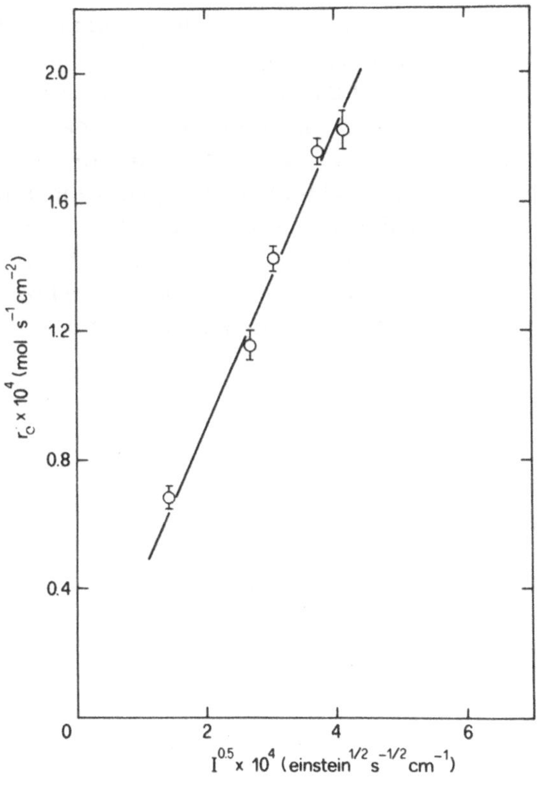

FIG. 5. Rate r_c of photocrosslinking (in moles of acrylate functions per square centimetre per second), as measured at the end of the photografting period by multiple internal reflection IR spectroscopic analysis, as a function of the square root of the absorbed radiation intensity $I^{0.5}$ per unit apparent cellulose surface, for 8·65 mg cm^{-2} of the same prepolymer as in Fig. 4, in the presence of 8·0 wt% of photoinitiator. Irradiations were carried out under a stream of nitrogen.[39]

favoured, probably because of a higher rate of radical formation from residual free monomer molecules than from less mobile pendant acrylate groups. That the latter reaction leads to photocrosslinking between reactive sites of grafted chains has been also demonstrated by modification of transport properties of the grafted layer through gas permeametric measurements.[40]

Besides, when operating in a liquid phase without any solvent, the second constant-rate period for photoinitiated grafting could not be observed,[37] unless the monomer adsorbed on the surface exceeded that which could be grafted at the end of the first constant-rate period, as shown in Fig. 6 for several diacrylate monomers. This quantity amounted to about 30 μmol cm^{-2}, which could be grafted,[37] in a certain range of experimental conditions, with a high rate corresponding to $\Phi_1 = 5{\cdot}3 \times 10^2 R$, where R is the ratio between molar concentration of the photoinitiator and that of monomer for the diacrylate monomers investigated, or with 100-fold greater values for triacrylates.[39]

The investigation of the influence of photochemical parameters in general is well established in the literature on grafting.[6,12] Among the more recent experimentation, a study[41] on unsensitized and photoinitiated polymerization reactions of glycidyl methacrylate from solu-

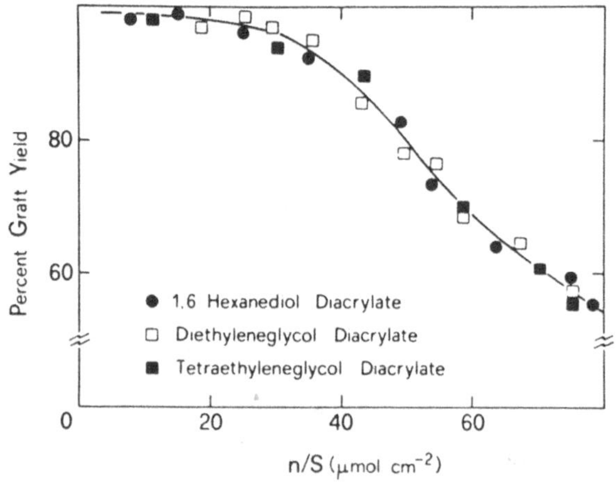

FIG. 6. Percentage graft yield measured at the end of the first constant-rate period as a function of n/S (moles of monomer adsorbed per unit apparent cellulose surface) for three diacrylate monomers, at various values of R (the ratio between molar concentration of photoinitiator and that of monomer) ranging from 0·025 to 0·095.[37]

tions of water and water–methanol with cotton cellulose fabrics deserves to be mentioned. If several layers of cotton fabric were immersed in solution, and only the first layer was exposed to irradiation, polymerization was also initiated in the inner layers. Chain transfer reactions were thought to be the main possible explanation. The effect of wavelength on polymer addition was: 300 nm > 350 nm > 254 nm, while on homopolymerization it was: 300 nm > 254 nm > 350 nm. The distribution of poly(glycidyl methacrylate) in the cotton fibres was independent of wavelength, and tended to be concentrated at the outer surface and to encapsulate them. Oxygen dissolved in the monomer solution is the most critical factor in reducing polymer addition, probably by reacting with the chain-propagating radical and terminating the reaction. Quantitative influences of oxygen on quantum yields of grafting have been pointed out in the literature[27,39] (cf. Fig. 4). In the study cited,[41] as the ratio of monomer solution to cotton cellulose increased, there was a small reduction in polymer addition at a given monomer concentration. Polymer addition increased with increasing water content of the methanol solution up to 50 vol.% water: then the solution separated into two phases, and polymer addition decreased. Polymer addition was directly related to reaction time and temperature. For reactivity of different monomers the order is as follows: methyl methacrylate > glycidyl methacrylate > 1,3-butylenedimethylacrylate > methacrylic acid > acrylonitrile > divinylbenzene.

2.4 Role of Photoinitiator System, of Temperature and of Radical Reactions

Initiation of graft copolymerization reaction of monomers with various backbone polymers, but to a great extent with cellulose, by ionic, charge-transfer, and free-radical processes has been widely reviewed.[6,12,14–16,42]

Some of the latest developments will be reported here. The influence of photoinitiator concentration on quantum yields in some grafting systems has already been discussed in the preceding section.

Classically, photografting has been dominated by studies involving organic or mixed organic/aqueous solvents. In view of the interest in the use of cellulosic copolymers in separation procedures by affinity chromatography, membrane fabrications and enzyme or whole-cell immobilization studies,[43] water-soluble photoinitiators were developed to operate in largely aqueous or polar media.[44–48] As the direct incorporation of a water-solubilizing group into an organic photoinitiator markedly reduces its efficiency, the chromophore was first protected by the use of bridging groups between it and the solubilizing function. By this method, however, unacceptably long cure times were the result. A second generation of photoactivators was then prepared, able to develop affinity with the substrate in terms of their anionic or cationic nature; two working examples are 4-(sulphomethyl)benzil sodium salt (SKS) and 4-(trimethylammoniummethyl)benzophenone chloride (BP2). Grafting of hydroxyethyl acrylate, hydroxypropyl acrylate, and hydroxyethyl methacrylate onto regenerated cellulose films was carried out, photoinitiated by BP2.[46] The monomer concentration of films at equilibrium uptake from monomer solutions was demonstrated to be quite different from bulk concentration: preferential adsorption depends on monomer concentration, being higher at lower concentration. The grafting efficiency depends on the availability of the monomer at the reactive site, its reactivity and monomer/solvent competition effects.

Diffusion of monomer to cellulose activated sites should occur within the lifetime of exposure, if UV is delivered as a series of regular bursts of radiation.[46] Radical inhibitors could reduce grafting and suppress homopolymerization. The photodecomposition of BP2 in aqueous solution was also investigated in this context[47] in the presence of alkanols and poly(vinyl alcohol) as hydrogen-donors. Various monomers were able to reduce the level of photodecomposition. The anionic water-soluble photoinitiator SKS, on the contrary, was rapidly decomposed when irradiated in aqueous solution, even in the absence of additional hydrogen-donating compounds.

The efficiency of SKS, moreover, was assessed, with or without dimethylaminoethanol as synergist, in the photoinitiated grafting of *N*-vinyl-2-pyrrolidone onto woollen substrates.[48] SKS is strongly adsorbed by woollen substrates, so there is the opportunity for radical sites to be generated on the surface of the wool. *N*-Vinyl-2-pyrrolidone has a low toxicity, improves flexibility of substrates to which it is cured, and, when grafted onto films, provides pervaporation membranes with preferential adsorption of particular components.[49] Solutions of *N*-vinyl-2-pyrrolidone were used to pad samples of wool, which were then irradiated. Without synergist, a maximum of grafting was seen at $pH = 4{\cdot}0$ (wool is known to be activated towards grafting at low pH): under acid conditions *N*-vinyl-2-pyrrolidone undergoes rapid hydrolysis to acetaldehyde and pyrrolidone, which combine with wool.

Grafting of *N*-vinyl-2-pyrrolidone onto cotton, using the vinyl sulphonyl dye Remazol Brilliant Blue as sensitizer, has also been investigated.[50] The maximal grafting was achieved with the greater level of association between the dye sensitizer and the cotton fabric. The sensitizer has the possibility of attachment to the fabric either by physical association or through the vinyl group. It acts only as a sensitizer, in two ways, depending on pH: at low pH through the quinone group, or involving hydrogen abstraction through the vinyl group, with subsequent homopolymerization and grafting. Electron micrographs show evidence of occluded homopolymer formation at pH = 10, and of fibre–fibre adhesion, thus giving support to the possibility of the two mechanisms for this system.

By a systematic series of studies,[51–61] photografting of methyl methacrylate from solutions in a variety of solvents onto various backbone polymers, in the presence of acriflavine as dye sensitizer, of charge-transfer complexes, *N*-bromosuccinimide, or peroxydiphosphate as photoinitiators, has been recently examined. The effects of monomer concentration, of amount (and consequently of surface area) as well as of the kind of substrate, of initiator concentration, of temperature, and of solvent have been investigated, with the aim of suggesting in each case a possible mechanism. Two main kinetic schemes have been used as a basis of interpretation of the experimental data:[51–61]

(1) a scheme compatible with a rate of copolymerization (calculated from graft yield kinetic curves) first-order-dependent on monomer concentration and one-half-order-dependent on photoinitiator (bimolecular termination of growing chains, both initially grafted onto the backbone polymer, similar to that occurring for classical photopolymerization in homogeneous phase);
(2) a second scheme distinguished from the first by a first-order dependence on initiator (termination by reaction of growing chains, grafted onto the substrate, with the solvent).

A type (1) mechanism is given for grafting onto silk in the presence of a lutidine–bromine charge-transfer complex,[51] onto nylon 6 fibres in the presence of γ-picoline–bromine,[52] onto cellulose fibres using the acid-catalysed decomposition of peroxydiphosphate as a photoinitiating system,[53] onto nylon 6 fibres using *N*-bromosuccinimide[54] or pyridine–bromine[58] as photoinitiator, onto cellulose using pyridine–bromine charge-transfer complex as initiator,[55] onto nylon 6 fibres using peroxydiphosphate,[56] onto poly(ethylene terephthalate) fibres using peroxydiphosphate[57] and onto nylon 6 fibres using acriflavine/ascorbic acid.[61]

The type (2) mechanism, on the contrary, is offered for graft copolymerization of methyl methacrylate onto wool using a peroxydiphosphate[60] or isoquinoline–sulphur dioxide charge-transfer complex[59] as photoinitiators.

The effect of solvent has been investigated by using various protic, dipolar aprotic, and non-polar solvents. The order of reactivity, in terms of percentage of grafting, generally follows the polarity of solvents. The effect of mixed solvents in the order of decreasing polarity (water–methanol > water–ethanol > water–*n*-propanol) was equally found in the photochemical graft copolymerization of methyl methacrylate onto nylon fibres.[62]

Fructose was also employed as a radical-producing system for photografting of acrylamide onto nylon fibres.[63,64] Potassium pervanadate has been found[65] to readily induce photograft copolymerization of acrylonitrile in neutral aqueous methanolic media onto starch, giving practically quantitative grafting efficiency and no formation of homopolymer. Termination in this system is considered to be mostly due to combination of growing chains with starch macroradicals. Redox initiators, such as oxalic acid–permanganate complex, were employed in photografting methyl methacrylate onto cellulose and oxycellulose, with limited amounts of water, with respect to the volume of monomer.[66] Potassium peroxydisulphate was, instead, used in the photograft copolymerization of methyl methacrylate on polyamine-modified oxycellulose in a similarly limited aqueous system.[67] Photograft copolymerization of methyl methacrylate onto natural rubber was, furthermore, carried out[68] using a quinoline-bromine charge-transfer complex as photoinitiator in benzene solution.

From the overall activation energy of the photografting process, an activation energy of initiation could be calculated,[51,52] not exceeding 8–30 kJ mol^{-1} for charge-transfer complex photoinitiation. The relatively large (45 kJ mol^{-1}) overall activation energy for photoinduced grafting on wool[59] was ascribed to its swellability, as well as to solubility and enhancement of the rate of diffusion of monomer.

Another very important aspect tightly correlated with the photoinitiator role is the elucidation of the nature and mechanisms of formation of free radicals on exposure to ultraviolet radiation.[69] Electron spin resonance (ESR) spectra in models or directly in the polymerizing systems are one of the most suitable techniques for providing this information. The ketone-photosensitized polymerization of *N*-methylol acrylamide with cotton[70] was investigated in this way. Similarly, ESR experiments showed that UV irradiation of starch, both in photosensitized and unphotosensitized conditions, resulted in chain scission and radical generation on the glucosidic ring by

TABLE 1
Proposed Structure[75] of Radicals Identified under UV Irradiation of Cellulose (1), Cellulose Acetate (2), Cellulose Triacetate (3), Glucose (4), Glucose Penta-acetate (5), Cellobiose (6), Cellobiose Octa-acetate (7), Maltose (8), Maltotriose (9), and Starch (10).

Radical	*Irradiated substrates*	*Recorded ESR signal*
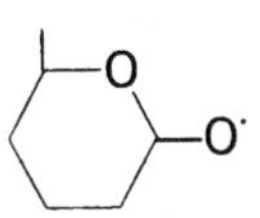	1, 2, 3, 6, 7, 8, 9, 10(?)	Singlet
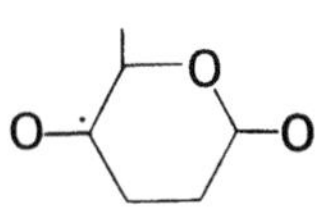	4,5 1, 2, 3, 6, 7, 8, 9, 10(?)	Singlet
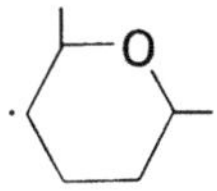	1, 2, 3, 6, 7, 8, 9, 10	Doublet
e.g. 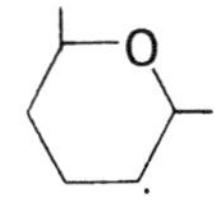	1, 2, 3, 8, 9, 10 6, 7(?)	Triplet
$CH_3^{\cdot}$	2, 3, 5, 7	Quartet

hydrogen abstraction.[71] It was thus shown that benzophenone, deoxybenzoin and its derivatives are efficient sensitizers. In the case of methyl methacrylate as monomer, 1,2-diphenyl-2,2-dimethoxyethanone appeared to be the most efficient initiator.

Detailed comparisons of grafting efficiencies measured with a large variety of vinyl monomers and a wide series of photoinitiators were also carried out,[72] showing the strong interrelations of the three factors: substrate, monomer, and initiator. In this group of experiments,[72–76] ESR spectroscopy was widely employed to detect free-radical generation by photosensitized excitation, in cellulose[73] as well as in cellulose acetate, cellulose triacetate and some oligosaccharide compounds (glucose, glucose penta-acetate, cellobiose, cellobiose octa-acetate, maltose and maltotriose).[75] These studies have clearly established that the UV irradiation of both photosensitized and unphotosensitized samples of cellulose derivatives and model compounds generates the same type of radicals (Table 1). Their subsequent evolution, however, is affected by the primary processes occurring after the absorption of photons. On cellulose acetate films,[75] for example, in the presence or absence of photosensitizers, yellowing resulted as a function of irradiation wavelength, and of oxygen concentration, as well as of type and concentration of the photosensitizer.

Finally, ESCA measurements, in the photografting study of acrylamide onto oriented polypropylene fibres,[77] enabled useful quantitative information to be obtained on the mechanism of grafting.

2.5 Vapour-Phase Photografting

Most photografting experiments have been carried out from solutions of monomers or from adsorbed monomers in liquid or solid phases. Operation in the adsorbed phase and the influence of solvent have already been discussed in Sections 2.1 and 2.4. One of the drawbacks when operating in solution, as has been pointed out, is the excessive formation of homopolymer, which is normally difficult to control. Little homopolymerization, on the contrary, occurs from adsorbed monomers. For this same purpose, and also to avoid uneven, spotty grafting and general loss of aesthetic properties of the substrate accompanied by extensive homopolymerization, photoinitiated vapour-phase grafting was studied.

An early investigation of this kind[78] has shown that methacrylic acid vapour can be grafted to the surface of nylon films when benzophenone vapour is used as photosensitizer. Similarly photografting of maleimide in the form of sublimation vapour has been found to proceed under UV irradiation[79,80] onto polyethylene and ethylcellulose films. Air retarded the grafting reaction,[80] but did not suppress it, nor did it alter the final extent of grafting. Equally, no effect by radical inhibitors was observed.[80] The grafted polymaleimide was found to be amorphous, even though UV irradiation produces a highly crystalline homopolymer in the solid as well as in the liquid phase.[81]

A series of acrylic monomers of different volatilities and reactivities including methyl acrylate, methyl methacrylate, acrylamide, acrylonitrile, and 2,2,2-trifluoroethyl methacrylate were successfully grafted onto several hydrophilic and hydrophobic textile fibres (wool, cotton, rayon, acrylics, polyester and polypropylene) and other polymeric surfaces in the vapour phase in the presence of biacetyl as initiator, and using convenient wetting agents to treat the surface (from water to alkanol, from benzene to *N,N*-dimethylformamide).[82] Following the nature of the substrate and/or the wetting agent, specially with water–organic solvent mixtures, vapour-phase photografting could be enhanced or reduced in comparison with grafting from solutions.[83]

More recently, several research articles have appeared on vapour-phase photografting.[84–89] Methyl methacrylate onto polypropylene and low-density polyethylene, in contrast to hydrophilic monomers, could be grafted with fairly high yield in the vapour phase,[84] provided the polyolefin films were coated beforehand with sensitizers.

The effects of mixed solvents on photografting were then investigated; a mixture of acrylic acid and solvent was supplied in the vapour phase.[85] Low- and high-density polyethylene, polypropylene, nylon and poly(ethylene terephthalate) films were previously coated with benzophenone, anthraquinone or benzoyl peroxide as initiators, by immersion in acetone or chloroform solutions of these compounds together with poly(vinyl acetate) as a binder. Water in the mixed solvent (a series of organic solvents was employed) exhibited an accelerating effect on the vapour-phase photografting.

ESR spectra of irradiated samples of cellulose suggested a negative effect of water on radical formation.[89] It is plausible, therefore, that water does not promote the formation of radicals from the polymer backbone, but can facilitate the diffusion of monomers in the polymer texture. If the latter is hydrophilic, this can lead to an effective utilization of substrate radicals capable of initiating grafting.

In another study[86] a similar technique for supplying the photoinitiator was employed, but grafting from aqueous liquid mixed solvents was carried out by means of a two-step method. By ESR examination, a thermally stable radical was found in the first step in which the sample was irradiated with light of wavelengths > 330 nm, in the presence of acrylic acid, methacrylic acid, acrylonitrile, or methyl methacrylate as monomers. The stable radical contributed to the second-step grafting with acrylamide quite efficiently. To the thermally stable radical a structure such as $—CH_2—C{\cdot}(CH_3)—COOH$ was assigned, corresponding to the active ends of the growing chain. The morphology of the liquid- and vapour-phase photografted structures were then compared.[87]

3 APPLICATIONS OF PHOTOGRAFTING PROCESSES

Even if the total sum of internal and external stimuli that act upon industrial applications of photografting is presently not very different from that existing when previous reviews[90,91] were written, new possibilities are now afforded by some of the latest developments of new processes, which look very promising.

Photografting processes for applying flame-retardant compounds to cotton textile fabrics continue to be of interest. Photoinitiated, unsensitized polymerization of *N*-methylolacrylamide and vinylphosphonate oligomers (molecular weight 500–1000) from aqueous solution to form graft copolymer gave a textile product that had a good flame resistance (FF 5–74 test).[92] The effects of wavelength of UV radiation (254, 300 and 350 nm), reaction time (up to 9000 s), solids in solution (10–60%), atmosphere (air, nitrogen, oxygen) and oligomer-to-monomer concentration (1:0 to 1:20 mole ratio) were studied. Grafted deposits appeared throughout the cross-sections of the fibres, even if copolymer and phosphorus tended to be more concentrated between and on the surfaces of the fibres. *N*-Methylolacrylamide was utilized because of its ready reactivity and amenability to textile finishing, as a previous examination has ascertained.[93] A continuous photocuring process to produce flame-resistant cotton textiles based on the *N*-methylolacrylamide–vinylphosphonate oligomer system was also set up.[94]

A wrinkle-resistant cotton fabric was produced by photografting glycidyl methacrylate from methanol–water solution, followed by crosslinking reactions of oxirane groups of the copolymer with cellulose.[95] The modified fibres were resistant to dissolution in cupriethylenediamine dihydroxide, had increased conditioned and wet recovery angles, and retained breaking and tearing strengths as well as stiffness values typical of wrinkle-resistant cotton.

UV-radiation-polymerized graft copolymers for leather have been recently developed,[96] ranging from epoxy to polyurethane oligomers, with or without the use of trimethylolpropane triacrylate as a crosslinkable diluent.

Photografting as a technique has become increasingly important in some applications in biotechnology. A new type of photoresponsive polymeric adsorbent which can regulate adhesion-dehesion of blood cells has been synthesized[97] by grafting an amphiphilic polymer, poly(2-hydroxyethyl methacrylate) onto the surface of controlled pore glass, followed by the introduction of an azobenzene moiety in the side chain of the amphiphilic polymer (azobenzene content 12–13%). The polarity of the surface of the polymeric adsorbent was increased by UV irradiation because of the *trans–cis* isomerization of the azobenzene moiety. Erythrocytes were thus adhered to the adsorbent in the dark and separated from the adsorbent by UV irradiation, as a consequence of the photoresponse of the polarity change on the surface of the polymeric adsorbent. These adsorbents with biospecific activity look very promising for the separation and purification of biosubstances, such as proteins and blood cells.

Another important aspect of these biotechnological applications of photografting is the use of graft copolymers for the immobilization of enzymes and whole cells. Many research papers have been published in this area, which give ground to expectation of fruitful developments.[43,48,50,98–106] Some workers are concerned primarily with chemical condensation reactions of bioactive materials,[43,48,50,98–100] such as bovine serum albumin, β-galactosidase, trypsin, pepsin, invertase and acid phosphatase, with poly(acrylic acid) grafts, the latter being obtained by hydrolysis of polyacrylonitrile branches of the graft copolymers onto various backbone polymers ranging from nylon to woollen substrates. In other articles, crosslinking between subunits of oligomeric proteins was directly induced by UV irradiation in the presence of furocoumarins,[103] or in the presence of iron(III) chloride and polysaccharide as support.[104]

The most interesting method from the practical point of view and probably the simplest one for entrapping enzymes, such as yeast invertase[101] or yeast microbodies containing FAD-dependent alcohol oxidase, catalase, and D-amino acid oxidase,[102] consists of irradiating mixtures containing the photoreactive and/or photocrosslinkable resin [e.g. poly(ethylene glycol dimethacrylate)] in the presence of photoinitiator and an enzyme solution or microbial cell suspension. Poly(vinyl alcohol) was also employed for similar studies[105,106] and the nature of the membrane structure obtained by this technique has been elucidated.

A new general method for the manufacture of membranes by photochemical grafting has been applied.[35,40,107–110] It was found to be quite suitable for the preparation of microfiltration and ultrafiltration membranes,[107,108] as well as that of membranes capable of separating gases.[35,40,109] The method consists of photoinduced grafting of appropriate thin layers of polyfunctional copolymers and/or their composites with sorbents or selective carriers onto macroporous substrates. Liquid membranes immobilizing Schiff base complexes of Co(III) have also been prepared by this technique; they were able to realize a very efficient O_2/N_2 separation, with the advantageous ratio of 50 between their fluxes.[109]

The membranes were characterized by the study of photochemical parameters in their kinetics of formation,[37,39] by morphological investigations,[39,107,108] by thermodynamic evaluation of sorption, such as of acetic acid from water for epoxy–diacrylate resin composites with zeolites,[107] by correlation of photocrosslinking with transport properties of gases (as evinced from permeametric measurements)[40] as well as by measurements of liquid permeability varying with pore diameter.[108,110] This latter type of investigation is particularly appropriate to throw light on the intrinsic behaviour of the photochemically synthesized membranes. Flow dynamics through porous media may be described by Poiseuille's laminar flow for relatively low fluxes. In Fig. 7, the logarithm of the mean pore diameter (R = mean pore radius) is reported as a function of the logarithm of flux (J) per unit applied pressure drop (ΔP). When the slope is close to 0·5 (see correlation line in the figure, extrapolated by the broken line down to the lowest fluxes which could be measured experimentally), Poiseuille's equation appears to hold and laminar flow is consequently operative. In this region, or near its higher borders, the gas separation and then the ultrafiltration ability of photosynthetic membranes may be exploited. The higher slope, approaching unity, in the range of more elevated fluxes suggests a turbulent flow with friction coefficients

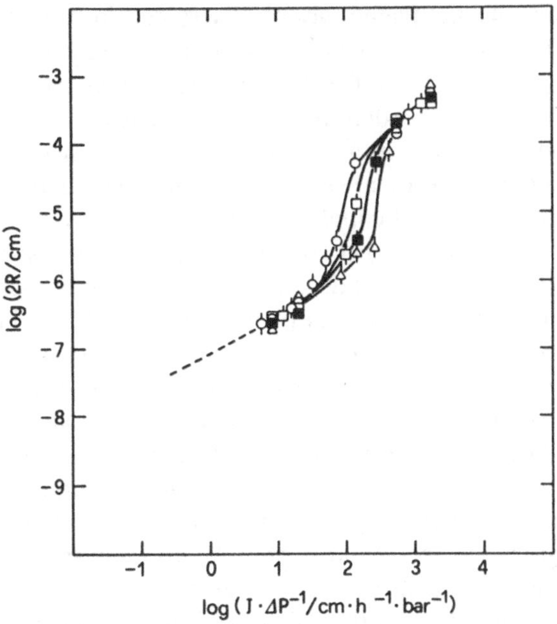

FIG. 7. Logarithm of mean pore diameter ($2R$) as a function of logarithm of water fluxes per unit applied pressure drop ($J \cdot \Delta P^{-1}$), measured at 20°C, for membranes synthesized by photografting of epoxy–diacrylate resin and its zeolite composites onto cellulose.[108] The polymeric membrane contained (○) no zeolite; (□) 20 wt% zeolite 3A; (■) 20 wt% zeolite 5A; (△) 20 wt% zeolite 13X.

(scarcely dependent on Reynolds number) which are allowed by the greater porosity values (microfiltration conditions).

Photografting of vinylsulphonic acid to allylated poly(vinyl alcohols) has been employed to prepare cation exchange membranes,[111] while diacrylated phospholipids were grafted as monolayers onto polyethylene films[112] (polymer-supported biomimetic membranes). Poly(4-vinylpyridine) grafts, introduced by gas-phase graft copolymerization onto polyethylene film with poly(vinyl acetate) as binder, were employed as ligands to absorb Cu(II) ions:[88] the catalytic activity of the complex, which was characterized thermodynamically by measurements of its instability constant, was investigated on the decomposition of hydrogen peroxide.

With a view to the manufacture of printing plates, photoinduced polymerization of diethylene glycol diacrylate in the presence of poly(methyl methacrylate), with or without 9-phenylacridine as the photoinitiator, has been studied:[113] the role of hydrogen abstraction reactions and of hydrogen donors has been discussed.

Finally, photografting continues to receive attention in recent years from the point of view of a tool towards macromolecular synthesis. Thus, in a study of hybrid copolymerization of alkyl vinyl ethers with glycidyl ethers,[114] the absence of copolymerization, as a result of incompatibility between epoxide and vinyl ether, was ascertained in conventional cationic as well as in photochemical cationic initiation by 4-chlorophenyldiazonium tetrafluoroborate. Acrylonitrile has been grafted successfully onto several copolycarbonates of bisphenol A and 1,1,1-trichloro-2,2-bis(4-hydroxyphenylethane), via the CCl_3 group, using $Mn_2(CO)_{10}$ as photoinitiator.[115] A novel monomeric photoinitiator, *p*-(*p*′-vinylbenzoyl)peroxybenzoic acid *t*-butyl ester,[116] was copolymerized with styrene and used to make graft copolymers with methyl methacrylate by means of UV irradiation.[117] Photograft copolymerization of methoxy poly(ethylene glycol methacrylate) and *N,N*-dimethylaminoethyl methacrylate to dithiocarbamated poly(vinyl chloride) in cyclohexanone was carried out and experimental results compared favourably with a model derived from photopolymerization kinetics.[118]

4 CONCLUDING REMARKS

From discussions in Section 2, it appears that no general mechanism can as yet be formulated for photochemically induced grafting onto polymers. Two main lines of interpretation of experimental data, however, may be clearly profiled:

(1) a mechanism for which the rate of grafting follows a one-half order with respect to initiator concentration (assuming that the photoinitiator is the dominant absorbing species) and hence more generally depends on absorbed radiation intensity. This kinetic scheme ensues the classical and well-known mechanism of photochemical polymerization of monomers in a homogeneous phase;
(2) another line of interpretation, according to which results are best fitted by a linear dependence of quantum yields on

absorbed radiation intensity (or photoinitiator concentration in specific cases), at least for a first grafting period, which may also result, in some conditions, as the prevailing process.

The kinetic distinction between these two types of mechanism is reflected by the termination reaction, being either [case (2)] of marked heterogeneous nature (reaction between a macroradical on the polymer backbone and the growing chain in the phase containing the reactive monomer), or [case (1)] represented by reaction of two growing polymeric chains, both already grafted onto the polymer substrate. Consequently, it is not unexpected that solvation of reactive sites in the polymer backbone as well as of adsorbed monomer, and physicochemical interactions between radical species produced during initiation on both, may largely determine the reaction path.

This determining aspect has been also discussed in the photograft polymerization of methyl methacrylate onto silica gel modified by the attachment of —CCl_3 groups, and using $Mn_2(CO)_{10}$ as initiator.[119] The choice of photoinitiator and/or photosensitizer, in view of their ability to promote direct hydrogen abstraction from the substrate and formation of copolymer in comparison with homopolymerization, is consequently of primary importance, both for the mechanism, and for the practical usefulness of the photografting technique.

Such facets of the problem have been known for a long time in the photografting of one polymer onto the backbone of some other different polymeric material,[120] a process which is akin to grafting of monomers or oligomers as well as to photocrosslinking. The study of primary photochemical events concerning the action of photoinitiator with techniques such as CIDNP[121] or laser photolysis[122] may undoubtedly contribute both to fundamental understanding and to applications. ESR and ESCA spectroscopies[123] are also useful tools, particularly in assessing the relative thermal stabilities of macroradicals and their mechanistic role in photografting processes.

Finally, it is easy to foresee the growing interest in photograft polymerization as a method of preparing highly filled composites, for a broad range of applications, from microelectronics to structural materials, from membranes to immobilization of sorbents and catalysts including biocatalysts, which will also lead research to examine more closely not only the physicochemical features of bonding of polymer to polymer, but also those related to the attachment of the propagating macroradicals to the surface of inorganic materials.

REFERENCES

1. Chapiro, A., *Radiation Chemistry of Polymeric Systems.* Interscience, New York, 1962.
2. Davison, W. H. T., Pinner, S. H. & Worrall, R., *Chem. Ind.*, (1957) 1274.
3. Williams, F. In *Fundamental Processes in Radiation Chemistry,* ed. P. Ausloos, Interscience, New York, 1968.
4. Stannett, V., *Pure Appl. Chem.*, **53** (1981) 673.
5. Liepins, R., Stannett, V. & Walsh, W. K., *Progr. Polym. Sci.*, **6** (1978) 1.
6. Hebeish, A. & Guthrie, J. T., *The Chemistry and Technology of Cellulosic Copolymers,* Springer-Verlag, Heidelberg, FRG, 1981, Ch. 3.
7. Witz, G., *Bull. Soc. Ind. Rouen,* **11** (1883) 188.
8. Oster, G. & Shibata, O., *J. Polym. Sci.*, **26** (1957) 233.
9. Oster, G., Oster, G. K. & Moroson, H., *J. Polym. Sci.*, **34** (1959) 671.
10. Geacintov, N., Stannett, V. & Abrahamson, E. W., *Makromol. Chem.*, **36** (1960) 52.
11. Geacintov, N., Stannett, V., Abrahamson, E. W. & Hermans, J. J., *J. Appl. Polym. Sci.*, **3** (1960) 54.
12. Arthur, J. C. Jr. In *Developments in Polymer Photochemistry—1,* ed. N. S. Allen, Applied Science, London, 1980, Ch. 3.
13. Hebeish, A. & Guthrie, J. T., *The Chemistry and Technology of Cellulosic Copolymers,* Springer-Verlag, Heidelberg, FRG, 1981, (a) p. 127; (b) p. 245
14. Phillips, D. In *Photochemistry,* Part IV, Polymer Photochemistry. Spec. Period. Rept., The Royal Society of Chemistry, London. (a) Vol. 1, 1970, p. 445; (b) Vol. 2, 1971, p. 776; (c) Vol. 3, 1972, p. 823; (d) Vol. 4, 1973, p. 894; (e) Vol. 5, 1974, p. 704; (f) Vol. 6, 1975, p. 659; (g) Vol. 7, 1976, p. 511; (h) Vol. 8, 1977, p. 545.
15. Allen, N. S. & McKellar, J. F. In *Photochemistry,* Part IV, Polymer Photochemistry. Spec. Period. Rept., The Royal Society of Chemistry, London. (a) Vol. 9, 1978, p. 563; (b) Vol. 10, 1979, p. 571.
16. Allen, N. S. In *Photochemistry,* Part IV, Polymer Photochemistry. Spec. Period. Rept., The Royal Society of Chemistry, London. (a) Vol. 11, 1981, p. 567; (b) Vol. 12, 1982, p. 488; (c) Vol. 13, 1983, p. 513; (d) Vol. 14, 1984, p. 460; (e) Vol. 15, 1985, p. 449; (f) Vol. 16, 1986, p. 487; (g) Vol. 17, 1987, p. 536.
17. Walsh, W. K., Makati, A. & Bittencourt, E., *Text. Chem. Color.*, **10** (1978) 220.
18. Park, K., Frame, R. L. & Bryant, G. M., *Text. Chem. Color.*, **11** (1979) 107.
19. Tsunooka, M., Tanaka, M. & Murata, I., *Kobunshi Kagaku,* **22** (1965) 107.
20. Ishibashi, H., *Kobunshi Kagaku,* **25** (1968) 481.
21. Kubota, H., Murata, Y. & Ogiwara, Y., *J. Polym. Sci., Polym. Chem. Ed.*, **11** (1973) 485.

22. Dilli, S., Garnett, J. L., Martin, E. C. & Phuoc, D. H. In *Proceedings of the Symposium on Graft Polymerization onto Cellulose* (*J. Polym. Sci. C*, **37**), ed J. C. Arthur, Jr. Interscience, New York, 1972, p. 57.
23. Davis, N. P., Garnett, J. L. & Urquhart, R. G., *J. Polym. Sci., Polym. Symp.*, **55** (1976) 287.
24. Ang, C. H., Garnett, J. L., Levot, R., Long, M. A. & Yen, N. T., *J. Polym. Sci., Polym. Lett. Ed.*, **18** (1980) 471.
25. Garnett, J. L., Jankiewicz, S. V., Long, M. A. & Sangster, D. F., *J. Polym. Sci., Polym. Lett. Ed.*, **24** (1986) 125.
26. Selli, E., Bellobono, I. R., Calgari, S. & Berlin, A., *J. Soc. Dyers Colour.*, **97** (1981) 438.
27. Selli, E., Bellobono, I. R., Tolusso, F. & Calgari, S., *Ann. Chim.* (*Rome*), **71** (1981) 147.
28. Calgari, S., Selli, E. & Bellobono, I. R., *J. Appl. Polym. Sci.*, **27** (1982) 527.
29. Ullman, E. F. & Baumann, N., *J. Am. Chem. Soc.*, **92** (1970) 5891.
30. Bellobono, I. R., Tolusso, F., Selli, E., Calgari, S. & Berlin, A., *J. Appl. Polym. Sci.*, **26** (1981) 619.
31. Bellobono, I. R., Calgari, S. & Selli, E. In *Polymer Additives*, ed. J. E. Kresta, Plenum Press, New York, 1984, p. 83.
32. Bellobono, I. R., Selli, E. & Marcandalli, B., *J. Appl. Polym. Sci.*, **32** (1986) 4323.
33. Decker, C., *J. Appl. Polym. Sci.*, **28** (1983) 97.
34. (a) Decker, C., *Polym. Photochem.*, **3** (1983) 131; cf. (b) Decker, C. & Bendaikha, T., *Eur. Polym. J.*, **20** (1984) 753.
35. Bellobono, I. R., Marcandalli, B., Selli, E., Comi, D. & Righetto, L., *Polym. Photochem.*, **6** (1985) 339.
36. Gaylord, N. G. & Tomono, T., *J. Polym. Sci., Polym. Lett. Ed.*, **13** (1975) 697.
37. Bellobono, I. R., Selli, E., Marcandalli, B., Comi, D. & Rastelli, E., *J. Photochem.*, **35** (1986) 99.
38. Merlin, A. & Fouassier, J. P., *Angew. Makromol. Chem.*, **86** (1980) 109.
39. Bellobono, I. R., Zeni, M., Selli, E. & Marcandalli, B., *J. Photochem.*, **35** (1986) 367.
40. Bellobono, I. R., Selli, E., Marcandalli, B., Comi, D. & Rastelli, E., *J. Photochem.*, **35** (1986) 231.
41. Harris, J. A., Arthur, J. C., Jr & Carra, J. H., *J. Appl. Polym. Sci.*, **22** (1978) 905.
42. Arthur, J. C., Jr, *Advan. Macromol. Chem.*, **2** (1970) 1.
43. Barker, P. S., Bottom, R. A., Guthrie, J. T. & Beddows, C. G., *Polym. Photochem.*, **2** (1982) 87.
44. Bottom, R. A., Guthrie, J. T. & Green, P. N., *Cellul. Its Deriv.*, ed. J. F. Kennedy. Horwood, Chichester, 1985, p. 475.
45. Guthrie, J. T., Ryder, M. & Abdel-Hay, F. I., *Polym. Bull.*, **1** (1979) 501.
46. Bottom, R. A., Guthrie, J. T. & Green, P. N., *Polym. Photochem.*, **6** (1985) 111.

47. Bottom, R. A., Guthrie, J. T. & Green, P. N., *Polym. Photochem.*, **6** (1985) 59.
48. Barker, P., Guthrie, J. T., Davis, M. J., Godfrey, A. A. & Green, P. N., *J. Appl. Polym. Sci.*, **26** (1981) 521.
49. Morel, G., Jozefonvicz, J. & Aptel, P., *J. Appl. Polym. Sci.*, **23** (1979) 2397.
50. Abdel-Hay, F. I., Barker, P. & Guthrie, J. T., *Makromol. Chem.*, **181** (1980) 2063.
51. Lenka, S., Nayak, P. L. & Mishra, M. K., *Angew. Makromol. Chem.*, **99** (1981) 45.
52. Lenka, S., Nayak, P. L. & Tripathy, K., *J. Appl. Polym. Sci.*, **27** (1982) 1853.
53. Lenka, S. & Nayak, P. L., *J. Appl. Polym. Sci.*, **27** (1982) 3787.
54. Lenka, S. & Nayak, P. L., *J. Appl. Polym. Sci.*, **27** (1982) 3625.
55. Lenka, S., Nayak, P. L. & Dash, M., *Polym. Photochem.*, **3** (1983) 109.
56. Lenka, S., Nayak, P. L. & Pradhan, A. K., *Polym. Photochem.*, **3** (1983) 357.
57. Lenka, S. & Nayak, P. L., *J. Polym. Sci., Polym. Chem. Ed.*, **21** (1983) 1871.
58. Lenka, S., Nayak, P. L. & Dash, M., *J. Appl. Polym. Sci.*, **28** (1983) 3665.
59. Samal, S., Lenka, S., Nayak, P. L. & Sahu, G., *Polym. Photochem.*, **6** (1985) 235.
60. Samal, S., Lenka, S., Nayak, P. L. & Sahu, G., *Polym. Photochem.*, **6** (1985) 293.
61. Lenka, S. & Mohanty, I. B., *Polym. Photochem.*, **7** (1986) 285.
62. Mukherjee, A. K. & Goel, H. R., *J. Macromol. Sci., Chem.*, **A17** (1982) 545.
63. Mukherjee, A. K., Sachdev, H. S. & Gupta, A., *J. Appl. Polym. Sci.*, **28** (1983) 2125.
64. Mukherjee, A. K., Sachdev, H. S. & Gupta, A., *J. Appl. Polym. Sci.*, **28** (1983) 2217.
65. Ghosh, P. & Paul, S. K., *Indian J. Technol.*, **24** (1986) 111.
66. Ghosh, P. & Ghosh, T. K., *J. Macromol. Sci., Chem.*, **A17** (1982) 847.
67. Ghosh, P. & Bandyopadhyay, A. R., *J. Appl. Polym. Sci.*, **31** (1986) 1499.
68. Ghosh, P. & Banerjee, M., *J. Macromol. Sci., Chem.*, **A17** (1982) 1273.
69. Arthur, J. C. Jr. In *Developments in Polymer Photochemistry—1*, ed. N. S. Allen, Applied Science, London, 1980, p. 74.
70. Reinhardt, R. M., Arthur, J. C., Jr & Hinojosa, O., *J. Appl. Polym. Sci.*, **24** (1979) 1691.
71. Merlin, A. & Fouassier, J. P., *Makromol. Chem.*, **182** (1981) 3053.
72. Hérold, R. & Fouassier, J. P., *Angew. Makromol. Chem.*, **97** (1981) 137.
73. Merlin, A. & Fouassier, J. P., *Angew. Makromol. Chem.*, **86** (1980) 109.
74. Hérold, R. & Fouassier, J. P., *Angew. Makromol. Chem.*, **86** (1980) 123.
75. Merlin, A. & Fouassier, J. P., *Angew. Makromol. Chem.*, **108** (1982) 185.

76. Hérold, R. & Fouassier, J. P., *Makromol. Chem., Rapid Commun.*, **2** (1981) 699.
77. (a) Tazuke, S. & Kimura, H., *Makromol. Chem.*, **179** (1978) 2603; cf. (b) Tazuke, S. & Kimura, H., *J. Polym. Sci., Polym. Lett. Ed.*, **16** (1978) 497.
78. Howard, G. J., Kim, S. R. & Peters, R. H., *J. Soc. Dyers Colour.*, **85** (1969) 468.
79. Hayakawa, K., Kawase, K. & Yamakita, H., *J. Polym. Sci. A-1,* **8** (1970) 1227.
80. Hayakawa, K., Kawase, K. & Yamakita, H., *J. Polym. Sci., Polym. Chem. Ed.*, **12** (1974) 2603.
81. Hayakawa, K., Yamakita, H. & Kawase, K., *J. Polym. Sci. A-1,* **10** (1972) 1363.
82. Seiber, R. P., & Needles, H. L., *J. Appl. Polym. Sci.*, **19** (1975) 2187.
83. Needles, H. L. & Alger, K. W., *J. Appl. Polym. Sci.*, **22** (1978) 3405.
84. Ogiwara, Y., Kanda, M., Takumi, M. & Kubota, H., *J. Polym. Sci., Polym. Lett. Ed.*, **19** (1981) 457.
85. Ogiwara, Y., Torikoshi, K. & Kubota, H., *J. Polym. Sci., Polym. Lett. Ed.*, **20** (1982) 17.
86. Ogiwara, Y., Takumi, M. & Kubota, H., *J. Appl. Polym. Sci.*, **27** (1982) 3743.
87. Ogiwara, Y., Kubota, H. & Hata, Y., *J. Polym. Sci., Polym. Lett. Ed.*, **23** (1985) 365.
88. Arai, K., Ogiwara, Y. & Sakamoto, T., *Makromol. Chem., Rapid Commun.*, **5** (1985) 837.
89. Ogiwara, Y. & Kubota, H., *J. Polym. Sci., Polym. Lett. Ed.*, **23** (1985) 15.
90. Hebeish, A. & Guthrie, J. T., *The Chemistry and Technology of Cellulosic Copolymers,* Springer-Verlag, Heidelberg, FRG, 1981, p. 326.
91. Arthur, J. C. Jr. In *Developments in Polymer Photochemistry—1,* ed. N. S. Allen. Applied Science, London, 1980, p. 85.
92. Harris, J. A., Arthur, J. C., Jr & Goynes, W. R., Jr, *J. Appl. Polym. Sci.*, **24** (1979) 201.
93. Reinhardt, R. M. & Arthur, J. C. Jr, *J. Appl. Polym. Sci.*, **24** (1979) 147.
94. Harris, J. A., Keating, E. J. & Goynes, W. R., Jr, *J. Appl. Polym. Sci.*, **25** (1980) 2295.
95. Reinhardt, R. M., Arthur, J. C., Jr & Muller, L. L., *J. Appl. Polym. Sci.*, **25** (1980) 933.
96. Buechler, P. R., Kronick, P. L. & Scholnick, F. In *Renewable Resource Materials,* ed. C. E. Carraher & L. H. Sperling, Plenum, New York, 1986, p. 251.
97. Ishihara, K., Kim, M., Shinohara, I., Okano, T., Kataoka, K. & Sakurai, Y., *J. Appl. Polym. Sci.*, **28** (1983) 1321.
98. Abdel-Hay, F. I., Beddows, C. G. & Guthrie, J. T., *Polym. Bull.*, **2** (1980) 607.

99. Beddows, C. G., Gil, M. H. & Guthrie, J. T., *Polym. Bull.*, **3** (1980) 645.
100. Beddows, C. G., Barker, P. S. & Guthrie, J. T., *Polym. Bull.*, **4** (1981) 149.
101. Fukui, S., Tanaka, A., Iida, T. & Hasegawa, E., *FEBS Lett.*, **66** (1976) 179.
102. Tanaka, A., Yasuhara, S., Osumi, M. & Fukui, S., *Eur. J. Biochem.*, **80** (1977) 193.
103. Schiavon, O. & Veronese, F. M., *Photochem. Photobiol.*, **43** (1986) 243.
104. D'Angiuro, L. & Cremonesi, P., *J. Appl. Biochem.*, **3** (1981) 459.
105. Imai, K., Shiomi, T., Sato, K. & Fujishima, A., *Biotechnol. Bioeng.*, **25** (1983) 613.
106. Imai, K., Shiomi, T., Uchida, K. & Miya, M., *Biotechnol. Bioeng.*, **28** (1986) (a) 198; (b) 1721.
107. Bellobono, I. R., Zeni, M. & Selli, E. In *Synthetic Polymeric Membranes,* ed. B. Sedláček & J. Kahovec. W. de Gruyter, Berlin, 1987, p. 113.
108. Zeni, M., Bellobono, I. R., Muffato, F., Polissi, A., Selli, E. & Rastelli, E., *J. Membrane Sci.*, **36** (1988) 277.
109. Bellobono, I. R., Muffato, F., Selli, E., Righetto, L. & Tacchi, R., *Gas Separ. Purif.*, **1** (1987) 103.
110. Bellobono, I. R., Selli, E., Righetto, L. & Muffato, F., *Mater. Chem. Phys.*, **19** (1988) 131.
111. Lehoussel, J., Wattiez, D. & Marechal, E., *Eur. Polym. J.*, **15** (1979) 915.
112. Regen, S. L., Kirszensztejn, P. & Singh, A., *Macromolecules,* **16** (1983) 335.
113. Braun, D. & Gerisch, R., *Polym. Photochem.*, **6** (1986) (a) 407; (b) 415.
114. Solaro, R., D'Antone, S., Orsini, M., Andruzzi, F. & Chiellini, E., *J. Appl. Polym. Sci.*, **28** (1983) 3651.
115. Ibrahim, T. M., Al-Lamee, K. G. & Adam, G. A., *Polym. Photochem.*, **4** (1984) 13.
116. Gupta, S. N., Thijs, L. & Neckers, D. C., *J. Polym. Sci., Polym. Chem. Ed.*, **19** (1981) 855.
117. Gupta, I., Gupta, S. N. & Neckers, D. C., *J. Polym. Sci., Polym. Chem. Ed.*, **20** (1982) 147.
118. Miyama, H., Fujii, N., Shimazaki, Y. & Ikeda, K., *Polym. Photochem.*, **3** (1983) 445.
119. Andrianov, A. K., Mislavskii, B. V., Olenin, A. V., Garina, E. S., Zubov, V. P. & Kabanov, V. A., *Doklady Akad. Nauk SSSR,* **280** (1985) 900.
120. Kachan, A. A., Lebo, Yu. G. & Shrubovich, V., *Vysokomol. Soedineniya,* **12A** (1970) 214.
121. Borer, A., Kirchmayr, R. & Rist, G., *Helv. Chim. Acta,* **61** (1978) 305.
122. Lougnot, D. J., Jacques, P., Fouassier, J. P., Casal, H. L., Kim-Thuan, N. & Scaiano, J. C., *Canad. J. Chem.*, **63** (1985) 3001.
123. Rånby, B. & Rabek, J. F. *ESR Spectroscopy in Polymer Research,* Springer Verlag, Berlin, 1977.

Chapter 5

PHOTOCHROMIC IMAGING

R. E. Jacobson

School of Communication, Polytechnic of Central London, London, UK

1 INTRODUCTION

Originally the term photochromism was applied to light-induced, reversible photochemical reactions which resulted in a visible colour change. However, this definition has been extended to cover reversible transformations between two states with differing absorption spectra which may be in the UV, visible or IR regions of the electromagnetic spectrum, the transformation being induced in at least one direction by the absorption of electromagnetic radiation and, in the other direction, spontaneously or thermally.[1] Equation (1) summarises photochromism according to the above definition.

$$\underset{(\lambda_1)}{A} \underset{h\nu' \atop \Delta T,\ \text{dark or } \lambda_2}{\overset{h\nu}{\rightleftharpoons}} \underset{(\lambda_2)}{B} \qquad (1)$$

More recently, however, photochromism has also been applied to systems which undergo refractive-index changes.[2] As mentioned above, photochromism is normally applied to reversible transformations which involve changes in spectral absorbance and hence appear to be a useful means for obtaining visible images. In transformations involving refractive-index changes the image read-out may involve radiation which is not absorbed by the species B in eqn (1). Such systems are of higher efficiency than those relying on absorption and involve, for example, reflection and interference phenomena which are

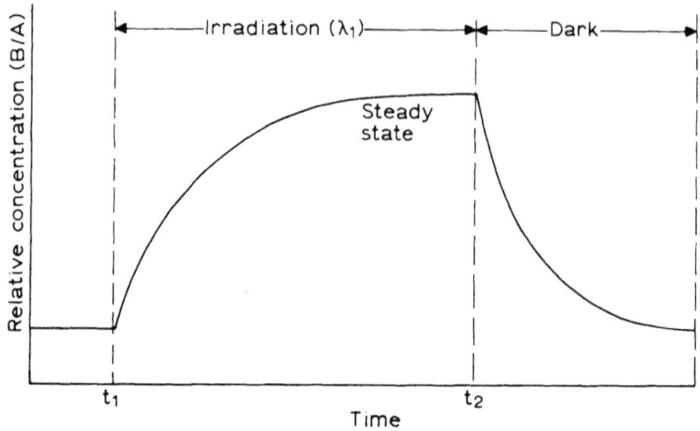

FIG. 1. Characteristics of a typical photochromic system; t_1 and t_2 represent times at which irradiation starts and stops.

able to handle higher power levels. Their main limitation is that most organic compounds have refractive indices in the range 1·3–1·8 whilst absorptions can vary by eight or more orders of magnitude.[2] This means that refractive-index systems are inherently less sensitive than absorption systems, which themselves are the least sensitive of all image-recording photochemical systems.

The behaviour of a typical photochromic system is shown in Fig. 1, in which the molar ratio (B/A) is plotted against time for the photochromic transformation of A to B. The absorption of the exciting

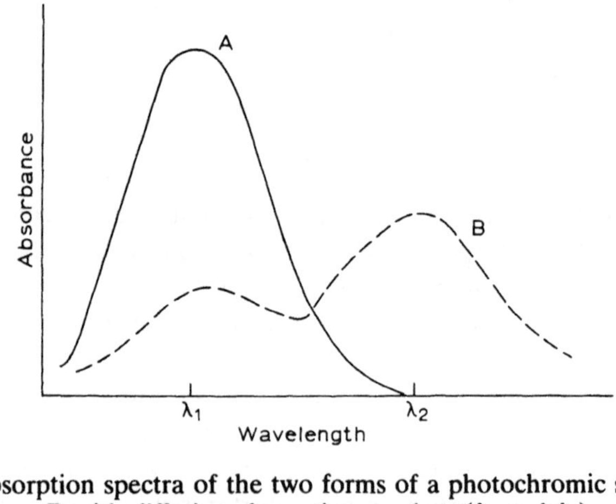

FIG. 2. Absorption spectra of the two forms of a photochromic system A and B with differing absorption maxima (λ_1 and λ_2).

radiation at time t_1 induces a change of A to B and builds up to a steady-state equilibrium value, the molar ratio at which depends upon the intensity of the radiation, temperature, solvent effects, the rate of the reverse reaction and inefficiencies. The inefficiencies are deactivations by collision, luminescence and any permanent side reactions. The quantum yields for photochromic reactions are usually substantially less than unity because of these inefficiencies and they are also affected by the inner filter effect which occurs if there is overlap in the absorption spectra of A and B as shown in Fig. 2. At time t_2 the exciting radiation has been turned off and the molar ratio (B/A) returns to the original value. The kinetics of the reverse reaction are usually first or second order and most are thermally induced or speeded up by absorption of radiation.

2 IMAGING

For imaging purposes a change in spectral absorption characteristics without the need for one or more chemical processing stages, as are required by virtually all photochemical imaging systems, is an attractive proposition. Figure 3 shows the basic steps involved in the production of an image.[3] Absorbed photons initiate a photochemical or photophysical reaction in a coated layer of the sensitive material. After exposure, additional energy is supplied to the system to amplify the effect of the absorbed photons of the primary phase of the reactions and to make the image stable. In conventional silver halide photographic processes, with overall quantum yields or amplification factors of approximately 10^9, these secondary stages are brought about by wet chemical processing techniques. In some processes, such as photopolymerisation, amplification occurs during exposure and involves a chain reaction with amplification factors of the order of 10^3. Since photochromic imaging involves no amplification stage, its sensitivity must be very low. It is a monophotonic process involving one event for the absorption of one photon with no subsequent amplification.

The most significant and unique features of photochromism for application in imaging are that image formation is direct and that it is possible to re-use the material. These and other features of photochromic systems have been exploited in a few imaging systems which involve the production of hard-copy, but for various reasons, which

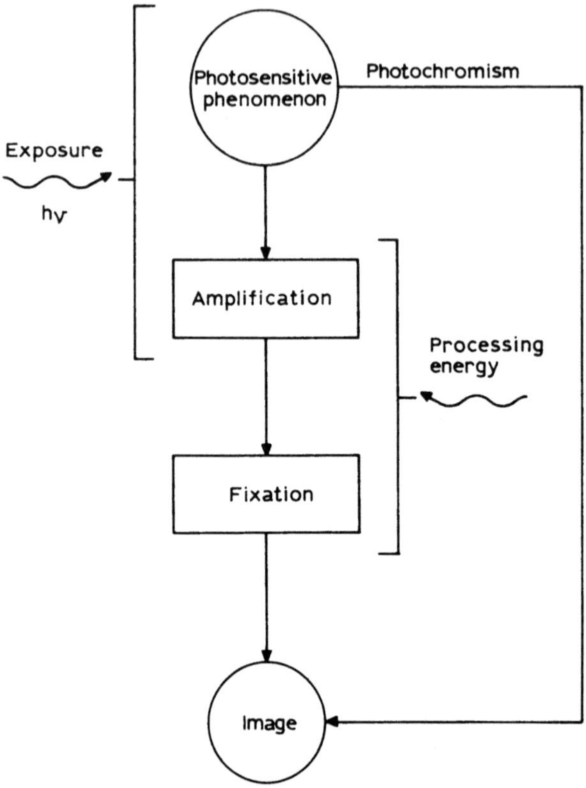

FIG. 3. Photochemical imaging system.[3]

will be discussed in this chapter, none has achieved commercial prominence. They have also been proposed for optical data storage in the form of optical discs,[4] for holographic recording and in integrated optics for refractive-index photochromic systems,[2] as photoresists[5] and in various information and data display devices.[6] These and other imaging applications of photochromism will be discussed further in Section 6.

The optimum properties for a photochromic absorption imaging system may be summarised as follows

(1) High quantum yield for transformation.
(2) High absorbance of form A.
(3) Absence of inner filter effect.

(4) Negligible rate of thermally induced reverse reaction (B to A), i.e. form B is stable.
(5) High absorbance of form B at a different wavelength from form A.
(6) Absence of fatigue if write–erase properties are required.

The thermally induced reverse reaction is a limitation to the use of photochromic systems for imaging but a number of approaches have been adopted to decrease the rate of this reaction so that images persist and are only erased by radiation. These approaches include chemical processing treatments which stabilise or fix the B form,[7] the use of polymeric matrices,[8,9] and the use of photochromic compounds with a thermally stable form B.[10] These approaches will be discussed further in Section 5 on the fixing and stability of photochromic images.

Until recently, organic photochromic systems suffered from fatigue which severely limited the number of possible recycles. Fatigue-free photochromic compounds which are capable of more than 8000 cycles are now available.[10] The characteristics of a fatigue-free photochromic system are shown in Fig. 4 for an inorganic system.

Some organic photochromic systems suffer from the 'yo-yo' effect, illustrated in Fig. 5, in which a coloured form when bleached with visible light rapidly recolours in the dark to an absorbance of one-ninth of that of the original absorbance. This process repeats itself to the limit of observable absorbance. The explanation involves the formation of a second colourless form and equilibria set up between

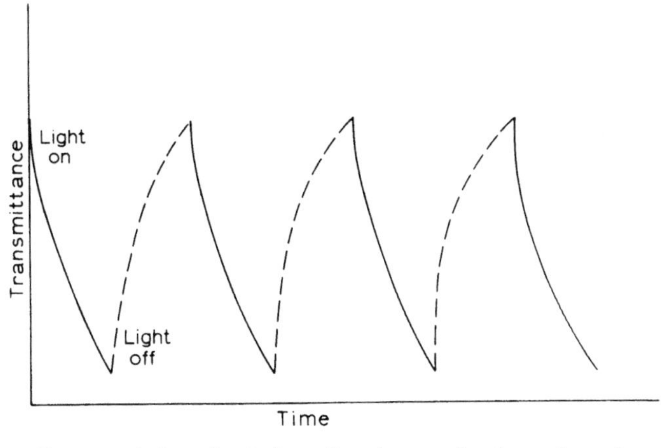

FIG. 4. Characteristics of a fatigue-free inorganic photochromic system.

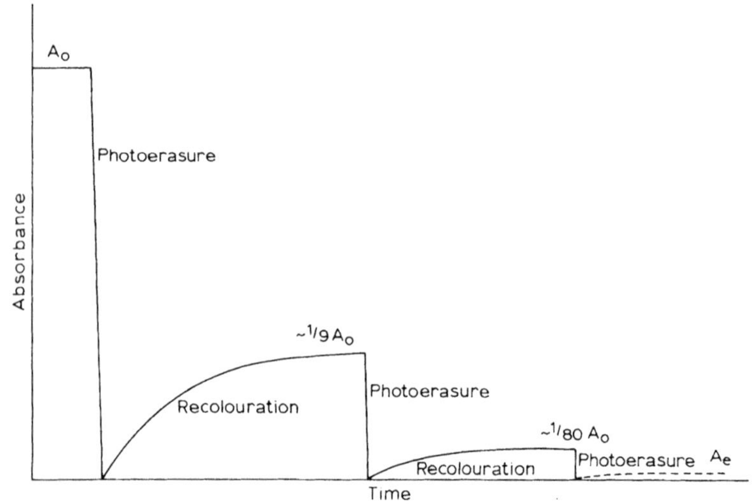

FIG. 5. The 'yo-yo' effect: thermal recolouration after erasure by light.[11]

the forms:[11]

$$\text{Colourless}_1 \rightleftarrows \text{Coloured} \rightleftarrows \text{Colourless}_2$$

The unique advantages of a photochromic imaging system may be summarised as follows.

(1) No wet processing required; image read-out is direct.
(2) Write–erase possibility, implying re-use of material; information may be added to existing images and materials need not be stored in the dark before use.
(3) High resolution; coatings of sensitive material involve molecules rather than suspensions or dispersions of large crystals or droplets.
(4) Possibility of multicolour images.

However, it is only recently that photochromic compounds with properties approaching the optimum properties given above have begun to become available.[10]

3 CALCULATION OF SENSITIVITY OF AN ORGANIC MONOPHOTONIC SYSTEM

A monophotonic imaging system such as one involving photochromism may be represented by the simple equation (2), for which A is colourless and absorbs radiation in the UV region while B is coloured and absorbs in the visible region.

$$\mathrm{A} \xrightarrow[\Phi=1]{h\nu} \mathrm{B} \tag{2}$$

For the purpose of determining the maximum sensitivity of the above system the following assumptions can be made.[3,12]

(a) That $\lambda_{max} = 365$ nm for A and $\lambda_{max} = 600$ nm for B, and there is no overlap in their spectral absorbances, i.e. no inner filter effect.
(b) That form B is stable; the rate of the thermally induced reverse reaction is slow relative to the forward reaction.
(c) That $\varepsilon_A = \varepsilon_B = 10^5\ \mathrm{l\ mol^{-1}\ cm^{-1}}$ at their respective λ_{max} values. This represents a high value, particularly for compounds absorbing in the visible region.
(d) That the absorbance of A and also of B is 2·5.

Sensitivity calculation

$$\text{Absorbance of A} = \varepsilon_A C_A d = 2{\cdot}5 \tag{3}$$

where C_A mol cm^{-3} is the concentration of A, and d cm is the layer thickness.

The reacting amount of A is given by $R_A = C_A d$, which from eqn (3) and the assumed value for ε_A gives:

$$R_A = 2{\cdot}5 \times 10^{-8}\ \mathrm{mol\ cm^{-2}} \tag{4}$$

If the complete transformation of A to B requires E mJ cm^{-2}, the maximum sensitivity of the system is given by:

$$E = E_{365} \times R_A \times \Phi^{-1} \tag{5}$$

For radiation of wavelength of 365 nm the energy of 1 einstein is given by eqn (6).

$$E_{365} = Nhc/\lambda \tag{6}$$

Substituting in eqn (6) for the known values for the wavelength and constants gives a value for E_{365} of $3{\cdot}28 \times 10^5\ \mathrm{J\ mol^{-1}}$. From eqns (4) and (5) the sensitivity, E, is $8{\cdot}2 \times 10^4\ \mathrm{mJ\ m^{-2}}$, assuming a quantum yield of unity.

Because of the assumptions made this value represents the maximum possible sensitivity for a monophotonic system. In practice the quantum efficiency (Φ), the molar extinction coefficient (ε) and the absorbance are all likely to have values very much lower than the assumed ones, which would result in higher values for E or a lower sensitivity to bring about the complete conversion of A to B.

A typical value for a practical organic monophotonic system such as the diazotype one is 3×10^5 mJ m^{-2}, which is an order of magnitude less sensitive than the maximum sensitivity calculated above. Photochromic materials have a sensitivity[13] of the order of 1×10^7 mJ m^{-2}. These values should be compared with those for a typical photographic system involving silver halides for which even a slow system such as a lithographic material has a sensitivity of 10 mJ m^{-2}. This system and some systems based on photopolymerisation have the advantage of the involvement of amplification, either during a processing technique (as used for photographic materials) or during exposure (as in photopolymerisation systems). This amplification stage is not available for photochromism and remains to be discovered, although the coloured form of certain photochromic compounds can be used to catalyse secondary ionic or thermal chain reactions.[14]

A photochromic system does have a number of advantages. These include the reversibility of the reaction, enabling the system to have a write, erase and re-write facility, together with the potential for high resolution because such materials comprise layers of molecules and the resolution is governed by the size of the molecules and not by the size of crystals as is the case for silver halide materials. These advantages, together with the ease of access to the image with no intrusion of wet chemical solutions, have been exploited to a limited extent in some practical imaging systems (see Section 6).

However, in organic photochromic systems involving conformation changes, rates of reaction are markedly influenced by the state of the system and polymeric matrices have a profound effect. These effects are discussed in Sections 4.1 and 4.5.

4 TYPES OF PHOTOCHROMIC SYSTEMS

For organic materials a variety of diverse reactions are involved in photochromism. These are grouped below according to the type of reaction involved. Inorganic photochromic systems have been excluded from this review.

4.1 *Cis–Trans* Isomerisation

A wide variety of compounds have been shown to exhibit *cis–trans* isomerisation according to the generalised scheme:

$$\mathit{Trans} \underset{\Delta T \text{ or } h\nu'}{\overset{h\nu}{\rightleftharpoons}} \mathit{Cis}$$

Generally the *trans* isomer is the more stable thermodynamically and there is a dark or thermal reverse reaction which may be very rapid (milliseconds) or may take months depending upon the nature of the compound, its environment and temperature. A significant disadvantage of *cis–trans* isomerisation from the imaging point of view is that there is usually little difference in the absorption maxima of the two forms. However, the quantum yields for the *trans* → *cis* isomerisation are often low whilst those for *cis* → *trans* can have values greater than 0·5 and are insensitive to temperature.

Compounds which exhibit *cis–trans* isomerisation include those with ethylenic double bonds (such as stilbenes and indigoid dyes), those with azo bonds (such as azobenzenes), and those with azomethine linkages. These photochromic systems have been extensively reviewed by Ross & Blanc,[15] although much of their review was concerned with reactions and mechanisms from studies in solution. Eisenbach[9] has investigated the influence of polymers on photochromism of azo compounds and spirobenzopyrans (see Section 4.5). He linked the photochromic compound to polyacrylates or polymethacrylates or dissolved them in the polymer matrix. These polymers enabled the glass transition temperature (T_g) to be varied by changing the ester side group of the polymer. He showed that isomerisation was possible in glassy solids at temperatures below T_g provided that the time of irradiation was extended. Generally *cis*/*trans* isomer ratios of 0·75 could be achieved in solution or in the solid state. This indicates that even in a glassy matrix isomerisation is possible due to there being sufficient motion of the polymer chains to allow enough free volume for conformational change of the photochromic entity.

The thermal reverse reaction of the *cis* → *trans* isomerisation in the glassy state was shown to depart from first-order kinetics. Eisenbach resolved it into two simultaneous first-order reactions: an anomalously fast reaction by a fraction of the *cis* isomer while the rest isomerise at a rate almost as fast as in solution. Overall the rate of reaction is faster in the glassy state than in solution, whilst the fraction of *cis* isomer with anomalously fast reaction increases as the temperature approaches T_g. From these considerations and evaluations of activation energies, Eisenbach proposed a model for isomerisation in bulk polymers which is illustrated in Fig. 6. The similarities between values for activation energies obtained for *cis* → *trans* isomerisation and those for 'crankshaft-like motions' of the polymer matrix of about 60 kJ mol^{-1} and a proportion of about 24 kJ mol^{-1} required to overcome van der Waals forces led to the model shown in Fig. 6: on the left is shown the rotation–relaxation mechanism in which the rotation of the chain segment provides free volume around the chromophore, whilst on the right is shown the translation–relaxation

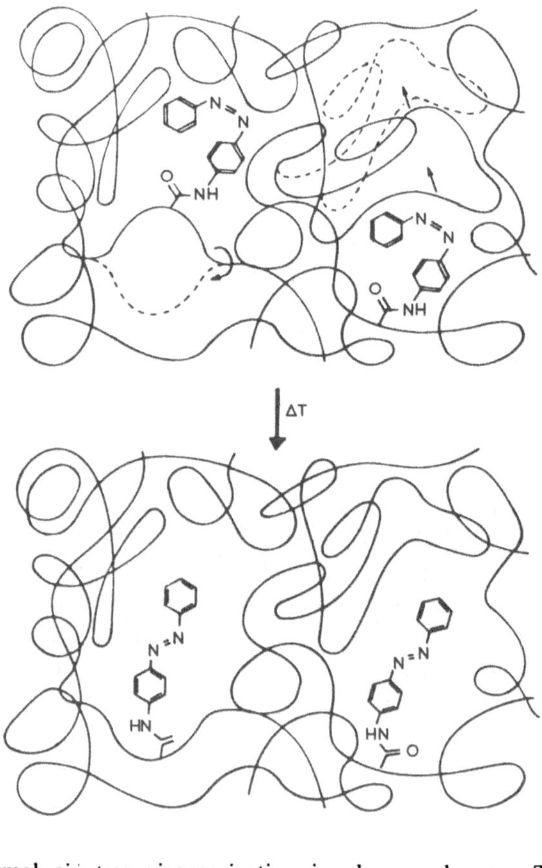

FIG. 6. Thermal *cis–trans* isomerisation in glassy polymers. The rotational mechanism is shown on the left and the translational mechanism on the right.[9]

mechanism in which only van der Waals forces between adjacent chains have to be overcome to provide sufficient free volume for isomerisation.

For application to imaging, Eisenbach[9] made the following statement: 'It is clear that photochromic processes in bulk polymers depend on the free volume and that they are controlled by molecular motions of the polymer matrix itself. It may be assumed that this is valid for photoreactions in general. As a consequence of these features it seems very unlikely that practical application of photochromism such as in photoimaging processes has a good future.' Despite this statement work continues on the application of photochromism in imaging, as will be considered in Section 6. Also, the limitations given by Eisenbach presumably apply to photochromism which involves conformational changes and it is worth noting that the mechanism of vision involves a photochromic *cis*→*trans* isomerisation of retinal as the first step in a complex series of reactions which occur with high efficiency.[16] Perhaps, in the future, biotechnology may provide an in-vitro photochromic imaging system based on this in-vivo reaction scheme.

4.2 Homolytic Cleavage

A number of organic compounds reversibly dissociate into free radicals by optical dissociation or by energy transfer from an absorbing species:

$$X{-}Y \underset{\Delta T \text{ or dark}}{\overset{h\nu}{\rightleftharpoons}} X\cdot + Y\cdot$$

These dissociations are accompanied by changes in spectral absorbance due to the formation of coloured radicals which may be stable and persist for long periods of time.

An example of this type of reaction is given in eqn (7) for the formation of naphthoxy radicals from tetrachloro-1(4*H*)-naphthalenone:[17]

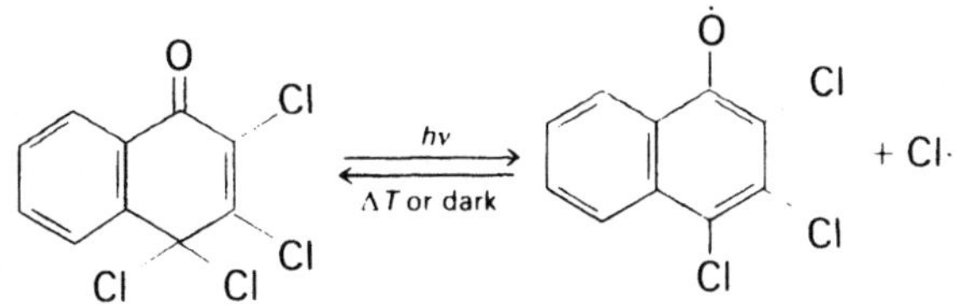

(7)

Hexa-aryl ethanes and bisimidazoles also exhibit photochromism by homolytic cleavage as shown in eqn (8) for hexaphenylbisimidazole.[18]

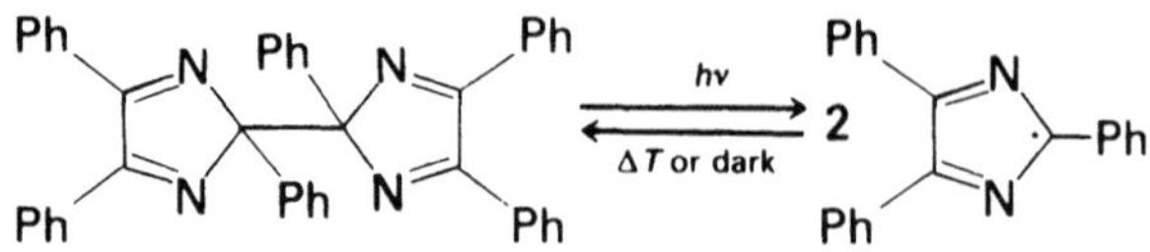

(8)

However, the radicals generated by the above reactions are highly reactive and combine with oxygen which destroys their photochromic properties after only a few cycles.[19,20] This, of course, makes them unsuitable candidates for use in photochromic imaging systems.

4.3 Redox Reactions

Photoredox reactions for most organic molecules involve two-electron transfers and are usually irreversible. However, certain dyes, such as Thionine and Methylene Blue, can exist in two stable forms: a colourless (leuco) reduced form and coloured oxidised form of the dye itself which are interconvertible by a photochemically induced two-electron redox reaction:[21]

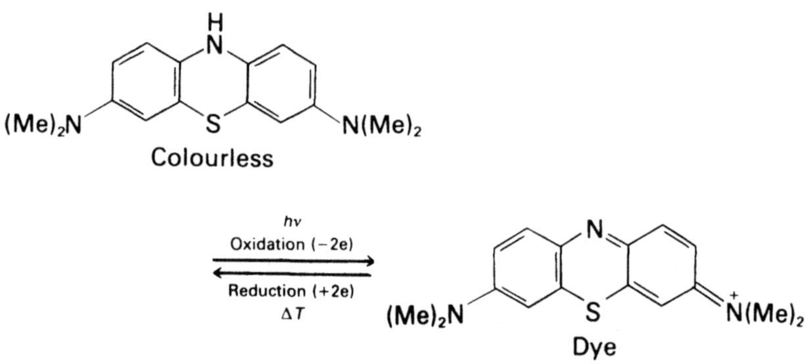

(9)

The forward reaction involves absorption of a photon by the colourless form, which in its excited state can be oxidised to the normal or coloured form of the dye. The reverse or thermal process, in the absence of radiation, involves the reduction back to the colourless form. An advantage of this system is that the dye has a high extinction coefficient and a small amount of photo-oxidation results in a large change in optical density. The rates of the forward and backward reactions and the intensity of the colour produced depend upon the temperature, pH, concentrations of dye, reducing agent and oxidising agent, and the matrix or solvent used for the system.

The system is susceptible to interference by oxygen, which destroys the reversibility when Fe(II)/Fe(III) is used as the redox system, and it becomes only partially reversible when ascorbic acid is used.[22]

The success of this photochromic system depends upon complete reversibility of the dye and the reducing agent between their different oxidation states. Generally these characteristics are achieved in solution in the absence of oxygen but good results have been achieved in a rigid silica gel matrix.[21]

4.4 Triplet Photochromism

Relatively long-lived triplet states can occur in polycyclic aromatic compounds such as picene or dibenzanthracene, in rigid matrices such as polymer films. This phenomenon is shown in Fig. 7.

Absorption of UV radiation populates excited singlet states (S_1) which by intersystem-crossing cross over to the triplet state (T_1). Triplet–triplet absorption occurs in the visible region of the spectrum. The triplet states can have lifetimes of the order of seconds before returning to the ground state by the paths shown in Fig. 7, involving delayed fluorescence or phosphorescence. This process is affected by oxygen, which is a very good quencher of the triplet state, and the lifetime of the triplet state is too short to be of any direct application in an imaging system. However, the triplet state is involved in the sensitization of other types of photochromism which is discussed in Section 4.6.

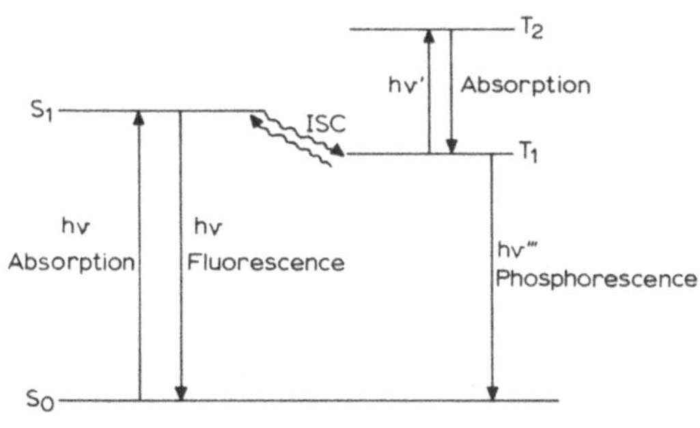

FIG. 7. Triplet photochromism of a polynuclear hydrocarbon in a polymer film.[26] S_0, singlet ground state; S_1, first excited singlet state; T_1, T_2, triplet states; ISC, intersystem crossing.

4.5 Heterolytic Cleavage

Under this heading comes the most thoroughly investigated class of photochromic compounds, the spiropyrans. Heterolytic cleavage causes the formation of charged ions with absorption characteristics differing from those of the parent compound. A typical representative member of the spiropyrans is 1′,3′,3′-trimethylspiro[2*H*-1-benzopyran-2′,2′-indoline] which has been given the acronym BIPS (**I**).

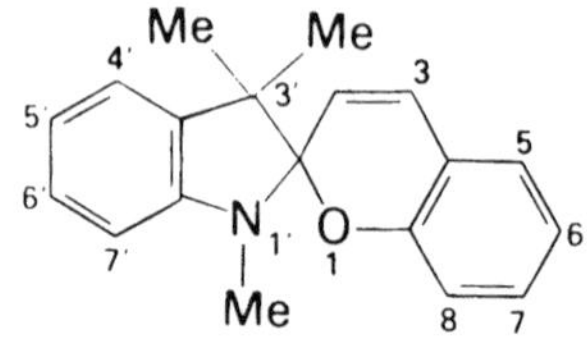

I

The general scheme for photochromism of BIPS and analogous compounds is given in eqn (10). It involves scission of the carbon–oxygen bond and the reverse reaction, a thermal or photocyclisation reaction.

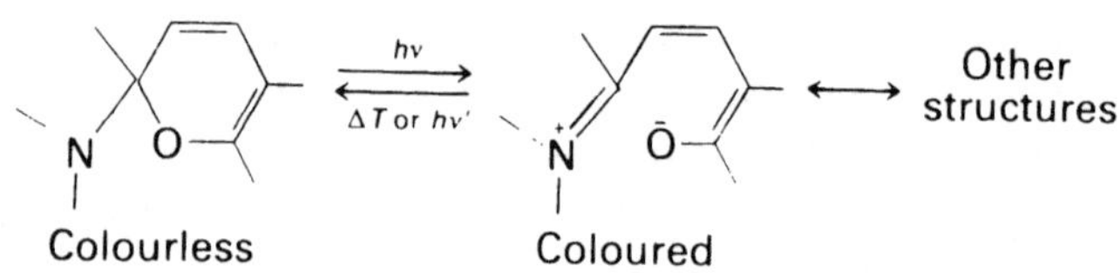

(10)

In addition to heterolytic carbon–oxygen bond scission, *cis–trans* isomerisation is also involved and there are a variety of isomers of the coloured form of BIPS.[23–25] The reaction can be regarded as the interconversion of a spiropyran (colourless form) into a merocyanine (coloured form). From studies of photobleaching of various BIPS compounds in polymeric binders on coated layers, Sahyun[25] showed that triplet states were involved and that *trans*→*cis* isomerisation of the merocyanine occurs. Also, photochromic reactions of spiropyrans can be made more efficient in solution by the use of triplet sensitisers, such as benzophenone which acts as the primary absorber which is able to transfer its energy to the triplet states of the spiropyran,[26,27] as shown in Fig. 8. In a rigid medium the rate of merocyanine formation was, however, substantially reduced, as is expected for diffusion-controlled energy-transfer processes.

As already discussed in Section 4.1, polymer matrices have a profound effect on photochromism. This is of great significance in

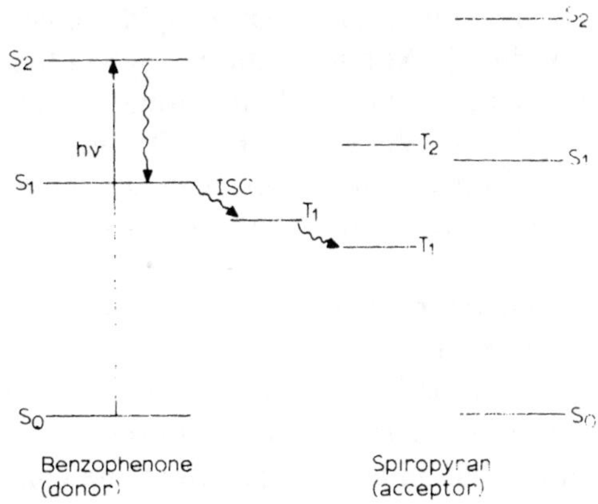

FIG. 8. Sensitized photochromism. Benzophenone (the donor) absorbs radiation and is converted to the triplet state by intersystem crossing (ISC) and then by energy transfer to the triplet state of the spiropyran (acceptor).

imaging systems in which the photochromic compound is either supported in a polymeric binder, or chemically bonded to a polymer to enable the system to be coated on a substrate.

Like the azo compounds mentioned in Section 4.1 the photochromic reaction of BIPS is affected by the polymer matrix in a similar way to that illustrated in Fig. 6 because conformational changes are also involved.[9,28,29] Delzenne[30] has shown that for a spiropyran polyester (**II**), irradiation at temperatures below the glass transition temperature

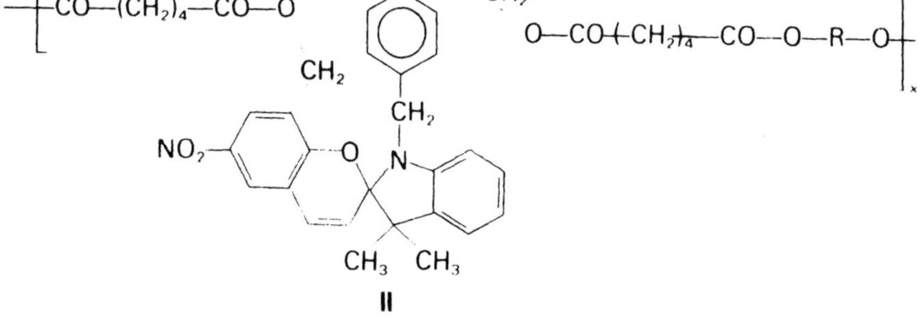

II

(T_g) does not cause a colour change but a rapid colour change occurs within the region of the T_g. At temperatures beyond the T_g the reverse thermal reaction competes with colour formation and the density decreases. These effects, shown in Fig. 9, were obtained from irradiation of the spiropyran polyester with an 80 W high-pressure mercury lamp at a distance of 10 cm for 60 s, followed by cooling to 30°C.[30]

The influence of polymeric matrices and other components on the reverse reaction is given in Section 5.

In most cases photochromism of spiropyrans results in a colourless form being converted into a metastable coloured form on absorption of short-wavelength radiation and the return to the colourless form by the absorption of longer-wavelength radiation or thermally. It would be expected that the environment will affect the rate of thermal reversion since the coloured, charged form is relatively more stable in a polar environment. In addition to this effect and the influences of polymeric matrices a phenomenon of 'inverse photochromism' has been shown to occur for 6-nitro-BIPS in a cyclohexane–silica gel matrix.[31] In this system the open or coloured form is the most stable: on irradiation with wavelengths greater than 350 nm it bleaches rapidly to the colourless form. The colourless spiropyran reverts to the highly coloured form in the dark with a half-life of about 40 min or in a few seconds on irradiation with wavelengths less than 350 nm. This inverse

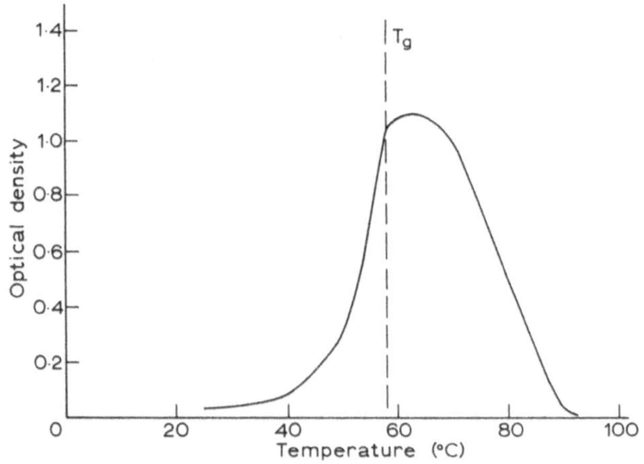

FIG. 9. The influence of temperature on photochromism of a spiropyran polyester.[30]

photochromism is shown in reaction scheme (11).

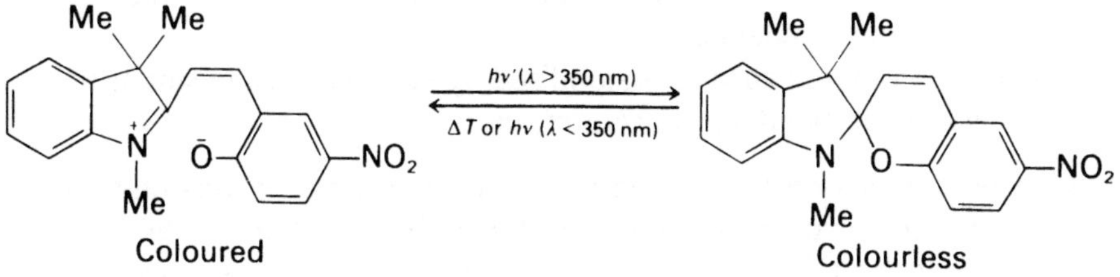

(11)

This is the reverse of the original equation for photochromism (eqn (10)). On highly polar, partially hydrated silica gel, the merocyanine charged coloured form becomes the more stable and 'inverse photochromism' results.

4.6 Tautomerism

Tautomerism by intramolecular hydrogen atom transfer is the mechanism by which a number of organic compounds are photochromic. Examples include *ortho*-substituted phenyl ketones, aromatic nitro compounds, and anils.[32,33]

Salicylidene anils are examples of compounds which are photochromic by tautomerism:

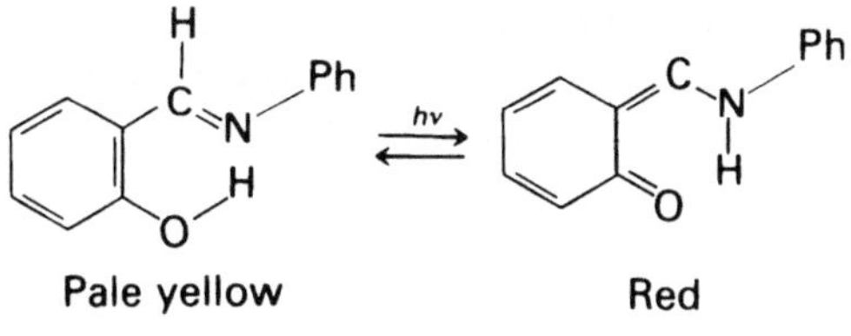

Margerum & Miller[32] have summarised the basic requirements for this type of photochromism as follows:

(a) The hydrogen transfer photolysis step occurs through a six-membered ring transition state.
(b) *Ortho*-substituted aromatic structures are the thermally stable forms.
(c) Quinoid structures are produced by photochemical hydrogen transfer.
(d) One of the *ortho* substituents is polar and its excited state is of greater activity than the other group. Either an excited-state

hydrogen abstraction becomes favourable, or proton transfer takes place because the acid–base balance between the *ortho* substituents changes.

(e) The coloured species is stabilised by various mechanisms such as hydrogen bonding, ionisation, etc.

However, hydrogen-transfer photochromism has poor reversibility due to competing irreversible side reactions, but crystalline anils show good photochromic reversibility. Their coloured forms are not stable thermally and their photochromic properties disappear in fluid solutions or plastic matrices at normal temperatures.[10,34]

In addition to photochromism by tautomeric hydrogen atom transfer, Margerum & Miller[32] have defined a second type of tautomerism which can also give rise to photochromism. This second type, termed valence tautomerism, is a dynamic isomerism involving a shift in the position of the valence bonds. This may also involve a change in the relative positions of the nuclei, like all types of tautomerism, but this change may be small and not appear in normal structural formulae.

Margerum & Miller[32] also drew attention to the possible overlap of valence tautomerism with other classifications of photochromic compounds such as spiropyrans which also undergo valence tautomerism. However, these have been classified under heterolytic Cleavage (Section 4.5) which is their prime mechanism.

Fulgides, which are one of the oldest groups of photochromic compounds,[35] come under the classification of photochromic valence tautomerism. They are derivatives (**III**) of dimethylenesuccinic anhy-

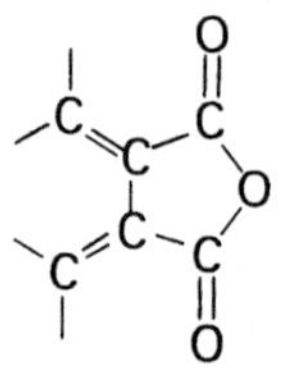

III

drides in which at least one of the substituents must be aromatic (or heterocyclic) for the compound to be photochromic. Photochromism of fulgides does not involve ions, radicals or dipolar intermediates and is inherently less susceptible to fatigue and side reactions than many other types of photochromism previously discussed. The only significant fatigue processes are thermal or photochemical hydrogen shifts[10] and some fulgide photochromic systems are not photodegraded or photo-oxidized and are thermally stable at temperatures below 100°C.[36]

The mechanism of photochromism in phenyl-substituted fulgides involves valence tautomerism and has been shown by Heller[37–39] to be a concerted photocyclisation by a conrotatory mode to form a coloured dihydronaphthalene. Irradiation of the coloured form causes the reverse reaction, also by a conrotatory mode as shown in reaction scheme (12). Thermal bleaching, on the other hand, involves ring-opening by a disrotatory mode [eqn (13)].

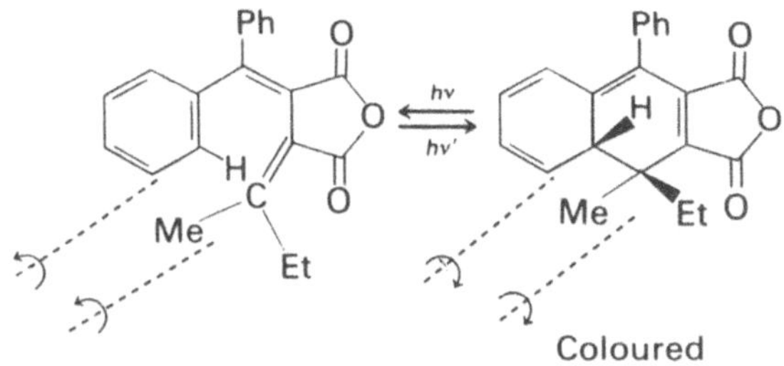

(12)

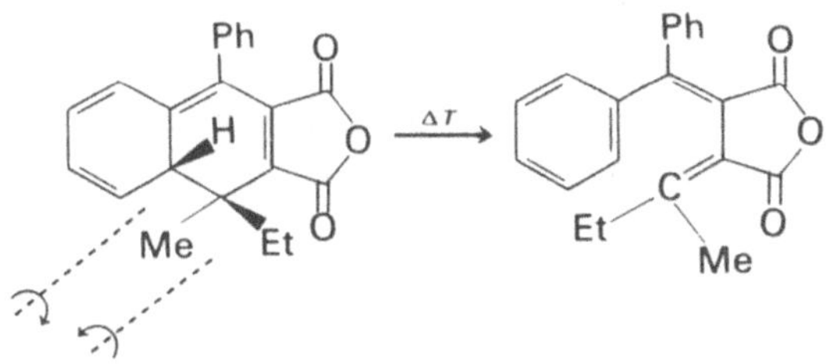

(13)

Heller has prepared and investigated a number of fulgides and has provided fulgides that are claimed to be fatigue-free by substituting the hydrogen atom that is involved in thermal and photochemical shifts by a methyl group, for example.[40] The general scheme (14), in which substituents in the benzene ring have been omitted, shows the two types of hydrogen shift which are avoided by substituting the hydrogen atom.

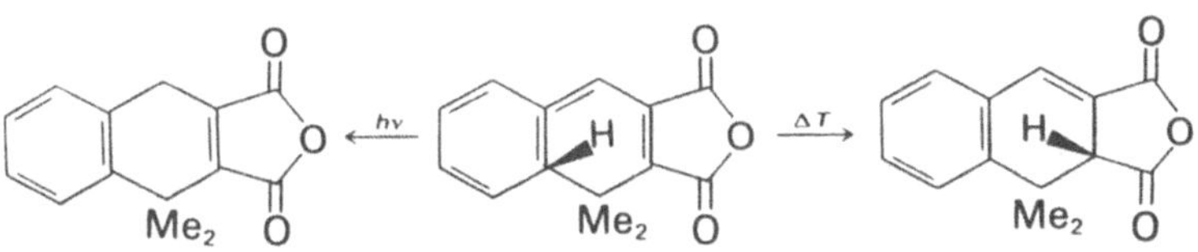

(14)

Substituting the hydrogen atom in the fulgide in eqn (14) eliminates fatigue reactions by having no hydrogen atom able to migrate and the thermal disrotatory mode of ring-opening is inhibited by steric effects of the methyl groups, whereas the conrotatory photochromic reverse reaction can proceed unhindered.

Heller[10] has termed these fatigue-resistant, stable, photochromic fulgides 'Aberchromes' from their discovery in Aberystwyth at the University College of Wales. Many different types are available with different photochromic properties. Some have a heterocyclic ring in place of the benzene ring. The general structural formula for Aberchromes is **IV**.

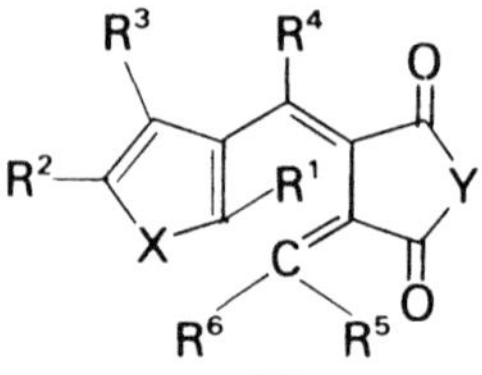

IV

The availability of stable, fatigue-resist, photochromic compounds has rekindled interest in photochromism as a means of data storage and for imaging, although their quantum yields are low. These applications are outlined in Section 6.

5 FIXING AND STABILITY OF PHOTOCHROMIC IMAGES

The main features of photochromic materials in conventional imaging are the direct access to the image and the ability to erase and re-record images. They can best be regarded as transient intermediates from which more permanent images can be reproduced by traditional photographic techniques. The scope of photochromic imaging systems could be extended if they could be rendered more stable, particularly to thermally induced reverse reactions. This can be achieved by a variety of methods which include chemical treatments and the use of appropriate polymeric matrices, or by the use of photochromic compounds which are inherently more stable. Some reference has been made to the last two methods in earlier sections but it is useful to discuss them further under individual headings.

5.1 Fixing of Photochromic Images by Chemical Treatment

A number of chemical treatments have been proposed for making photochromic images permanent. However, these techniques remove one of the most significant advantages of photochromic imaging, that of direct access to the image without a processing stage, whilst some of the proposed methods also destroy the reversibility of the system which converts a photochromic system into a more conventional imaging system.

An example of this type of process is a dry process which involves subjecting the irradiated material (in the form of a coated layer) to an electron-accepting gas, such as sulphur dioxide, iodine or boron trichloride, followed by heating to cause an acid–base reaction between the complex and a colour-forming compound also present in the coated photochromic layer.[41] The reaction scheme, together with the colour changes for this type of process, are shown in eqn (15), where BIPS is a spiropyran and MC is the merocyanine formed on irradiation (see Section 4.5). The coloured form of BIPS, the merocyanine image, persists for about 1 h, during which time the image can be erased or fixed by the above process.

$$\underset{\text{Colourless}}{\text{BIPS}} \overset{h\nu}{\rightleftharpoons} \underset{\text{Blue}}{\text{MC}} \xrightarrow{SO_2} \underset{\text{Yellow}}{\text{MC/SO}_2} \xrightarrow[\text{Chromogen}]{\Delta T} \underset{\text{Blue}}{\text{Dye}} \quad (15)$$

Alternative 'wet' processes can be used if the chromogenic material is omitted from the original coating and the irradiated material immersed in a solution of the chromogen after treatment with sulphur dioxide.

Photochromic images have also been fixed by converting the coloured forms of spiropyrans to silver images by a process similar to that described above.[42] This involves treating the MC/SO_2 complex with HCl vapour, followed by immersion in a silver nitrate solution. This converts the irradiated areas to silver halide, which can then be treated with a conventional photographic developer solution (reduction) to convert the silver halide to metallic silver. If a colour developer is used then a dye image could be formed.

Other chemical methods for fixing photochromic spiropyrans include the use of metal salts which form a stable coloured chelate image[43] and the conversion to a complex by reaction with a metal halide, such as zinc bromide.[44] This complex is photodissociable so that separated entities are accessible to various reagents and solvents.

5.2 Decreasing Oxygen-Induced Fatigue Reactions

A number of photochromic systems suffer from unwanted side reactions which induce fatigue into the system. One example of the effect of this type of reaction, which occurs with spiropyrans, was mentioned in Section 2 and can give rise to the 'yo-yo' effect (see Fig. 5). Many photochromic systems are susceptible to oxidative degradation, i.e. quenching of the triplet state by oxygen or reaction of reactive free radicals with oxygen. The effect of oxygen on a photochromic aziridine system[45] is shown in Fig. 10; it can be seen that the introduction of oxygen causes rapid decolouration of the coloured form, which is relatively stable when kept in an inert atmosphere.

The influence of oxygen on this photochromic system is also shown in Fig. 11. Irradiation at 60°C in the presence of oxygen induces instantaneous colouration followed by rapid thermal bleaching. This cycle is repeated six times and when nitrogen is introduced a stable coloured form results which persists for more than 3 h.

For practical systems Fanslow[45] has used oxygen-impermeable barrier layers of poly(vinyl alcohol) or gelatin coated on top of the photochromic layer. It is claimed that images are maintained for more than one year by this technique. A wide variety of oxygen-impermeable barrier layers have been proposed.[46] These include water-insoluble polymers such as cellulose esters, poly(*n*-butyl methacrylate), polyurethanes, polyamides, polyesters, etc. These barrier

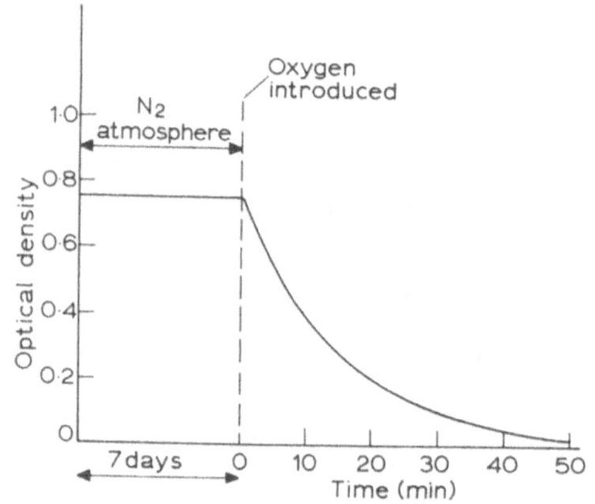

FIG. 10. The effect of oxygen on bleaching of an aziridine photochromic system.[45]

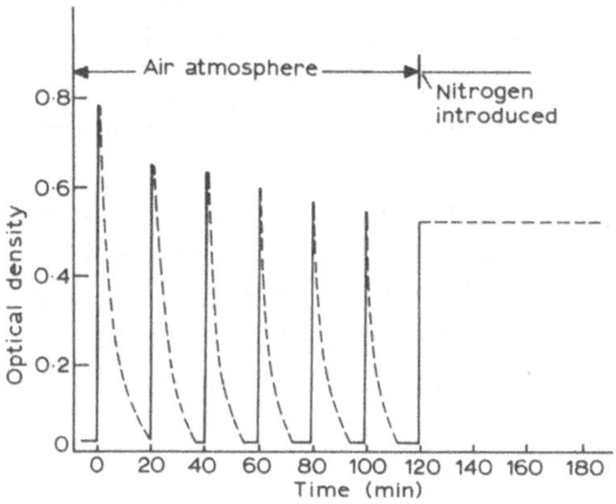

FIG. 11. The effect of nitrogen on bleaching of an aziridine photochromic system.[45]

layers may be applied from polymer solutions or dispersions, by lamination or adhesion of polymer sheets, or by polymerising monomers *in situ* on the photochromic layer.

5.3 Decreasing Thermal Self-bleaching

In Sections 4.1 and 4.5, the influences of polymeric matrices on photochromism was discussed and a model proposed for explaining the effects of polymers on conformational changes (see Fig. 6). Also, Fig. 9 shows the effect of the temperature on photochromism of spiropyrans linked to a polymer chain. These investigations have been

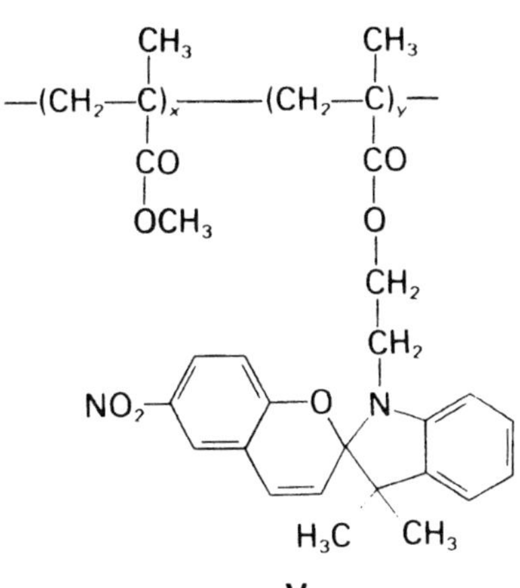

V

extended by Delzenne,[8] who, from kinetic studies, showed that the thermal self-bleaching reaction was slowed down when spiropyrans were linked to polymers. This technique makes the spiropyran photochromic system more useful in practical imaging systems.

Delzenne[8] used a copolymer of general structure **V**. He investigated the kinetics of photochromism of films cast from solvents of this polymer and for the corresponding spiropyran in a poly(methyl methacrylate) matrix. Kinetic curves for the thermal self-bleaching for these systems are shown in Fig. 12. The deviations from a first-order relationship have been interpreted as being due to the existence of

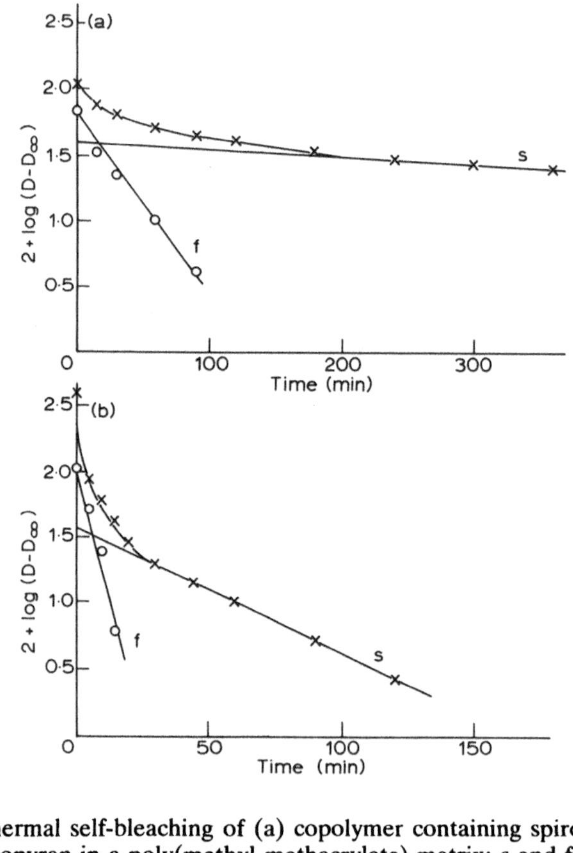

FIG. 12. Thermal self-bleaching of (a) copolymer containing spiropyran units and (b) spiropyran in a poly(methyl methacrylate) matrix: s and f refer to the slow and fast bleaching curves, respectively. D and D_∞ are the optical densities at times t and t_∞, respectively.[8]

several coloured isomeric merocyanines in slow equilibrium with each other.[8,29,47] The kinetics of bleaching in polymer matrices are characterised by the superimposition of two first-order processes which relate to different isomers. The isomers of the coloured merocyanines in equilibrium with each other are illustrated in Fig. 13.

When spiropryans are incorporated in photocrosslinkable films there occurs an even more significant reduction in the rates of self-bleaching.[8] An example is the azide group, which forms crosslinks under the influence of UV radiation.[48]

Generally polymer matrices exert a restriction on the mobility of the

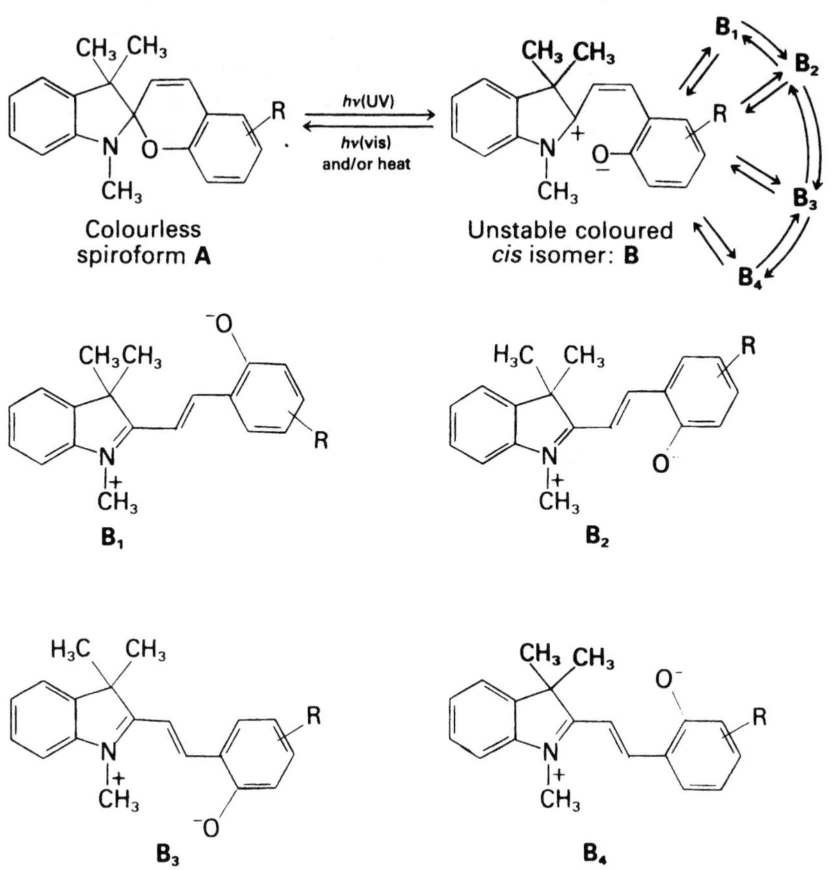

FIG. 13. Photochromism of spiropyrans. $\mathbf{B}_1$–$\mathbf{B}_4$ are the stable *trans* isomers of **B**.[8]

merocyanine isomers and this restriction has the effect of decreasing the thermal self-bleaching rate.

5.4 Stable Fatigue-Resistant Photochromic Systems

The techniques outlined in the preceding paragraphs are all ingenious ways of overcoming the major shortcomings of photochromic systems and extend the applicability of such systems for imaging purposes. A fundamental approach by Heller[10] to structural considerations, which was described in Section 4.6, appears to have provided stable and fatigue-resistant photochromic systems. These photochromic fulgides or 'Aberchromes' are available with widely differing properties depending upon their structure. They can be synthesised with high or low quantum yields for colouring or bleaching, thermal stability, or fast thermal fade, and a wide range of colours is possible. The absorption spectra of some typical Aberchromes[10] are shown in Fig. 14 as an example of the wide variety of possible colours obtainable after

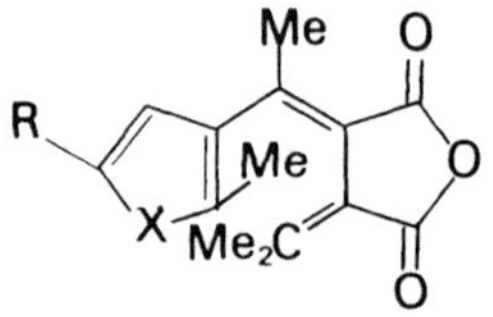

VI

Aberchrome 540: R = Me, X = 0
Aberchrome 850: R = H, X = NMe
Aberchrome 999: R = Ph, X = S

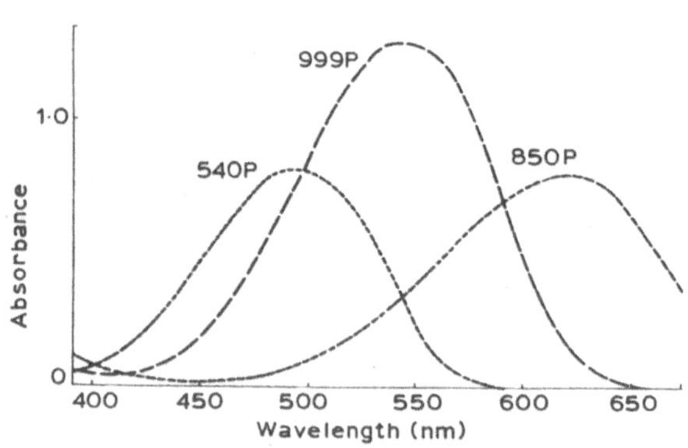

FIG. 14. Absorption spectra of some Aberchromes.[10] For a key to the numbers on the curves, see the text.

irradiation at 366 nm, depending upon the structure of the Aberchrome. The numbers on the spectral absorbance curves refer to structure **VI**.

6 APPLICATIONS OF PHOTOCHROMIC SYSTEMS IN IMAGING

For applications of photochromic systems in imaging, the term 'imaging' is used in its broadest sense to embrace areas of conventional imaging or the production of hard copy as well as to include its use in optical data storage and display. The term 'hard copy' applies to copies of images that can be physically handled and evaluated directly (e.g. a photograph), as opposed to projected or displayed images or information in a coded form.[49] In this context the term 'image' applies to the optical representation of the external form of an object produced by focusing rays of light from the object onto a two-dimensional surface of recording material.[49] The potential application of photochromic systems in the storage of optical (and other) data and in dynamic display devices has been recognised for more than two decades.[6,50,51] More recently their potential for optical data processing has also been considered.[52]

6.1 Direct Image-Forming Applications

In principle, photochromic systems have the advantage over other photochemical recording systems in not requiring chemical processing stages (see Section 2) but their sensitivity is low in comparison with the more usual systems (Section 3). Also, problems of fatigue and lack of stability have limited their application in image recording. A further limitation of photochromic systems in direct image recording is that the non-image areas still remain photosensitive. Their application has been restricted to the preparation of photographic prints or proofs of limited life from negatives produced by conventional silver halide processes,[53] masking in photographic printing as a means of controlling and modifying the contrast of the image,[54] and as intermediates in the production of microimages.[55]

The recent discovery by Heller[10] of fatigue-resistant stable photochromic systems, which were described in Section 4.6, is likely to extend the applications of photochromic systems to direct image-forming processes.

Photochromic recording paper[53] which was commercially available more than a decade ago required an exposure of 10^5 mJ m^{-2} at wavelengths between 300 and 400 nm. It was of high resolution (greater than 1000 lines mm^{-1}) and could be used for recording from display tubes with a P-16 phosphor and a fibre-optic face-plate, Cd, Ar and N_2 lasers, flashlamps, etc. Also it could be used in the reverse mode (decolouration) with He–Ne and CO_2 lasers. Because the exposed material still contained the photosensitive photochromic material, storage in transparent UV-absorbing envelopes was recommended. The image could be erased by heating to 100–140°C for less than 1 s. For permanent copies of images it was recommended that the photochromic image be photocopied in an office copying machine.

The application of photochromic systems in modifying the tone-reproduction characteristics in photographic printing have been extensively reviewed by Dorion & Wiebe.[56] In outline, this procedure involves printing the negative through a transparent sheet of photochromic film (or mask) onto the print material. For contrast modification the least dense areas of the negative cause maximum colouration of the photochromic mask and less exposure needs to be given to the print material than without the mask being present. Contrast of the image is reduced and photochromic masking is equivalent to printing a negative image sandwiched with a weak positive image. The characteristics of the photochromic mask material have been summarised by Dorion & Wiebe[56] as follows.

(1) The absorption spectrum of the coloured mask must overlap with the spectral sensitivity of the print material.
(2) The contrast and maximum density must be compatible with the density scale of the negative.
(3) The mask should be transparent in unexposed areas.
(4) It should be sufficiently long-lived to allow printing.
(5) It should preferably be unaffected by the light source used for printing.

Holography is another application of direct image recording in which appropriate photochromic materials could be of great benefit because of their extreme resolution and instant access to the image, but their lack in sensitivity is somewhat of a limitation. Use of photochromic recording materials would avoid the need for removing the material for processing and returning it to the system. Some holographic applications require precise repositioning of the developed hologram;[57] this would be avoided if the material formed an image directly and also the hologram would not be subjected to dimensional changes that occur with wet processing of silver halide holographic materials.

Photochromic materials are used as intermediates in the reduction of documents (up to 40 000:1 in area).[55] This is a stage in the production of microimages which enables inspection of the images and additions or deletions to be made. This PCMI (PhotoChromic MicroImage) system (National Cash Register Company) starts with the original document being transferred to a microfilm negative at a reduction of about 10:1. A second reduction stage is then carried out by printing onto an organic photochromic layer, using near-UV radiation in a step-and-repeat mode, to build up a matrix of microimages. The photochromic microimages, after inspection and correction, are contact printed onto a high-resolution photographic material which after processing becomes the master for subsequent printing onto high-resolution film. The PCMI system[58] is shown in Fig. 15. Typically, approximately 3000 average-sized pages can be recorded in an area 105 mm × 148 mm with a linear reduction of 150:1. For protection against damage the microform is laminated on both sides with a polyester film.

Photochromic materials have certain advantages in the preparation of photoresists for the production of integrated circuits. One of the main reasons for using a photochromic system in the above technique for the production of microimages is that corrections can readily and simply be made part-way through a complicated and exacting imaging process. The same reason applies in the mask-making process in the preparation of photoresists. Thus if a photochromic material is used in place of the photographic material, mistakes could be corrected during their production, rather than having to wait to the end of the process.[59] Although solubility differences between the irradiated and non-irradiated amorphous spiropyran layers have been proposed as a means of preparing photoresists,[5] there exist many other photochemical processes which have properties more appropriate for this application.[60] The exposed coloured merocyanine form of spiropyrans is less soluble in certain solvents than the unexposed form. Thus after exposure the non-image areas can be removed by the solvent. However, the remaining spiropyran resist image is not particularly resistant to strong etchants and their application is restricted to substrates which only require mild etchants, such as iron(III)

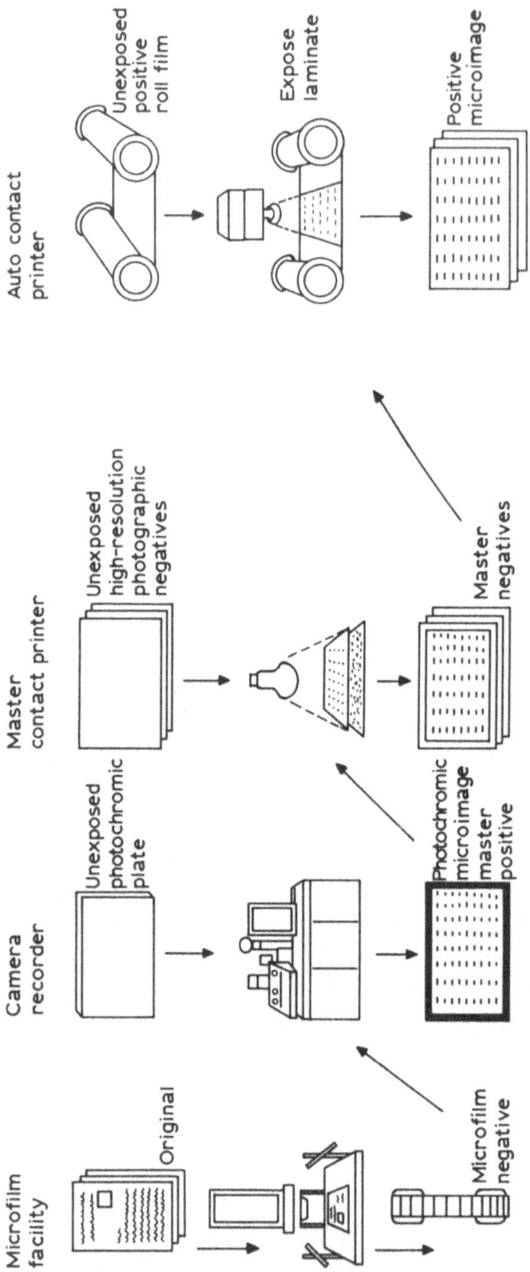

FIG. 15. The NCR PCMI microform process.[58]

chloride.[59] It remains to be seen whether the more recently introduced fulgides[10] are more appropriate for this application or can be molecularly tailored to be so.

6.2 Optical Data Storage

The ability of photochemical imaging materials to store large quantities of data is well known.[61,62] The theoretical limit of 10^{23} bits cm^{-3} has been calculated by Camras[61] based on the assumptions that a solid contains 10^{23} atoms cm^{-3} and that each atom represents one bit of information. This theoretical limit of storage density, together with storage densities of various media, is shown in Fig. 16.

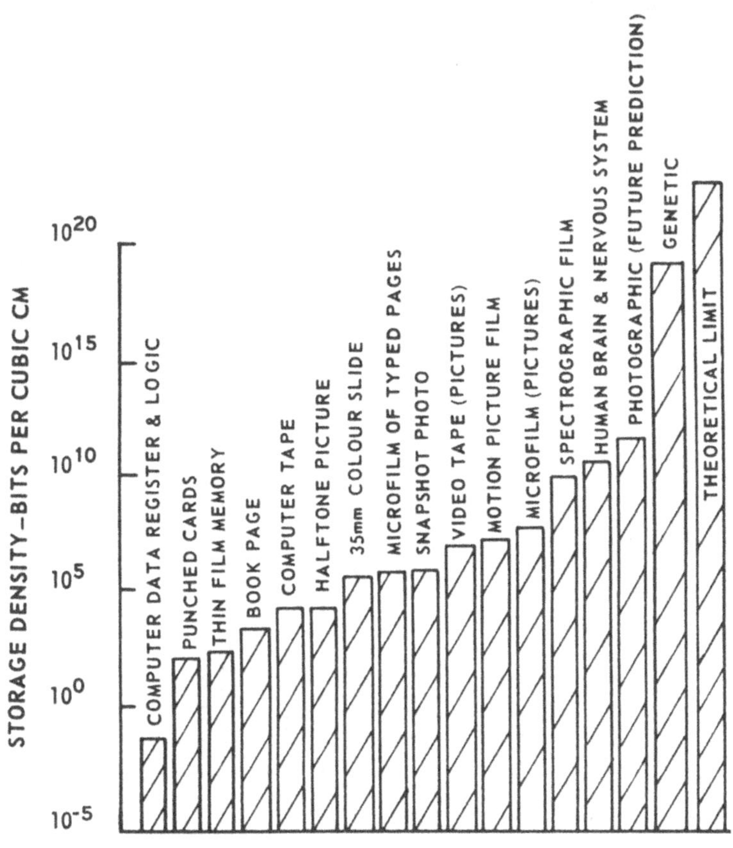

FIG. 16. Volume storage densities.[61]

The limiting value becomes 10^{21}–10^{23} bits cm^{-2} for recording layers 0·1–0·01 cm thick.[61] The gross area information densities for some typical storage media, according to Camras,[61] are given in Fig. 17.

The large difference between the theoretical limit and realisable values given in Figs 16 and 17 has been attributed to additional fundamental considerations. Atkinson *et al.*,[63] using the principles of quantum statistics, the Heisenberg Uncertainty Principle and relativity, calculated the limit for information density to be $2{\cdot}5 \times 10^9$ bits cm^{-2} (5000 lines mm^{-1}) for green light (500 nm). An experimental value[62] for a high-resolution silver halide photographic material is $1{\cdot}6 \times 10^8$ bits cm^{-2}, which is within two orders of magnitude of the limiting value. The conventional photographic material suffers from limitations due to image spread within the layer and the particulate nature of the recording layer.

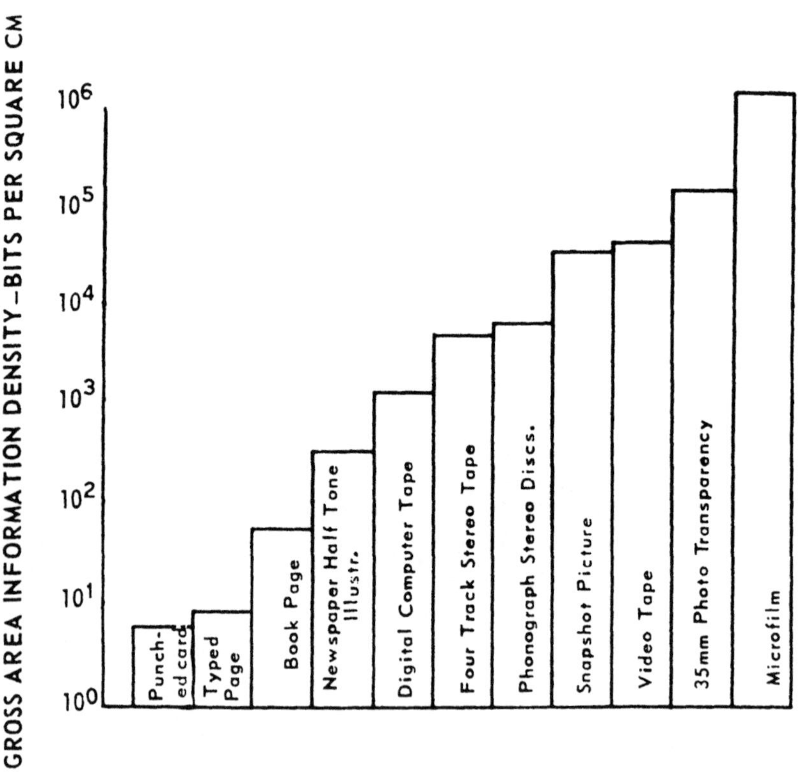

FIG. 17. Gross area information density including blank margins.[61]

Layers of organic photochromic materials, whose coloured forms have high extinction coefficients, permit thin coatings to be used which reduce image spread within the layer, and the layers themselves are molecular rather than particulate in nature. For an optical density of 2, layer thicknesses of 3–10 μm are required, which limits the resolution to 100–200 line pairs mm^{-1}.[13] The dimensions of the recorded bit is determined by the modulation transfer function (MTF) of the material and by the diffraction-limited spot size of the recording beam, according to the following equation:[4]

$$\text{Diameter} = (2 \times \lambda)/(\pi \times \text{Numerical aperture})$$

Current technology makes use of magnetic recording for storage of information which allows efficient erasing and re-recording. At present, optical recording techniques are restricted to read-only systems, in the form of optical and audio discs which use prerecorded data replicated from a master, or to Direct Read After Write (DRAW) systems. DRAW systems mainly use pit formation in tellurium layers although other techniques such as magneto-optic effects, phase changes, surface texture changes and bubble formation are being investigated in inorganic and organic layers.[4] The properties required for materials for re-usable erasable materials have been summarised by Wilson:[52]

(1) High-sensitivity or low-energy requirements for writing and erasing (less than 500 mJ cm^{-2} to produce an optical density change of 1).
(2) High storage capacity (10^6–10^8 bits cm^{-2}).
(3) Non-destructive read-out.
(4) Many recycles possible without loss in performance.
(5) Storage life in excess of 10 years.
(6) Instant access to the recorded image without processing.
(7) Low cost and ease of fabrication of large areas.

The majority of the above criteria have been met and the stable fatigue-resistant fulgides (see Section 4.6) are claimed to be capable of recording 30 million bits cm^{-2} on a photochromic layer with more than 8000 fatigue-free write–erase cycles.[10] Colour changes are preferable for direct spot recording as complex read-out systems are not required.[52] Phase changes are also useful in optical data recording and read-out, in optical data processing,[13] and in optical waveguide components.[64]

Figure 18 illustrates typical absorbance and refractive-index changes for a photochromic fulgide.[13]

Evaluation of photochromic layers for use as erasable optical discs by Ralston[65] has shown them to be capable of recording a 0·5 μm spot. Ralston used an encapsulated photochromic layer 10 mils (0·254 mm) thick which reached a maximum density of 2·7 and a density change of 1 required 170 mJ cm^{-2} when recorded with an Ar laser at a wavelength of 514·5 nm. This wavelength was used for both recording and read-out. Complete erasure required approximately 2 J cm^{-2} of UV radiation. Accelerated ageing tests, using high humidities and elevated temperatures, showed that the photochromic material was unaffected until 60°C and 90% RH, when degradation occurred after 4 h.

Kirkby & Bennion[13] have recently drawn attention to problems in cycle times associated with currently available UV lasers. With many fulgides the sensitivity to colouring is the same or greater than the sensitivity to bleaching but the power output of UV lasers at the wavelengths required for erasing is in milliwatts and erase rates are likely to be 10^3-fold slower than writing rates.

For convenience, the reading of data is carried out by using light

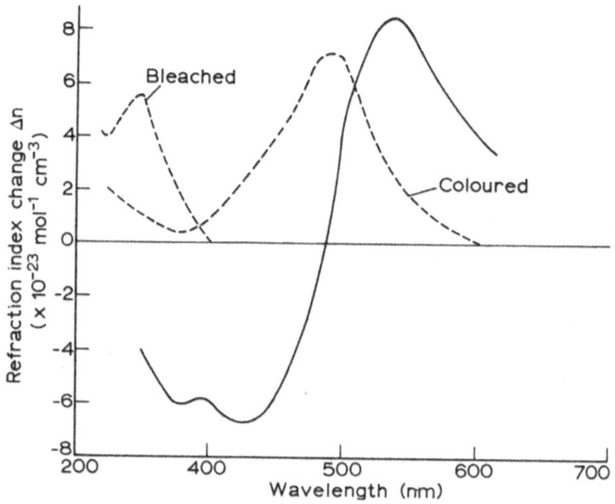

FIG. 18. Absorption spectra and refractive index changes for a photochromic fulgide.[13] ----, Absorption spectra of the two forms; —— refractive index change.

from the same laser at the same wavelength as that used for recording but at a much lower intensity. However, this must lead to degradations each time the data are read and the number of reads per write will be restricted. This destructive effect can be avoided by reading at wavelengths beyond the visible absorption range, using refractive-index changes (see Fig. 18).[52] For example, a He–Ne laser at 632·8 nm could be used. Also, the refractive-index changes which occur for some photochromic compounds at wavelengths beyond the visible region (700 nm) make possible the use of semiconductor lasers.[13]

6.3 Other Imaging Applications

Photochromic layers have been used in various display devices for large-area bright-screen displays.[6] These require UV sources and optics for writing, and projection systems for viewing. The hardware required for these applications has been extensively reviewed elsewhere.[6,66] However, these systems suffer from problems of image persistance, writing speed and visual contrast.[52]

Despite photochromic materials having been known for more than a century their imaging applications are still in their infancy. Much research has been done in this area but photochromic imaging products have so far not penetrated the commercial marketplace to any significant extent. In the past the inherent limitations of photochromic systems have outweighed their advantages. It is hoped that new discoveries, particularly of the stable fatigue-resistant fulgides,[10] will eventually provide photochromic imaging systems of significant practical usefulness. However, their major limitation of low photosensitivity will restrict their applications to those systems where sensitivity is not of prime importance and where high-intensity sources are available and appropriate.

At the end of a paper, much cited in this review, there appears the following statement by Professor Heller:[10] 'Why has no advantage been taken of these new and versatile fatigue-resistant photochromic organic compounds? There may be many reasons. Perhaps one is simple disbelief that, after three decades of extensive research, unsubstantiated claims, great expectations and repeated expensive failures, the long-sought-for photochromic organic compounds are finally readily available.'

It is hoped that these compounds will pave the way for new applications of photochromic systems and that we will at last see their successful application to imaging systems.

REFERENCES

1. Brown, G. H., *Techniques of Chemistry,* Vol. III, Photochromism, ed. G. H. Brown. Wiley–Interscience, New York, 1971, Ch. 1.
2. Tomlinson, W. & Chandros, E. A., *Adv. Photochem.,* **12** (1980) 240.
3. Jacobson, R. E., *J. Photogr. Sci.,* **31** (1983) 1.
4. Kuder, J. E., *J. Imaging Technol.,* **12** (1986) 140.
5. Orlovic, M., Stone, E. & Pearson, I. M., *J. Electrochem. Soc.,* **116** (1969) 1464.
6. Dorion, G. H. & Wiebe, A. F., *Photochromism: Optical and Photographic Applications.* Focal Press, London, 1970, Ch. VI.
7. Jacobson, K. I. & Jacobson, R. E., *Imaging Systems.* Focal Press, London, 1976, pp. 153–5.
8. Delzenne, G. A. In *Proc. 4th Int. Congress on Reprography, Hanover,* 1975, pp. 169–76.
9. Eisenbach, C. D., *Photogr. Sci. Engng,* **23** (1979) 183.
10. Heller, H. G., *IEE Proc.,* **130J** (1983) 209.
11. Bertelson, R. C. In *Techniques of Chemistry,* Vol. III, Photochromism, ed. G. H. Brown. Wiley–Interscience, New York, 1971, pp. 130–3.
12. Brinkman, E., Delzenne, G., Poot, A. & Willems, J. *Unconventional Imaging Processes,* Focal Press, London, 1978, p. 78.
13. Kirkby, C. J. G. & Bennion, I., *IEE Proc.,* **133J** (1986) 98.
14. Metzger, J. In *Non-silver Photographic Processes,* ed. R. J. Cox. Academic Press, London, 1975, pp. 155–9.
15. Ross, D. L. & Blanc, J. In *Techniques of Chemistry,* Vol. III, Photochromism, ed. G. H. Brown. Wiley–Interscience, New York, 1971, Ch. V.
16. Wayne, R. P. *Photochemistry.* Butterworths, London, 1970, pp. 237–42.
17. Eigenmann, E. In *Techniques of Chemistry,* Vol. III, Photochromism, ed. G. H. Brown. Wiley–Interscience, New York, 1971, pp. 449–54.
18. Hayashi, T., Maeda, K., Shida, S. & Nakada, K., *J. Chem. Phys.,* **32** (1960) 1568.
19. Hayashi, T. & Maeda, K., *Bull. Chem. Soc. Japan,* **33** (1960) 565.
20. Hayashi, T. & Maeda, K., *Bull. Chem. Soc. Japan,* **38** (1965) 2202.
21. Hoffmann, D. O. & Hovey, R. J. In *SPSE Symposium, Unconventional Photographic Systems, Washington, DC,* 1964, p. 109.
22. Dessauer, R. & Paris, J. P. *Advances in Photochemistry,* Vol. 1. Wiley–Interscience, New York, 1963, pp. 293–7.
23. Heiligmann-Rim, R., Hirshberg, Y. & Fischer, E., *J. Phys. Chem.,* **66** (1962) 2465.
24. Flannery, J. B., *J. Am. Chem. Soc.,* **90** (1968) 5660.
25. Sahyun, M. R. V., *Photogr. Sci. Engng,* **16** (1972) 63.
26. Bercovici, T. & Fischer, E., *J. Am. Chem. Soc.,* **86** (1964) 5687.
27. Becker, S. R. & Roy, J. K. *J. Phys. Chem.,* **69** (1965) 1435.
28. Gardlund, Z. G., *J. Polym. Sci., Part B,* **6** (1968) 57.
29. Gardlund, Z. G. & Laverty, I., *J. Polym. Sci., Part B,* **7** (1969) 719.
30. Delzenne, G. A. In *Proc. 4th Int. Congress on Reprography, Hanover,* 1975, p. 105.

31. Evans, T. R., Toth, A. F. & Leermakers, P. A., *J. Am. Chem. Soc.*, **89** (1967) 5060.
32. Margerum, J. D. & Miller, L. J. In *Techniques of Chemistry*, Vol. III, Photochromism, ed. G. H. Brown. Wiley–Interscience, New York, 1971, Ch. VI.
33. Exelby, R. & Grinter, R., *Chem. Rev.*, **65,** (1965) 247.
34. Cohen, M. D., Schmidt, G. M. & Flavian, S., *J. Chem. Soc.* (1964) 2041.
35. Stobbe, H., *Annalen,* **380** (1911) 1.
36. Heller, H. G. & Langan, J. R., *J. Chem. Soc., Perkin Trans. 2,* (1981) 341.
37. Hart, R. J. & Heller, H. G., *J. Chem. Soc., Perkin Trans. 1,* (1972) 1321.
38. Heller, H. G. & Szewczyk, M., *J. Chem. Soc., Perkin Trans. 1,* (1974) 1487.
39. Darcy, P. J., Hart, R. J. & Heller, H. G., *J. Chem. Soc., Perkin Trans. 1,* (1978) 571.
40. Heller, H. G. & Megit, R. M., *J. Chem. Soc., Perkin Trans. 1,* (1974) 923.
41. Becker, W. J. & Foris, P. L., US Patent 3 359 103 (1967).
42. Becker, W. J. & Foris, P. L., US Patent 3 364 023 (1968).
43. Taylor, L. D., Nicholson, J. & Davis, R. B., *Tetrahedron Lett.*, **17,** (1967) 1585.
44. Schwab, H., Foris, P. L., Becker, W. J. & Bowman, R. E. In *SPSE Symposium, Unconventional Photographic Systems, Washington, DC,* 1967, p. 106.
45. Fanslow, D. L. British Patent Application GB 2 006 808A (1978).
46. Leenders, L., Durr, H. & Gross, H., Research Disclosure No. 20532. **205** (1981) 188.
47. Verborgt, J. & Smets, G., *J. Polym. Sci., Pol. Chem. Ed.*, **12** (1974) 2511.
48. Jacobson, R. E. In *Light, Chemical Change and Life,* ed. J. D. Coyle, R. R. Hill & D. R. Roberts. Open University Press, Milton Keynes, 1982, p. 319.
49. Spencer, D. A. *The Focal Dictionary of Photographic Technologies.* Focal Press, London, 1973.
50. Exelby, R., *J. Soc. Dye Col.,* **75** (1967) 450.
51. Dorion, G. H. & Wiebe, A. F., *Photochromism: Optical and Photographic Applications.* Focal Press, London, (1970) Ch. V.
52. Wilson, A. E. J., *Phys. Technol.*, **15** (1984) 232.
53. ISSEC, *Photochromic Recording Paper. Laboratoires de Physiochimie,* Ferney Voltaire, France, 1974, pp. 13–22.
54. Nachtigall, G. W. & Dorion, G. H., *Photogr. Sci. Engng,* **9** (1965) 67.
55. Myers, W. C., *PCMI: Technology and Potential Applications.* The National Cash Register Co., Hawthorne, CA, 1964, pp. 9–27.
56. Dorion, G. H. & Wiebe, A. F., *Photochromism: Optical and Photographic Applications.* Focal Press, London, 1970, Ch. III.
57. Smith, H. M., *Principles of Holography,* Wiley–Interscience, New York, 1969, p. 190.

58. Tauber, A. S. & Myers, W. C., American Documentation, **13** (1962) 403.
59. Bertelson, R. C. In *Techniques of Chemistry,* Vol. III, Photochromism, ed. G. H. Brown. Wiley–Interscience, New York, 1971, p. 793.
60. Jacobson, R. E. In *Light Chemical Change and Life,* ed J. D. Coyle, R. R. Hill & D. R. Roberts. Open University Press, Milton Keynes, 1982, Ch. 5.13.
61. Camras, M., *IEEE Spectrum,* **2**(7) (1965) 98.
62. Higgins, G. C. In *Photographic Systems for Engineers,* SPSE, Washington, DC, 1966, p. 202.
63. Atkinson, J. H., McManigal, P. G., Taber-Zadeh, M. & Calvert, J. D., *Soc. Inf. Display–Soc. Photogr. Opt. Inst. Eng., Proc. I Joint Workshop, Los Angeles, California,* 1964, VIII–1.
64. Bennion, I., Hallam, A. G. & Stewart, W. J., *Radio Electron. Engn.,* **53** (1983) 313.
65. Ralston, L. M., *SPIE Proc.,* **299** (1981) 53.
66. Bertelson, R. C. In *Techniques of Chemistry,* Vol. III, Photochromism, ed. G. H. Brown. Wiley–Interscience, New York, 1971, Ch. 10.

Chapter 6

FOUR-CENTER TYPE PHOTOPOLYMERIZATION OF DIOLEFIN CRYSTALS

MASAKI HASEGAWA
The University of Tokyo, Japan

1 TOPOCHEMICAL PHOTOPOLYMERIZATION OF 2,5-DISTYRYLPYRAZINE

Crystalline-state organic reactions proceed with a minimum of atomic and molecular motion, indicating strict control of the crystal lattice. Therefore, many of the reaction conversions are quantitative and the products have an extremely stereospecific configuration which can be attained only through a crystalline-state reaction.

The crystalline-state polymerization of vinyl and cyclic ether compounds by a chain reaction mechanism has been extensively investigated since the 1950s. Among these reactions, certain correlations between the molecular arrangement of the monomer and the growth direction of the polymer chain were demonstrated in the polymerization of trioxane crystals.[1] The term 'topochemical polymerization' was introduced for the first time for this polymerization. However, any definite and clear correlation had not been demonstrated between the monomer and the polymer in the dimensions of the crystal unit cell in these polymerizations.

[2 + 2]-Photodimerization of two types of cinnamic acid crystals has been known for many years. Bernstein and Quimby had predicted the correlation between crystal structures and configuration of photoproducts for the two crystals of cinnamic acid, assuming a crystal lattice controlled process.[2] In 1964 Schmidt and coworkers correlated the crystal structures of several olefinic derivatives, including two kinds of cinnamic acids, with their photoreactivity and the configuration of their photoproducts.[3] The crystal structures of these derivatives were divided into α-, β- and γ-type crystals. In α-type crystals the neighboring molecules are inversionally related and the double bonds are arranged in parallel and make contact at a distance of 3·6–4·1 Å across a center of symmetry. The β-type crystal is characterized by a lattice having one axial length of 3·9–4·1 Å between translationally related molecules. In the γ-type crystal no double bonds of neighboring molecules are within 4·7 Å.

After establishment of the topochemical concept, its extension into the topochemical polymerization was attempted and little reported.[4] However, the first clear-cut example of topochemical polymerization was the reaction of a 2,5-distyrylpyrazine (DSP) crystal which was reported in 1967.[5]

The photoreaction of DSP crystals takes place under exposure to sunlight, repeating cyclobutane formation to result in an extremely highly crystalline linear polymer in quantitative yield:

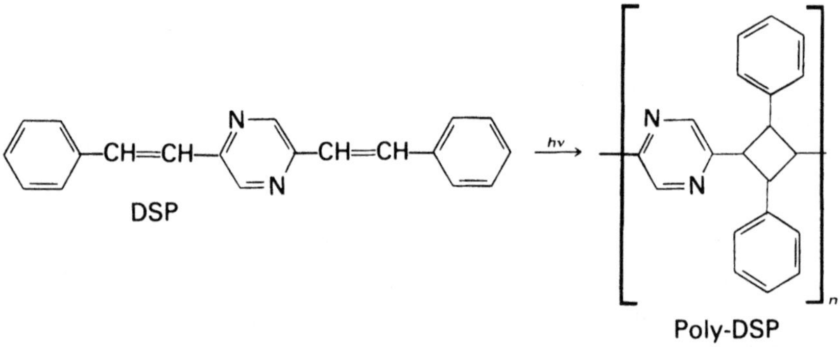

The reactivity of the DSP crystal was fully interpreted in terms of the topochemical concept.[6] In the DSP crystal the double bonds are related to the center of symmetry and are separated by 3·939 Å. Monomer and polymer crystals are closely related at the level of unit cell dimensions; the reaction proceeds with retention of the space group of the crystal. During the reaction the center of gravity in the monomer moves approximately 1·8% in the direction of chain growth (in the direction of the *c*-axis) and the crystal volume shrinks very slightly (approx. 1%); see Table 1.

The crystallographic correlation between the starting substance and the final product was demonstrated between DSP and poly-DSP for

TABLE 1
Crystal Data of DSP and Poly-DSP[a]

	Space group	*a* (Å)	*b* (Å)	*c* (Å)	*Z*	*Density*	*C=C distance* (Å)
DSP	*Pbca*	20·638	9·599	7·655	4	1·244	3·939 Å
Poly-DSP	*Pbca*	18·36	10·88	7·52	4	1·257	

[a] After our first reports on the crystal structure correlations between DSP and poly-DSP crystals, a different crystallographic result was reported on the poly-DSP crystal.[7] We reconfirmed, however, that our first structural analysis was correct, reporting this in Ref. 6(c).

the first time in a chemical reaction. In addition the polymerization of the DSP crystal was a first extension of a topochemical [2 + 2]-photocyclodimerization to polymerization from an α-type olefinic crystal to an α-type diolefinic crystal. The polymerization was named 'four-center type photopolymerization'.

Furthermore it deserves special emphasis that the four-center type polymerization of DSP was the first example of photopolymerization by a step-by-step mechanism, not limited to the crystalline state but occurring widely in all the reaction phases. Also in 1967, a patent announced the crystalline-state photopolymerization of dibenzylidene-benzene diacetonitrile derivatives.[8]

2 DEVELOPMENT OF FOUR-CENTER TYPE PHOTOPOLYMERIZATION

Since the discovery of the photopolymerization of a DSP crystal, research work in this field entered into a new stage of development and a great number of photoreactive diolefinic crystals have now been reported.[9] Most of these monomers are prepared by aldol condensation between an appropriate aromatic aldehyde and methyl groups. Wittig's reaction is more convenient for the preparation of several monomers.

Almost all of the reaction behavior of diolefin crystals and configurations of the resulting products are clearly interpreted in terms of the topochemical principle of [2 + 2]-photocyclodimerization enunciated by Schmidt. Most of the photoreactive symmetric diolefins have the crystal structure in which the local arrangement of double bonds is the same as in the α-packing of cinnamic acid, and give a highly crystalline linear polymer having a 1,3-*trans* type configuration of cyclobutane in the main chain. The molecular shape of all these diolefinic derivatives is rigid rod-like, presumably because this shape may be easy for potentially reactive orientation of the double bonds. However, several examples of exceptional behavior, which are related to the distance and parallelism of reactive double bonds, have been recently reported for diolefin as well as mono-olefin crystals. For example, one of the polymorphs of DSP which is obtained by sublimation is photostable although the distance between neighboring double bonds (4·187 Å) is within that of normal reactive bonds.[10] The crystal of the unsymmetric diolefin derivative, methyl 4-[2-(4-pyridyl)ethenyl]cinnamate, is photoreactive and it is converted, on photoirradiation, into a hetero-adduct type linear polymer (cf. Scheme 2). In this crystal, however, the intermolecular double bonds are aligned not exactly in parallel but separated by 4·023(9) Å for one pair of reactive two-carbon atoms and 4·097(10) Å for the other.[11] Several other examples which deviate from the topochemical rule are described in recent review articles.[12,13]

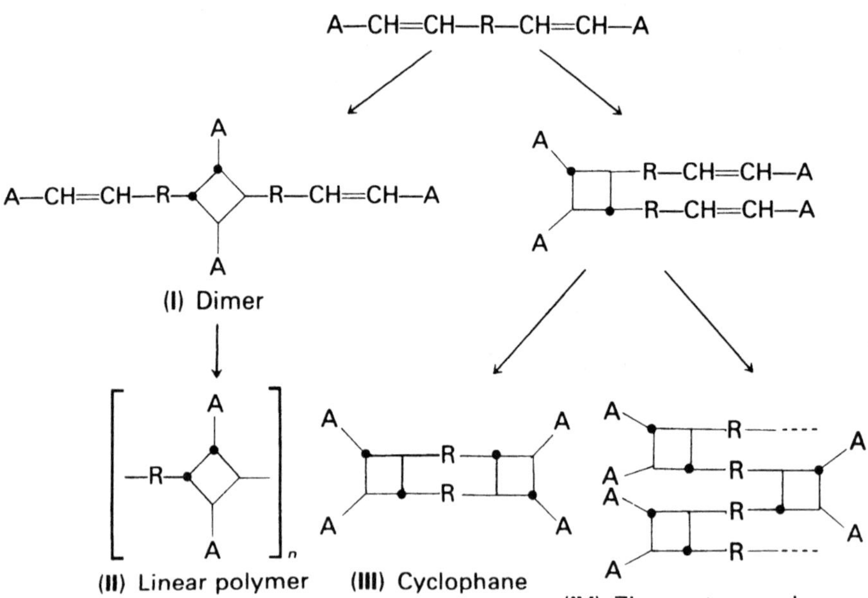

SCHEME 1. Structures of the photoproducts isolated by topochemical reaction of symmetric diolefin crystals.

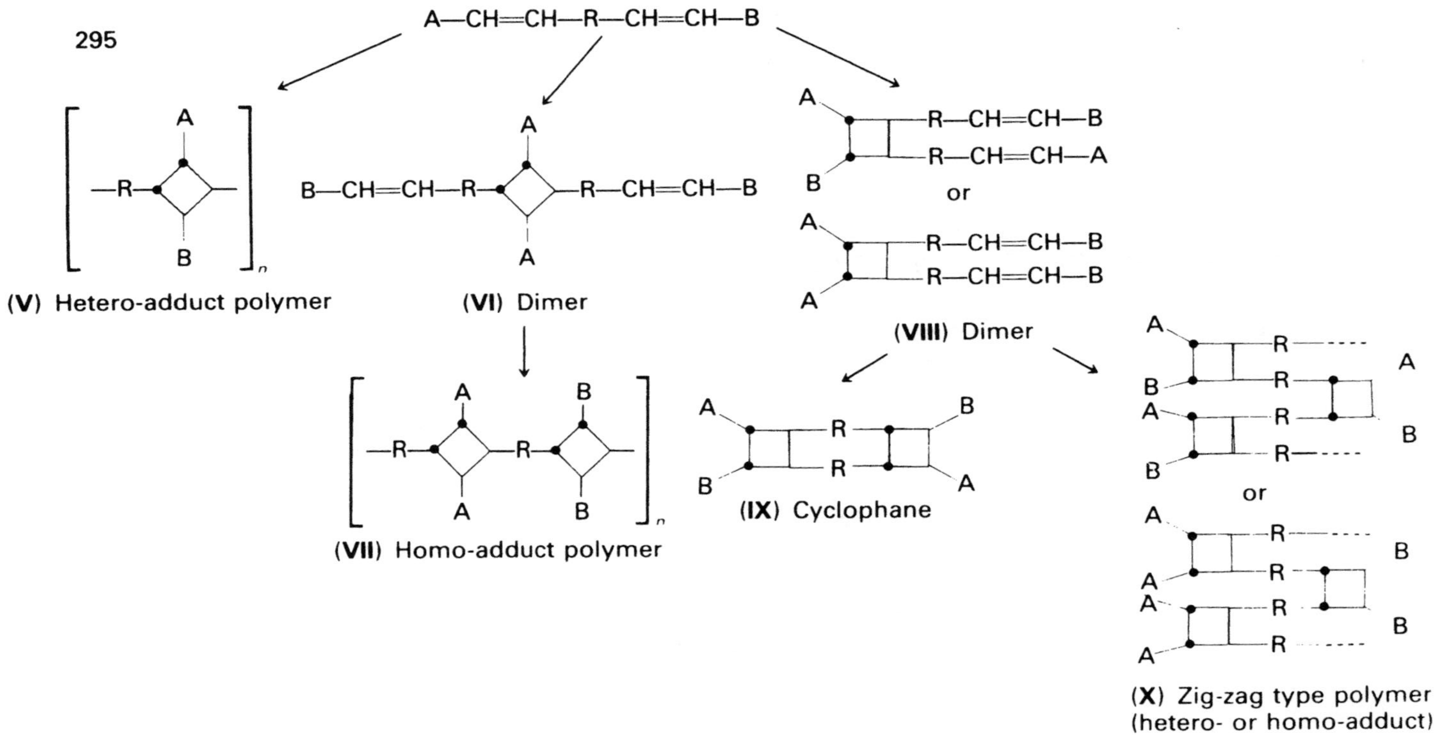

SCHEME 2. Structures of the photoproducts isolated by topochemical reaction of unsymmetric diolefin crystals.

TABLE 2
Topochemical Behavior of Typical Diolefin Crystals

Structure of diolefin	*Photoproduct*			
	Structure[a]	*Yield (%)*	*Morphology*	η_{inh}[b]
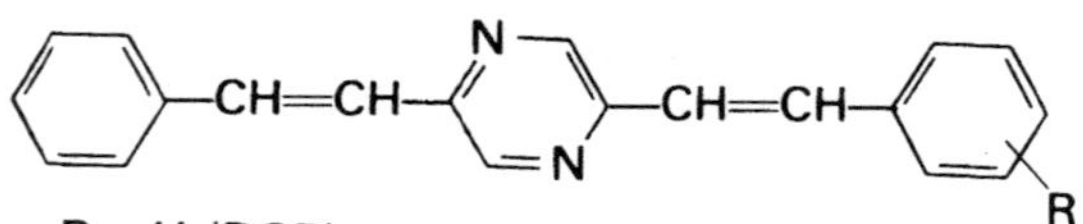				
R = H (DSP)	**II**	Quant.[c]	Cryst.	3
R = 4-COO*n*Pr	**V**	Quant.	Cryst.	2·4
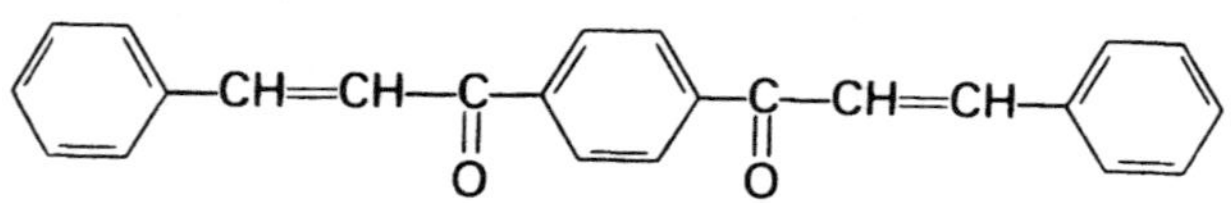	**III**, Major	80–90	Amorph.	Cyclophane
	IV, Minor	3–10		Oligomer
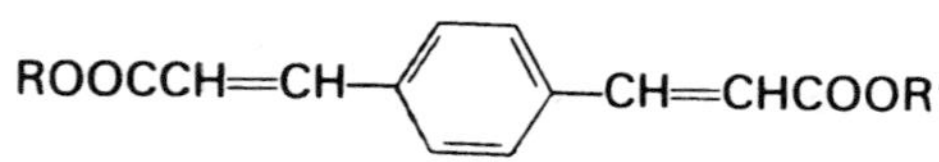				
R = R′ = Et	**II**	Quant.	Cryst.	>2
R = R′ = Et	**I**	30[d]	Cryst.	Dimer
R = Me, R′ = Et	**VII**	Nearly quant.	Amorph.	0·2–3·1

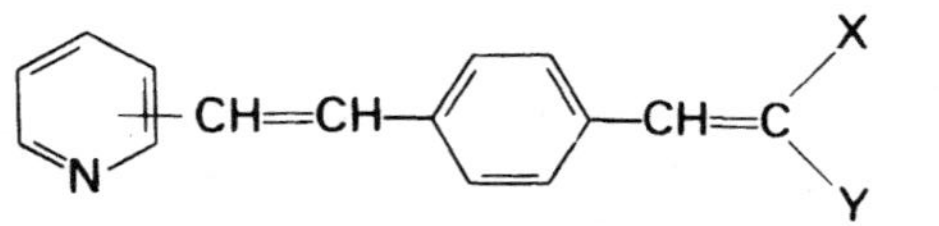

2-Pyridyl, X = CN, Y = COOMe	**VIII**	Quant.	Amorph.	Dimer
4-Pyridyl, X = H, Y = COOMe	**V**	Quant.	Amorph.	1
4-Pyridyl, X = CN, Y = COO*n*Pr	**VI**	Quant.	Cryst.[e]	Dimer
4-Pyridyl, X = CN, Y = **COOEt/COO*n*Pr**[f]	**IX**	Quant.	Cryst.	Cyclophane

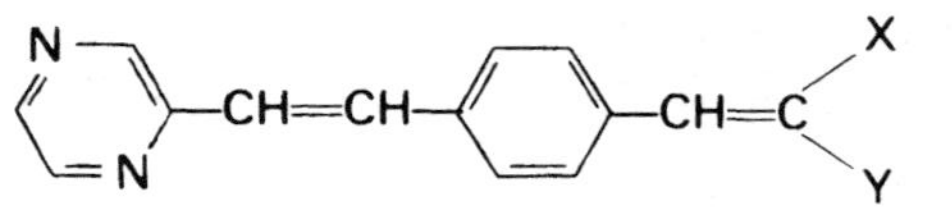

X = H, Y = COOEt	**V**	Quant.	Cryst.	8·2
X = CN, Y = COOMe	**X**	>80	Amorph.	Oligomer

[a] Refer to Schemes 1 and 2.
[b] 2–3 g/litre solvent, 30°C.
[c] Quantitative.
[d] Photoirradiated at 60–65°C.
[e] The crystal as prepared is photostable while the crystal recrystallized from propan-1-ol is highly photoreactive and gives a high polymer (**VII**).
[f] Mixed crystal of ethyl and *n*-propyl esters (1 : 1).

The first photoreaction of unsymmetric diolefin crystals with cinnamic and α-cyanocinnamic units in a single molecule was reported by Addadi & Lahav.[14] They succeeded in obtaining the optically active hetero-adduct type dimer and oligomers by means of the four-center type photopolymerization of the chiral crystal of achiral unsymmetric molecules.

Recent topochemical studies of unsymmetric diolefin crystals have revealed a great variety of reactivities and photoproducts, which are markedly influenced by a slight modification of the chemical structure or crystallization process of the monomer.[15] Topochemical behavior of diolefinic derivative crystals is classified according to their product configurations and these are schematically drawn in Schemes 1 and 2 for the reactions of symmetric and unsymmetric derivatives, respectively. All of the products in these Schemes, except the zigzag type high polymers (**IV** and **X**), were not only identified spectroscopically but also isolated on a preparative scale. These are linear and zig-zag type polymers (hetero- or homo-adduct type), dimers (hetero- or homo-adduct type, and photostable or photoreactive) and cyclophanes.

Many hetero-adduct polymers from the unsymmetric diolefin crystals, as well as those derived from the symmetric diolefins, are highly crystalline with a high molecular weight. For example, an ethyl 4-[2-(2-pyrazyl)ethenyl]cinnamate crystal gives a linear high-polymer crystal (η_{inh} 0·82 l/g, 3·0 g/l in 2-hexafluoropropanol at 30°C) on photoirradiation by a 100 W high-pressure mercury lamp for 5 h, or even under sunlight for a day.[16] In contrast with the symmetric diolefins, however, the unsymmetric crystals often afford amorphous polymers with a fairly low degree of polymerization. Even in reactions that afford the amorphous products, the cyclobutane ring in the products consists of a single configuration, with a few exceptions, and therefore the crystal lattice has an important influence on the whole reaction process. All of the zig-zag type polymers obtained are amorphous and oligomeric, as is expected from the molecular arrangement.

Reaction behavior of typical monomers is illustrated in Table 2.

Very exceptional examples are in the photopolymerization of pentaerythritol tetracinnamate[17] and of poly(methylene biscinnamate) crystals[18] which are not rigid conjugated molecules. On photoirradiation, the former crystal is converted into insoluble amorphous network polymers having a repeating cyclobutane structure without any apparent change of crystal shape. Of the latter crystal any indication of cyclobutane formation has not been confirmed in the author's laboratory.

3 POLYMERIZATION MECHANISMS

3.1 Kinetic Behavior

The four-center type photopolymerization of diolefin crystals proceeds step-by-step by repeating the intermolecular [2 + 2]-photocyclodimerization, as long as the reaction temperature is within the range in which the rigid crystal lattice is maintained.[9] Therefore, in the typical topochemical reaction where the crystal-to-crystal transformation occurs, the apparent reaction rate decreases with a reduction in temperature only by a small extent. On the other hand, the lower the reaction temperature, the higher the molecular weight of the final polymer. However, many of the topochemical polymerizations of unsymmetric diolefin crystals in which monomer crystals are transformed into the amorphous polymers are considerably depressed by reducing the reaction temperature, suggesting a rather large thermal motion during the polymerization. In contrast to the reaction at the lower temperature approaching the crystal melting point, all the topochemical photoreactions are so deteriorated that both the apparent reaction rate and the molecular weight of the final polymer markedly decrease due to enhanced thermal vibration of the crystal lattice.[19] In such a high temperature range, the final conversion, rather than an apparent reaction rate, would characterize the topochemical kinetic behavior more accurately.[20]

For example, the diethyl 1,4-phenylenediacrylate crystal (1,4-PDAEt, m.p. 96°C) photopolymerizes quantitatively even when the temperature is as low as that of liquid helium (4·2 K).[21] On the other hand, above room temperature the same crystal does not polymerize quantitatively but gives a fairly low-molecular-weight polymer with a certain amount of by-products.[22] The dimer of 1,4-PDAEt was prepared (30% conversion) on irradiation at 60–65°C and isolated as fine crystals from the oligomeric photoproducts.

The time versus conversion and time versus reduced viscosity curves of 1,4-PDAEt at various temperatures are shown in Figs 1 and 2.

As a result the optimum temperature (T_{opt}) at which the apparent rate is a maximum, appears several tens of degrees lower than the

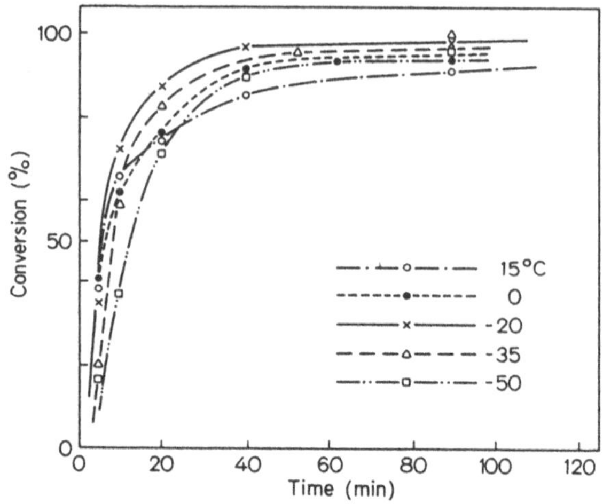

FIG. 1. Plot of time versus conversion for polymerization of 1,4-PDAEt at various temperatures.

melting point for individual crystals (approximately −20°C for the 1,4-PDAEt crystal).

With reference to the above kinetic behavior of 1,4-PDAEt and crystallographic information on several photoreactive diolefin crystals,

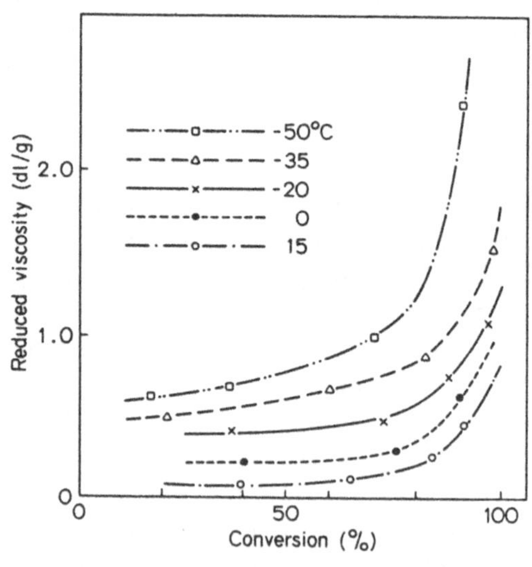

FIG. 2. Plot of conversion versus reduced viscosity for polymerization of 1,4-PDAEt at various temperatures.

the equation for the apparent reaction rate, k, has been derived for one- and two-dimensional thermal motion of molecules in the crystal. In order to derive the equation it is assumed that k is proportional to the reaction probability, which is predominantly dependent on the extent of thermal deviation of the two reactive olefin bonds from the optimal position for the topochemical reaction in the crystal.[19]

The rate equation for the two-dimensional case is derived as follows:

$$k = A \exp \{-(RT_{opt} + E_a)/RT\}$$

where $A = \beta T_{opt}/T$ and β is a constant related to the individual reaction. Although the equation is formally very similar to the Arrhenius equation, $k = A \exp \{-E_a/RT\}$, the temperature, T, in the present equation is correlated not with the 'activation energy', but with the thermal energy for molecular motion in the topochemical reaction.

3.2 Quantum Yield

Since the π-electron conjugation of the monomer is interrupted by the formation of a cyclobutane ring, the π–π^* electronic transition of molecules larger than the dimer is shifted to a higher energy level than that of the monomer. Therefore, on photoirradiation at the long-wavelength edge of photoabsorption of the monomer ($h\nu$), only dimerization and oligomerization (the reaction between excited monomer and the terminal olefin unit of a growing chain in the ground state) proceed, to result in oligomeric crystals. A high polymer is produced on successive irradiation of the oligomer with the wavelength of light that excites the terminal olefin unit ($h\nu'$). By such a selective excitation technique, in parallel with control of the irradiation time and temperature, average molecular weight and molecular weight distribution of the final product are broadly controllable:

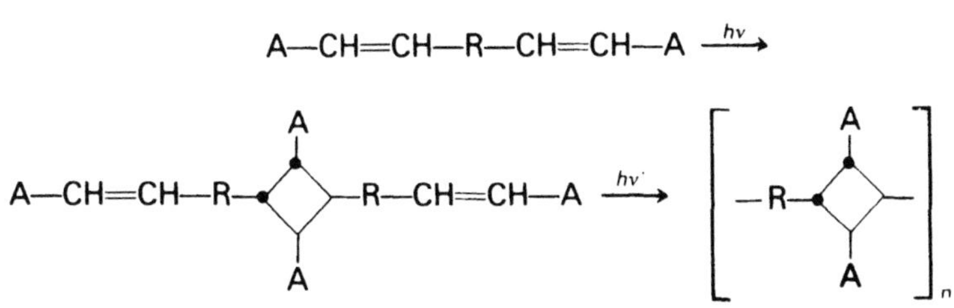

$(\nu < \nu')$

TABLE 3
Quantum Yields of Oligomerization and Subsequent Polymerization of DSP, P2VB[a] and 1,4-PDAMe[b]

	Oligomerization: Wavelength used in irradiation (nm)	Φ	*Polymerization: Wavelength used in irradiation (nm)*	Φ
DSP	436	1·2	365	1·6
P2VB	405	0·04	—	—
1,4-PDAMe	365	1·2	313	0·7

[a] 1,4-Bis-[2-(2-pyridyl)ethenyl]benzene.
[b] Dimethyl 1,4-phenylenediacrylate.

For example, the average degree of polymerization of the final product corresponds to the trimer-pentamer if the DSP crystal is irradiated with wavelengths longer than 400 nm at room temperature.[23,24]

The quantum yields of the oligomerization and polymerization of several symmetric diolefin monomers have been measured with respect to the number of olefinic double bonds consumed to form cyclobutane per absorbed quantum, and the results are summarized in Table 3.

These quantum yields support the hypothesis that the four-center type photopolymerization belongs to a single-photon reaction in which the theoretical maximum value is equal to 2. The reactions of the DSP and 1,4-PDAMe crystals, but not of the P2VB crystal, proceed very efficiently. Such high quantum yields of DSP and 1,4-PDAMe reflect a high probability of effective collisions in the photopolymerization. The crystal structures of DSP and P2VB, and those of the corresponding polymers, are extremely similar to each other, but during the photopolymerization the unit cell volume of the DSP crystal shrinks whereas the volume of the P2VB crystal expands.[25] However, the marked difference in the quantum yields of these two reactions has not, as yet, been explained.

In addition to the two excitation species, i.e. those of the conjugated diolefin monomer and the terminal olefin unit, it should be noticed that the topological environment must vary step-by-step in the reacting crystal, resulting in a variation in the quantum yield of cyclobutane formation in each step of the elementary processes.

Such a step-by-step variation of the quantum yield is particularly exaggerated in the reaction of several unsymmetric diolefin crystals.

3.3 Characteristic Behavior of Unsymmetric Diolefin Crystals

In many of the unsymmetric diolefin crystals, two pairs of intermolecularly facing double bonds are placed differently in relative positions and orientations; accordingly, topochemical behavior is markedly different even when the two pairs of intermolecular reactive double bonds have extremely similar chemical structures to each other.[15]

For example, an ethyl methyl 1,4-phenylenediacrylate (1,4-PDAEtMe) crystal photopolymerizes into a linear polymer (Table 2). However, a large amount of the dimer is formed by the exclusive photodimerization between two methoxycarbonylethenyl groups and is accumulated at the intermediate state of polymerization even without any control of the irradiating wavelength. The intermediate photoreactive dimer crystal is easily isolated from the reacting crystals, and none of the other types of dimers is detected throughout the whole reaction process:[26]

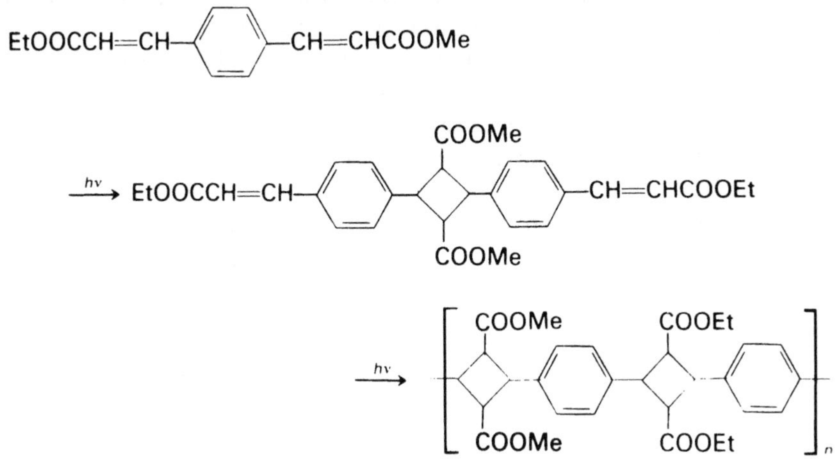

In the crystal of 1,4-PDAEtMe, every molecule is related to its neighboring molecules by two different centers of inversion along the stack direction. The distances of the intermolecular double bonds are 3·891(6) Å between two methoxycarbonylethenyl groups (A in Fig. 3), 4·917(6) Å between two ethoxycarbonylethenyl groups (B in Fig. 3),

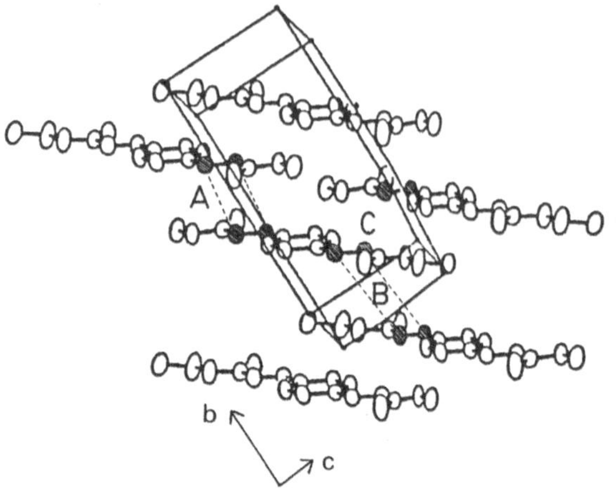

FIG. 3. Crystal structure of 1,4-PDAEtMe. A = 3·891(6) Å; B = 4·917(6) Å; C ≈ 6 Å.

and approximately 6 Å between ethoxy- and methoxy-carbonylethenyl groups (C in Fig. 3). From the crystallographic result, the observed exclusive photodimerization between two methoxycarbonylethenyl groups is quite reasonable because in the 1,4-PDAEtMe crystal only the distance 3·891(6) Å is within the topochemically reactive distance. It is assumed that during the first photodimerization, two ethoxycarbonylethenyl groups come closer, from 4·917(6) Å in the pure monomer crystal to the normal reactive distance in the reacting intermediate crystals.[16]

In the topochemical photopolymerization of ethyl α-cyano-4-[2-(4-pyridyl)ethenyl]cinnamate, which is similar to that of 1,4-PDAEtMe, one type of photoreactive dimer is accumulated at the intermediate stage and, on further photoirradiation, is converted into a linear homo-adduct polymer.[15]

In the crystal the two pairs of double bonds are separated by distances of 3·758(8) Å between the two α-cyanoethoxycarbonylethenyl groups and 4·878(10) Å between the two 4-pyridylethenyl groups, respectively, as is schematically shown in Fig. 4.

As the former distance is within the normal reactive distance whilst the latter is longer than the reactive distance, the dimerization occurs exclusively between former double bonds. No growing species with an odd-numbered degree of polymerization have been detected in GPC curves during the reaction, as is illustrated in Fig. 5.

The GPC study demonstrates that the monomer reacts only with

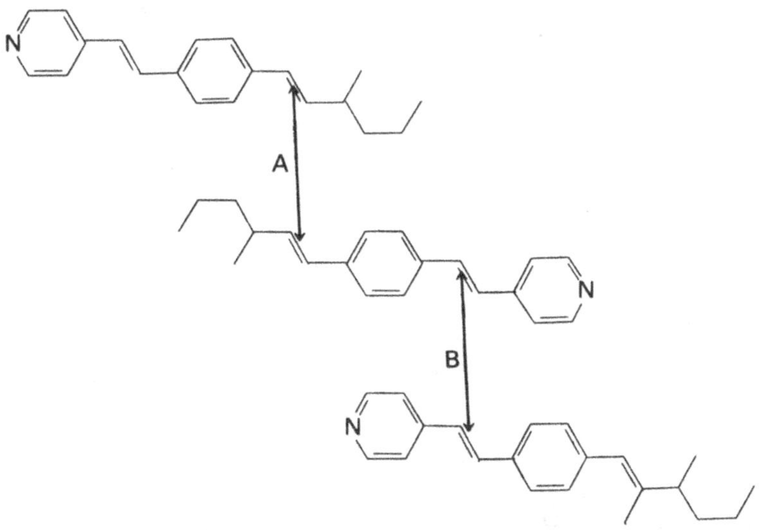

FIG. 4. Schematic drawing of exclusive dimer formation from ethyl α-cyano-4-[2-(4-pyridyl)ethenyl]cinnamate crystal. A = 3·758(8) Å; B = 4·878(10).

other monomers, and that after then all the reacting species have an even-numbered degree of polymerization ('even-numbered polymerization mechanism'):

$$M + M \xrightarrow{h\nu} \text{Dimer}(M\text{—}M) \xrightarrow{h\nu} \text{Polymer}(M\text{—}M)_n$$

$$M + \text{Dimer} \overset{h\nu}{\nrightarrow} \text{Trimer}$$

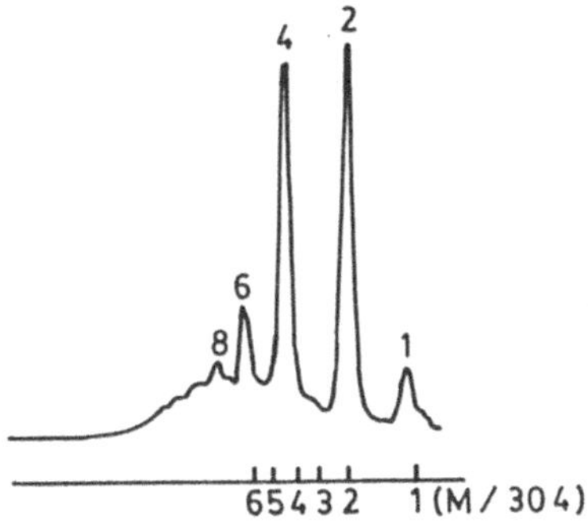

FIG. 5. GPC curve at the intermediate stage of polymerization of an ethyl α-cyano-4-[2-(4-pyridyl)ethenyl]cinnamate crystal.

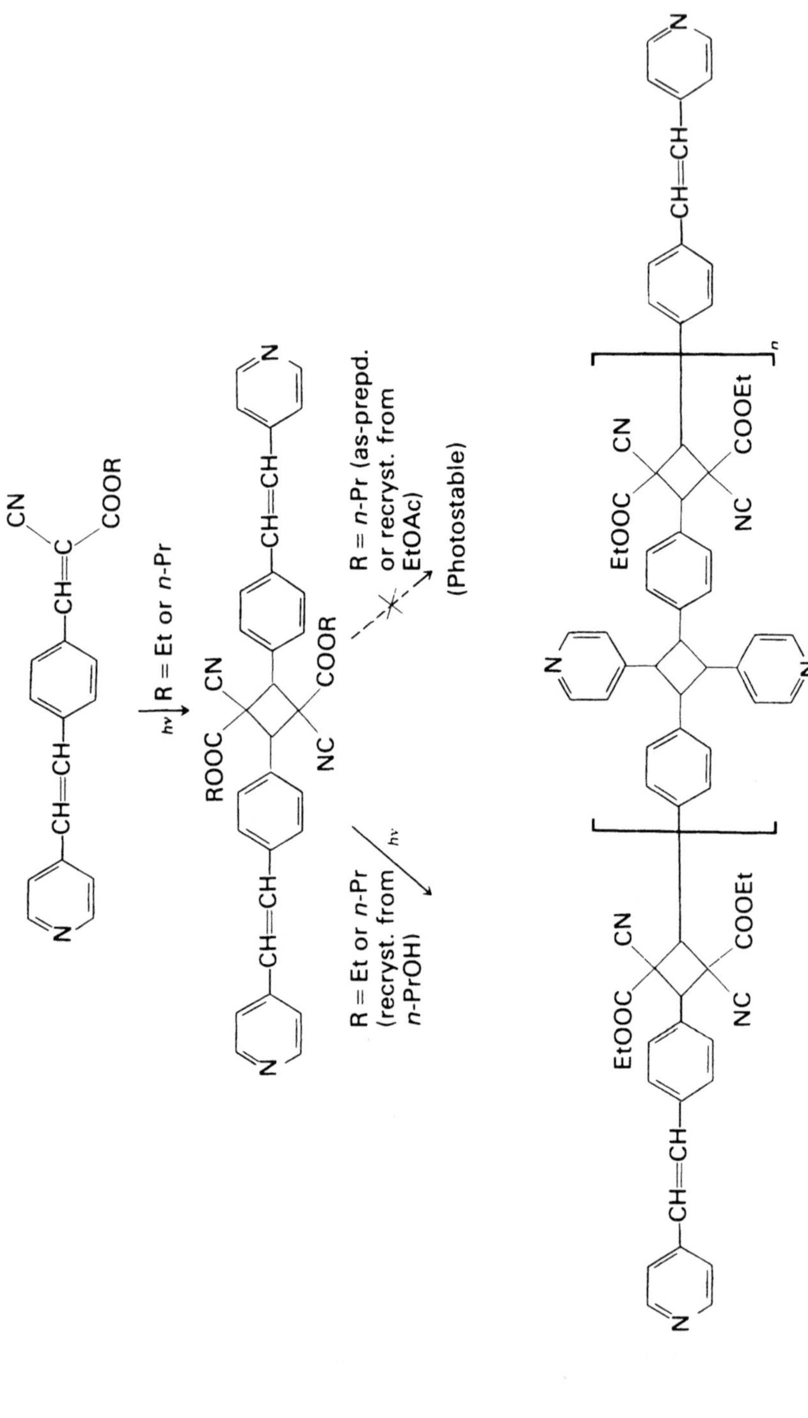
CN
COOR
CH=CH
CH=C
hν
R = Et or n-Pr
ROOC
CN
NC
COOR
R = n-Pr (as-prepd. or recryst. from EtOAc)
(Photostable)
R = Et or n-Pr (recryst. from n-PrOH)
hν
EtOOC
CN
NC
COOEt
n

SCHEME 3

In addition to the even-numbered polymerization mechanism, all the reactive species in the reaction of ethyl α-cyano-4-[2-(4-pyridyl)ethenyl]cinnamate should have terminal 4-pyridylethenyl groups.

In contrast to the ethyl ester derivative, photoirradiation of the same *n*-propyl ester crystal gives only the crystalline dimer in a quantitative yield.[15,27] Not only the configurations but also the X-ray diffraction patterns of the two dimers derived from the ethyl and the *n*-propyl ester crystals are essentially the same. In the monomer crystal of the *n*-propyl ester, only the distance between two *n*-propoxycarbonylethenyl groups is within a normally reactive one (3·729 Å). As expected, the distance of the intermolecular residual double bonds in the dimer crystal is still too great for them to react with each other, even though the distance becomes shorter from 5·088 Å in the monomer to 5·047 Å in the dimer. The dimer crystal, as prepared or recrystallized from ethyl acetate, is photostable whereas the same dimer recrystallized from propan-1-ol is highly photoreactive and on photoirradiation is converted into a linear high polymer having a homo-adduct type structure.

The topochemical behavior of the ethyl and the *n*-propyl ester crystals is shown in Scheme 3.

4 SPECIAL TOPICS ON [2+2]-TOPOCHEMICAL PHOTOREACTION

All the linear polymers derived from conjugated diolefin crystals have an extended rigid configuration with an alternating 1,4-phenylene and 1,3-*trans*-cyclobutane ring structure. As the cyclobutane derivatives having chromophere group(s) are generally photocleaved into two olefin moieties, almost all of the polymers prepared by the four-center type photopolymerization depolymerize photochemically into the monomer in solution. On heating, the linear polymer crystals as prepared do not melt, but thermally depolymerize topochemically into the oligomer crystals. At the oligomer stage these crystals begin to melt and, on further heating, the molten oligomers depolymerize into the monomer.[28]

$$\text{Monomer crystal} \underset{\Delta(2)}{\overset{h\nu(1)}{\rightleftarrows}} \text{Oligomer crystal} \underset{\Delta(3)}{\overset{h\nu'(1)}{\rightleftarrows}} \text{Polymer crystal}$$

(1) Topochemical photopolymerization.

(2) Thermal depolymerization in the molten state.
(3) Topochemical thermal depolymerization.

Unusual thermal stability of these polymers, which is markedly dependent on their molecular weight, has been observed in poly-DSP and a few other linear polymers both in the crystalline state[29] and in solution.[30] The observed phenomenon has been interpreted by means of the model of a narrow and slender stick. According to the model, it has been predicted that the molecular chain cannot grow infinitely via the topochemical process even in the ideal crystal with no structural defects, and that an upper limit of molecular length should exist at the specified temperature. Quite recently this hypothesis was rationalized by theoretical calculation.[31] The concept is generally applicable not only for the present cyclobutane polymers but for any of the rigid rod-like linear polymers.

From the unsymmetric diolefin crystals two types of new polymers have been prepared in which either one side of the rigid polymer chain is hydrophobic and the other side is hydrophilic (hetero-adduct polymer), or there is an alternating hydrophobic and hydrophilic structure in the polymer chain (homo-adduct polymer), as shown in Fig. 6. The skeleton of these polymers is unique and may be promising as an extreme structural model for LB-film technology.

The mixed crystal consisting of an equimolar amount of ethyl and *n*-propyl α-cyano-4-[2-(4-pyridyl)ethenyl]cinnamates is very photoreactive and gives the [2.2]paracyclophane crystal quantitatively as shown:[32]

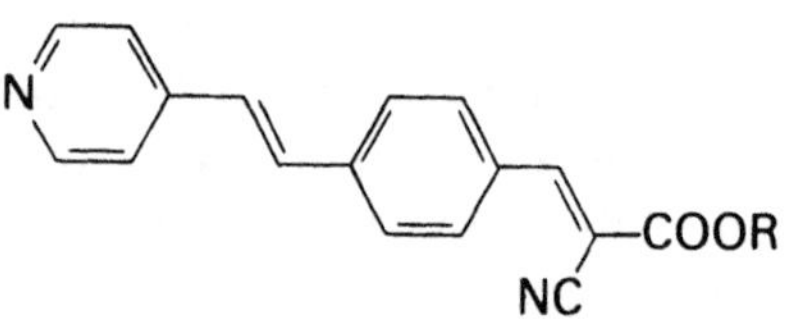

I: R = Et
II: R = Pr^n

1 : 1 Complex of I and II $\xrightarrow[\text{solid state}]{h\nu}$

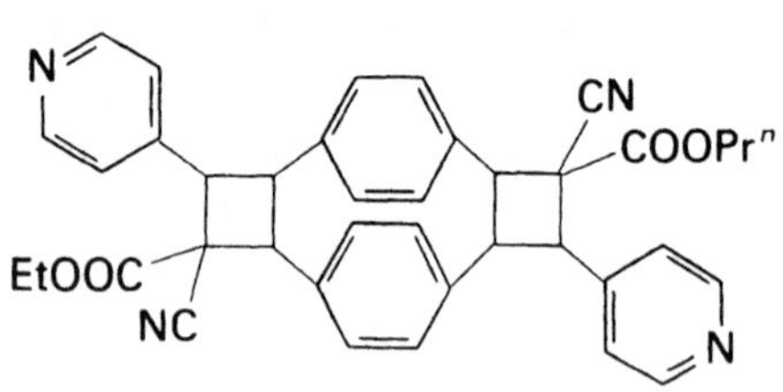

It should be remembered that the pure ethyl ester crystal gives a homo-adduct linear polymer whereas the pure *n*-propyl ester crystal

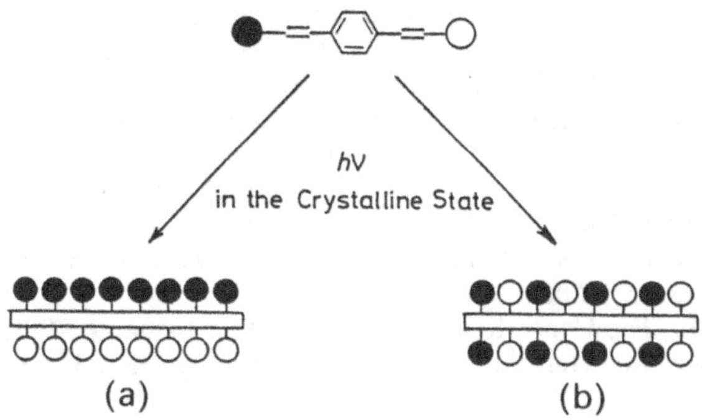

FIG. 6. (a) Hetero-adduct polymer. (b) Homo-adduct polymer. ●, Hydrophobic; ○, hydrophilic.

gives a photostable dimer crystal. A few papers have described the photoreactive mixed diolefin crystals.[33] However, quantitative formation of the above cyclophane which is a clear-cut topochemical reaction is reported here for the mixed crystal for the first time. Topochemical reaction of the mixed diolefin crystals will offer a new topic in the field of copolymerization.

For the 1,4-dicinnamoylbenzene crystal, again contrary to the expected topochemical polymerization, the major product is the cyclophane having a δ-type cyclobutane configuration.[34] As illustrated in the behavior of cyclophane formation, a correlation between chemical and crystal structures is so subtle that further investigations are required, especially on the unsymmetric diolefin crystals in order to establish the 'crystal engineering' concept.

Several recent articles describe topics on [2 + 2] and some other types of topochemical reactions.[13,35]

REFERENCES

1. Okamura, S., Hayashi, K. & Nishii, M., *J. Polym. Sci., C4,* (1969) 839.
2. Bernstein, H. I. & Quimby, W. C., *J. Am. Chem. Soc.*, **65** (1943) 1845.
3. (a) Cohen, M. D. & Schmidt, G. M. J., *J. Chem. Soc.* (1964) 1969; for a review see (b) Schmidt, G. M. J., *Pure Appl. Chem.*, **27** (1971) 647.
4. (a) Hirshfeld, F. L. & Schmidt, G. M. J., *J. Polym. Sci., A2* (1964) 2181; (b) Lahav, M. & Schmidt, G. M. J., *Tetrahedron Lett.*, **26** (1966) 2957.
5. Hasegawa, M. & Suzuki, Y., *J. Polym. Sci.*, **B5** (1967) 813.
6. (a) Sasada, Y., Shimanouchi, H., Nakanishi, H. & Hasegawa, M., *Bull.*

Chem. Soc. Jpn, **44** (1971) 1262; (b) Nakanishi, H., Hasegawa, M. & Sasada, Y., *J. Polym. Sci.*, *A2* (1972) 1537; (c) Nakanishi, H., Sasada, Y. & Hasegawa, M., *Polym. Lett.*, **17** (1979) 459.
7. Meyer, W., Lieser, G. & Wegner, G., *J. Polym. Sci., Polym. Phys. Ed.*, **16** (1978) 1365.
8. (a) Holm, M. & Zienty, F., US Patent 3 312 688; *Chem. Abstr.*, **67** (1967) 12151; (b) *J. Polym. Sci.*, *A1* **10** (1972) 1311.
9. For the reviews see (a) Hasegawa, M., *Adv. Polym. Sci.*, **42** (1982) 1; (b) Hasegawa, M., *Chem. Rev.*, **83** (1983) 507.
10. Nakanishi, H., Ueno, K. & Sasada, Y., *Acta Cryst.*, **B32** (1976) 3352.
11. Hasegawa, M., Harashina, H., Kato, S. & Saigo, K., *Macromolecules*, **19** (1986) 1276.
12. Hasegawa, M., Saigo, K. & Mori, T. In *Materials for Microlithography*, ed. L. G. Thompson, C. G. Wilson & J. M. J. Fréchet, ACS Symp. Series, (1984) 255.
13. Ramamurthy, V. & Venkatesan, K., *Chem. Rev.*, **87** (1987) 433.
14. Addadi, L. & Lahav, M., *Pure Appl. Chem.*, **51** (1979) 1269.
15. Hasegawa, M., *Pure Appl. Chem.*, **58** (1986) 1179.
16. Hasegawa, M., Saigo, K., Kato, S. & Harashina, H. In *Crystallographically Ordered Polymers*, ed. D. J. Sandman, ACS Symp. Series, **337** (1987).
17. (a) Watanabe, S. & Hasegawa, M., *Kenkyu-Hokoku Seni Kobunshi Zairyo Kenkyusho*, **134** (1982) 33; for the crystal structure see (b) Danusso, F., Ferruti, P., Moro, A., Tieghi, G. & Zocchi, M., *Polymer*, **18** (1977) 161.
18. Miura, M., Kitami, T. & Nagakubo, K., *Polym. Lett.*, **6** (1968) 463.
19. Hasegawa, M. & Shiba, S., *J. Phys. Chem.*, **86** (1982) 1490.
20. Desiraju, G. R. & Kannan, V., *Proc. Indian Acad. Sci. (Chem. Soc.)*, **96** (1986) 351.
21. Gerasimov, G. N., Mikova, O. B., Kotin, E. B., Nekhoroshev, N. S. & Abkin, A. D., *Dokl. Akad. Nauk SSSR*, **216** (1974) 1051.
22. (a) Nakanishi, H., Nakanishi, F., Suzuki, Y. & Hasegawa, M., *J. Polym. Chem., Polym. Chem. Ed.*, **11** (1973) 2501: (b) Nakanishi, F., Nakanishi, H. & Hasegawa, M., *Kenkyu-Hokoku Seni Kobunshi Zairyo Kenkyusho*, **103** (1974) 49.
23. Tamaki, T., Suzuki, Y. & Hasegawa, M., *Bull. Chem. Soc. Jpn*, **45** (1972) 1988.
24. (a) Braun, H.-G. & Wegner, G., *Makromol. Chem.*, **184** (1983) 1103; (b) Braun, H.-G. & Wegner, G., *Mol. Cryst. Liq. Cryst.*, **96** (1983) 121.
25. Nakanishi, H., Hasegawa, M. & Sasada, Y., *J. Polym. Sci., Polym. Phys. Ed.*, **15** (1977) 173.
26. Hasegawa, M., Kato, S., Yonezawa, N. & Saigo, K., *J. Polym. Sci. Part C, Polym. Lett. Ed.*, **4** (1986) 513.
27. Hasegawa, M., Kato, S., Saigo, K., Wilson, S. R., Stern, C. L. & Paul, I. C., *J. Photochem. Photobiol. Chem.*, **41** (1988) 385.
28. Hasegawa, M., Nakanishi, H. & Yurugi, T., *J. Polym. Sci., Polym. Chem. Ed.*, **16** (1978) 2113.

29. Hasegawa, M., Nakanishi, H. & Yurugi, T., *Chem. Lett.* (1975) 497.
30. Hasegawa, M., Nakanishi, H. & Yurugi, T., *J. Polym. Sci., Polym. Lett. Ed.*, **14** (1976) 47.
31. Hanamura, E., *Solid State Commun.*, **63** (1987) 1097.
32. Hasegawa, M., Maekawa, Y., Kato, S. & Saigo, K., *Chem. Lett.* (1987) 907.
33. Nakanishi, H., Jones, W. & Parkinson, G. M., *Acta Cryst.*, **B35** (1979) 3103.
34. Hasegawa, M., Saigo, K., Mori, T., Uno, H., Nohara, M. & Nakanishi, H., *J. Am. Chem. Soc.* **107** (1985) 2788.
35. Scheffer, J. ed. Tetrahedron Symposia-in-Print No. 29, Organic Chemistry in Anisotropic Media. *Tetrahedron*, **43** (1987) 1197.

Chapter 7

EXCITED-STATE PROPERTIES OF PHOTOINITIATORS: LASERS AND THEIR APPLICATIONS

J. P. FOUASSIER

Laboratoire de Photochimie Générale, Ecole Nationale Supérieure de Chimie, Mulhouse, France

1 INTRODUCTION

The applications of lasers in polymer photochemistry spring from the meeting point of two fascinating research fields.

Firstly, light-curable monomer and resin formulations are being increasingly developed because of the rapid growth of their industrial applications in various areas such as adhesives, dental and medical equipment, inks, metal decoration, flooring, coatings and varnishes (for optical fibres, wood, plastics, papers, etc.), printing plates and printed circuits. Improvement of the systems requires (i) further investigation on the (photo)physical and (photo)chemical processes involved, in particular the photoreactivity of sensitizers used to start the reaction; and (ii) new ideas concerned with the use of powerful irradiation devices in the case of specific applications.

Secondly, during the past decade laser sources appeared as very convenient tools in photochemistry and photophysics. One of their most attractive applications lies in using a short pulse of laser light as an excitation source to generate transient states whose absorption is monitored, in real time, with a colourimetric analysis device. This technique has been successfully applied to the study of a large variety of processes in molecules and macromolecules. Other applications emerging range across the field of laser-induced chemistry: besides the use of lasers in photopolymer processing (photopolymerization, crosslinking or modification), monitoring the laser-induced reaction can be of importance for analysing the excited-state processes.

The present chapter deals with both aspects and stresses the fact that lasers used as irradiation sources and probes can improve knowledge of the primary processes involved in the excited states of polymerization photoinitiators.

2 OUTLINE OF THEORY

The basic idea[1] is to subject a monomer M to UV irradiation, in order to generate an active species such as a free radical (M·) or an ion (generally a cation M^+), which in turn leads to a polymer chain (P) by addition of other M units (Fig. 1). It is known that the direct absorption of UV energy by the monomer is not very effective and, therefore, one adds light-absorbing substances (photoinitiators: A),

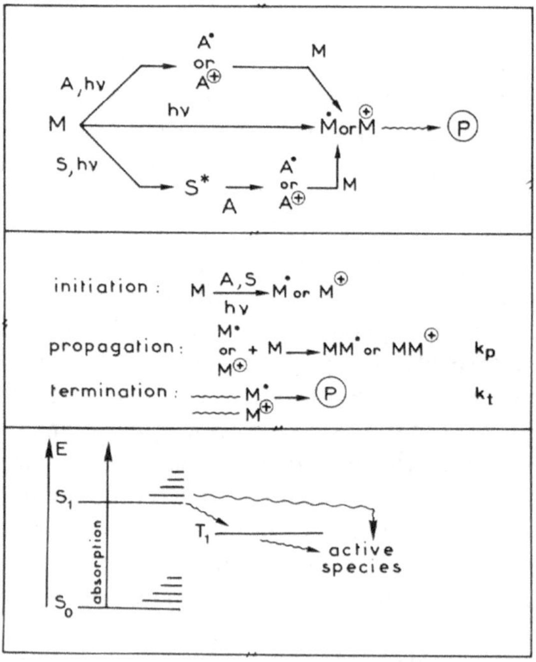

FIG. 1. General representation of the processes involved in a photoinduced polymerization reaction.

which decompose into active radical (or ionic) species or chemically react with additives. In some cases, one also adds compounds (photosensitizers: S) which first absorb and then transfer energy to the molecules used as photoinitiators. One of the problems is, therefore, the adaptation of the absorption spectra of these substances (Fig. 2) to the wavelength emitted by the equipment (a polychromatic conventional mercury lamp or monochromatic laser source).

When a molecule is exposed to UV light (Fig. 1), it first passes to

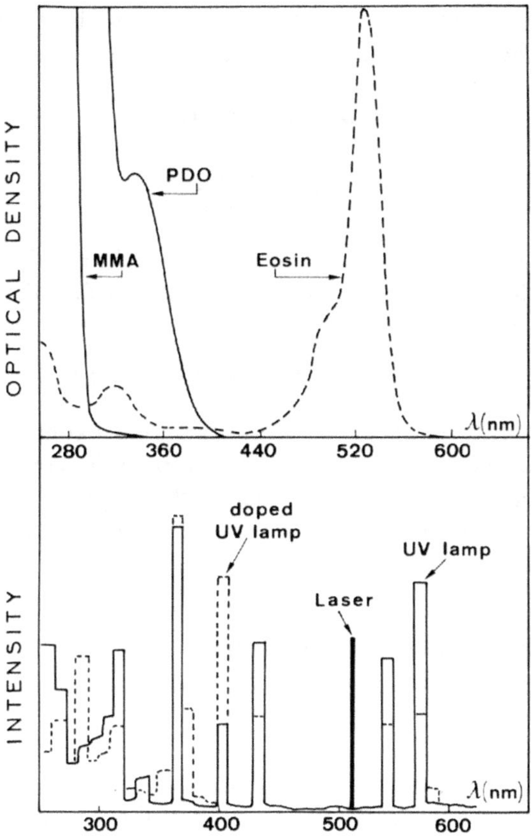

FIG. 2. Typical absorption spectra of an acrylate monomer (MMA), a UV photoinitiator (PDO) and a visible photosensitizer (eosin), showing the matching with the emission spectrum of a usual conventional polychromatic mercury lamp and an Ar^+ laser.

the singlet excited state S_1, then to a triplet excited state T_1 (change in spin multiplicity) via an intersystem crossing which is often very fast ($\sim 10^8$–$10^{10}\,s^{-1}$). The T_1 state generally produces very reactive species which attack the monomer unit and initiate the polymerization reaction. Some reactions can also originate from the singlet state S_1.

Let us consider a monofunctional monomer whose solution polymerization takes place only in the presence of a photoinitiator. The various steps of the free-radical polymerization are shown in Fig. 1.

It is possible to establish kinetic equations which are related to the measurable experimental variables (rate, R_p, and degree of polymerization, DP_n), to the rate constants (k_p, k_t) of the propagation and termination steps and to the rate of initiation (R_i).

$$R_p = k_p\left(\frac{R_i}{k_t}\right)^{1/2}[M]$$

$$DP_n = \frac{k_p[M]}{(k_t R_i)^{1/2}}$$

This is a general expression for the free-radical polymerization. In the case of thermal initiation, R_i is expressed by fk_d. For photochemical initiation it becomes:

$$R_i = \phi_i I_{abs} = \phi_i I_0[1 - \exp(-2{\cdot}3\varepsilon l[A])]$$

with I_0 = intensity of incident light, ε = molar extinction coefficient, [A] = initiator concentration, l = optical path length, ϕ_i = quantum yield of initiation (number of initiated chains per absorbed photon) and I_{abs} = number of photons absorbed by the solution at a given wavelength of irradiation.

2.1 Free-Radical and Ionic Initiation

The main types of initiation fall into two categories.

2.1.1 Free-Radical Type

This process, which is by far the most used, involves the creation of free radicals in the medium, according to well-established mechanisms, of the Norrish type I cleavage, hydrogen abstraction and electron transfer followed by a proton transfer (Scheme 1). Some of these reaction mechanisms have been elucidated in model systems by resorting to techniques used in steady-state photolysis[2] (analysis of photolysis products, measurements of rates and degrees of polymeri-

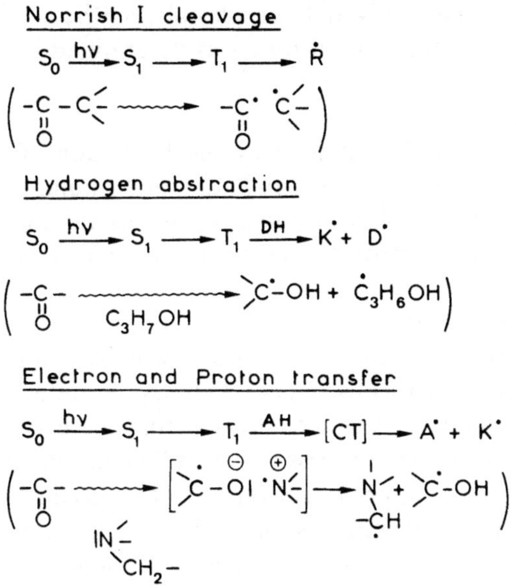

SCHEME 1. Free-radical initiation: usual primary processes.

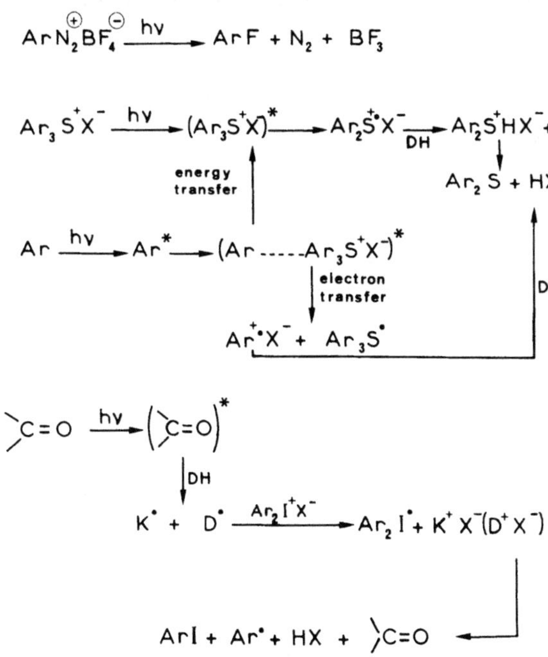

SCHEME 2. Cationic initiation: direct and sensitized mechanisms.

zation), and in time-resolved laser photolysis (the evolution of the excited states[3,4] as a function of time was determined in this way).

2.1.2 Cationic Type

This process is generally obtained by the formation of active species (e.g. strong acids), produced by irradiating diazonium, onium or arene salts (Scheme 2). If photosensitizers are used, the efficiency is enhanced, owing to electron or/and energy transfer reactions.[5,6]

2.2 Initiation Efficiency

The excited triplet state can also be deactivated by non-radiative and radiative relaxation processes. Figure 3 shows the different routes with reference to competitive reactions arising from the excited state. With the exception of typical processes depending on the nature of the spectroscopic states (internal conversion, fluorescence, phosphorescence, formation of complexes or photoproducts), in the case of free-radical initiation the excited states of photoinitiators deactivate mostly via two processes:

(a) *Deactivation by oxygen:* While the inhibiting effect of oxygen on

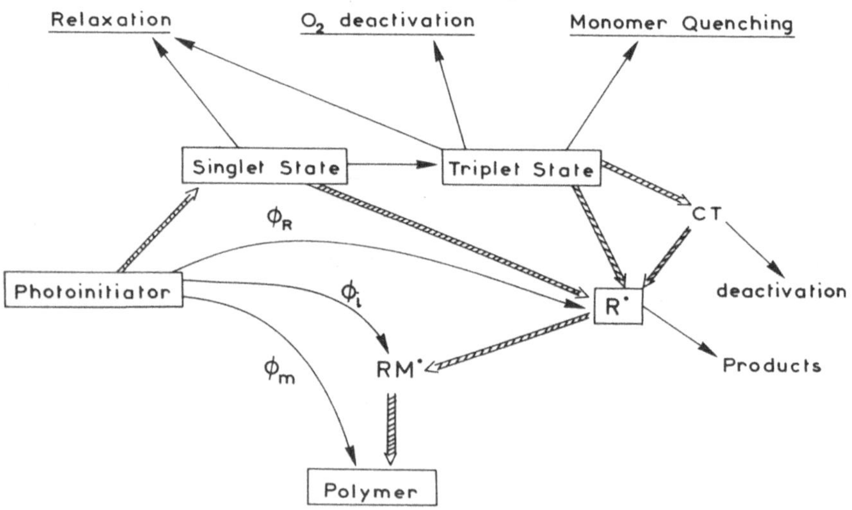

FIG. 3. General diagram showing the modification of the excited states of a radical photoinitiator.

generated free radicals is very well known (oxygen acts as a trap for initiating radicals), it also more often appears that oxygen acts as a quencher of the triplet state, thus being able to reduce the efficiency with which the initiating species R· is produced.

(b) *Deactivation by the monomer:* The monomer reacts on the excited triplet state of the initiator and with the initiating radical. The first reaction is unfavourable, the second leads to initiation (it should be noted that the chemical nature of R· plays a role in the attack on the monomer); it is obvious that the longer the lifetime of the triplet state, the less efficient is free-radical generation and the more efficient is the deactivation:

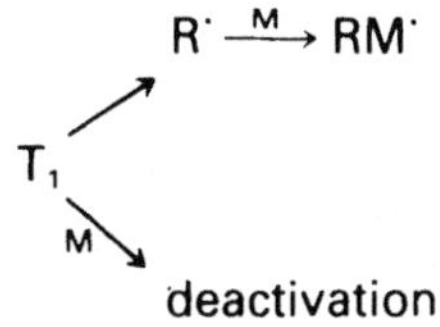

The efficiency of the photoinitiator reaction can, thus, be visualized by the quantum yield of initiation ϕ_i defined above which shows the extent of the reaction R· + M → RM· in comparison with the others, e.g. in the schematic and simplified sequence of reactions shown in Scheme 3.

According to the values of the different rate constants, a large variety of cases may be expected for the dependence of R_p and DP_n on [M], [RH], I_0 and OD. Thus, it is apparent that the rate and the degree of polymerization will be governed by several types of factors:

(a) *The experimental conditions:* the characteristics of the light source (irradiation wavelength, number of photons emitted), the monomer and initiator concentrations, the presence of additives and oxygen, the environment.
(b) *The nature of the initiator:* extinction coefficient, quantum yield of intersystem crossing, rate constants of the processes involved, efficiency of the initiating radicals, role of the photolysis products, side reactions;
(c) *The nature of the monomer:* rate constant of the bimolecular quenching, efficiency of the initiating reaction, rate constants of propagation (k_p) and termination (k_t).

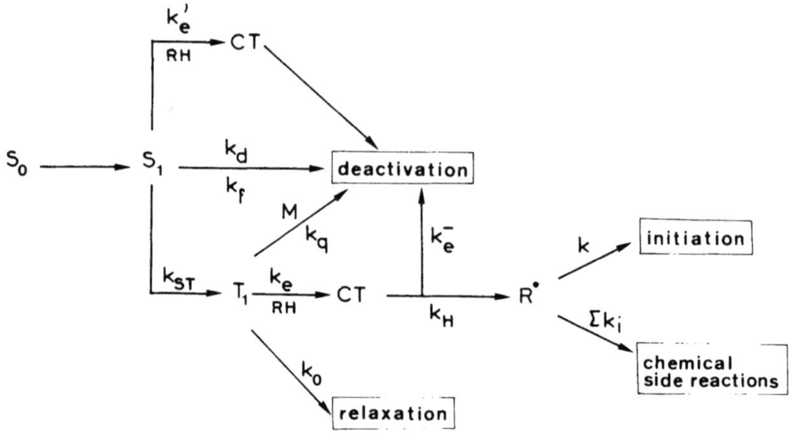

$$\Phi_i = \frac{k_{ST}}{k_{ST}+k_d+k_f+k'_e[RH]} \times \frac{k_e[RH]}{k_0+k_q[M]+k_e[RH]} \times \frac{k_H}{k_H+k_e^-} \times \frac{k[M]}{k[M]+\Sigma k_i}$$

SCHEME 3. Example of a typical diagram showing the initiation processes and allowing the initiation quantum yield to be calculated.

2.3 The Role of the Photoinitiator

The choice of photoinitiators is partly governed by the requirement of a high curing speed. Many reactive monomers and efficient photoinitiators have been synthesized by now and are commercially available. They make it possible to formulate a broad range of appealing photocurable coatings. However, improved photoinitiator systems exhibiting increasingly faster curing speeds or/and increased photosensitivity are unceasingly required.

The intrinsic reactivity in the excited states determines the interest of a given photoinitiator with reference to its efficiency; however, other factors, e.g. high molecular absorption coefficients and a broad spectral absorption range (limiting the inner filter reabsorption in pigmented media), synergistic effects, yellowing, extractability and the effect on the long-term stability of the material, must be considered as important items.

By way of examples, typical compounds of the main classes of radical photoinitiators are listed in Table 1. Their photochemical behaviour corresponds to α-cleavage (**I**, **II**, **III**, **IV**, **IX**, **XI**, **XII**) electron transfer (**V**, **VI**, **VII**, **XIII**), α- or γ-cleavage (**VIII**), C—S cleavage (**XIV**), or β-cleavage (**X**). Recent review papers have been published on one or other of these mechanisms.[3,4,7–12]

Very attractive fields of research and development are found in the role played by decisive parameters affecting the reactivity of a given structure, such as chemical substitution in well chosen positions,[13] the character of water solubility,[14] acid releasing ability,[15–17] and copolymerization availability.[18,19]

Although the reactivity of photoinitiators can be successfully investigated through steady-state experiments, laser spectroscopy proved a most convenient tool for the real-time investigation of excited-state dynamics and the processes involved.

3 TIME-RESOLVED LASER SPECTROSCOPY

Several types of time-resolved laser spectroscopy techniques are now available. Although single photon counting fluorescence is less used (most photoinitiators do not emit light) and sophisticated devices such as CIDNP laser photolysis[20] or picosecond Raman spectroscopy[21] are not extensively used, they could provide a better insight into the field of photoinitiators in conjunction with nanosecond (or picosecond) laser absorption spectroscopy, which is most commonly used.

3.1 Nanosecond and Picosecond Laser Spectroscopy

3.1.1 Principles

A typical device for laser spectroscopy with nanosecond time resolution is shown in Fig. 4.[22]

The excitation source consists of a neodymium-YAG laser Q-switched by a Pockels cell which produces a 3 ns pulse of up to 300 mJ at 1·06 μm. This emission can be converted to secondary emissions at 532 nm, 355 nm and 265 nm. Other usual excitation sources are ruby,[3,23] nitrogen or excimer lasers.[24]

The monitoring system consists of a xenon arc, a monochromator and a photomultiplier at the detection end. The intensity of the monitoring source is increased during the measurement by discharging a capacitor bank through its power supply during a short and adjustable period of time. The monitoring and excitation beams are arranged orthogonally and the excited volume is limited by a set of rectangular diaphragms. The electric signal delivered by the photo-

TABLE 1
Typical UV Photoinitiators for a Radical Initiation

Compd	*Structure*	*Compd*	*Structure*
I	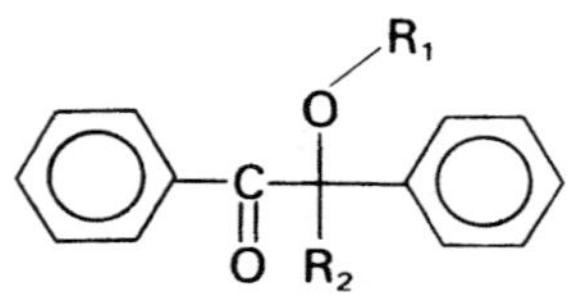	VIII	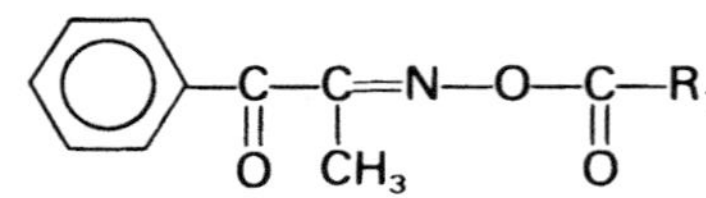
II	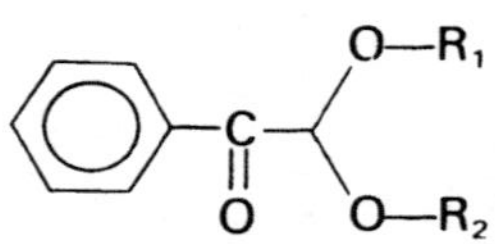	IX	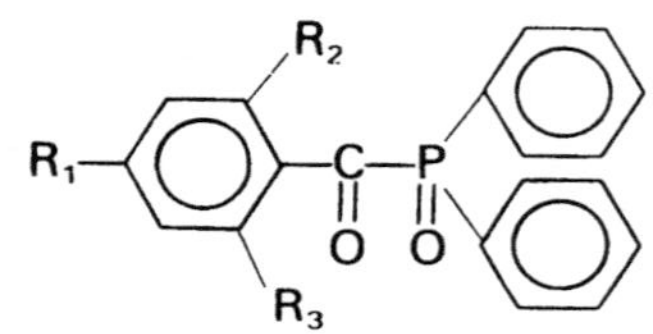
III	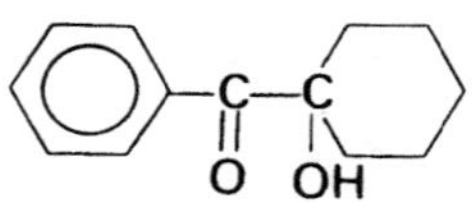	X	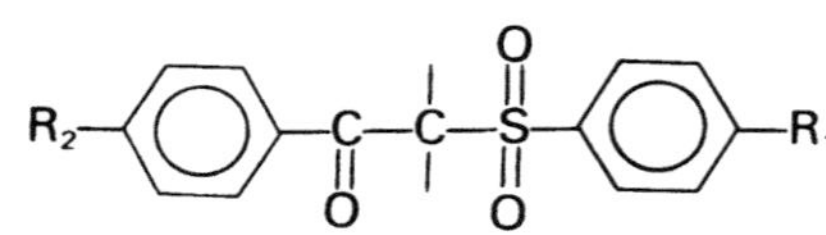

IV
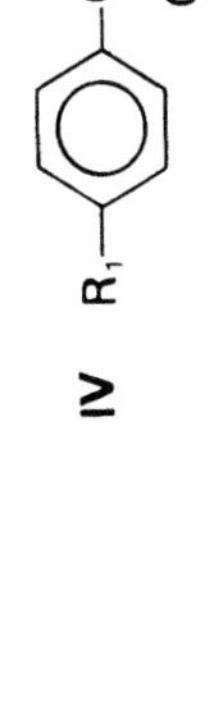

V
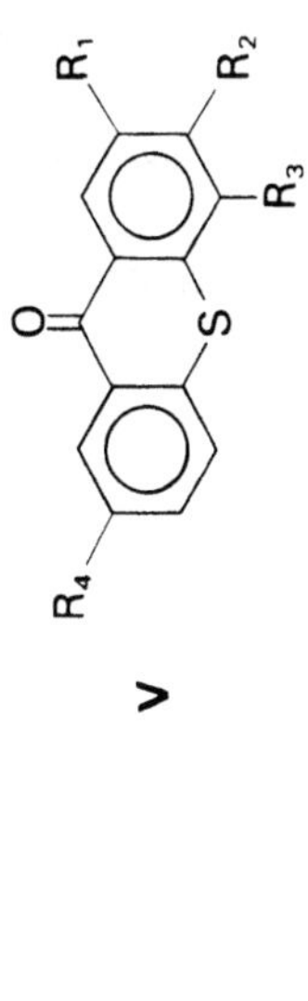

VI
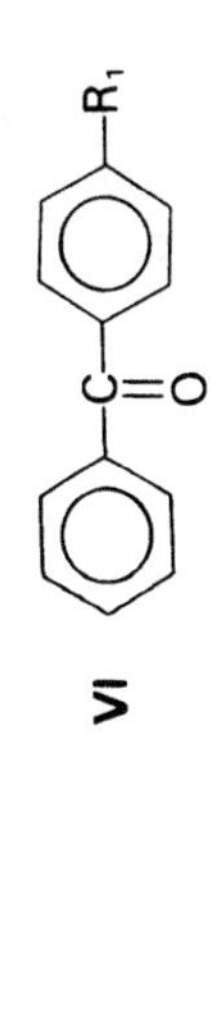

VII
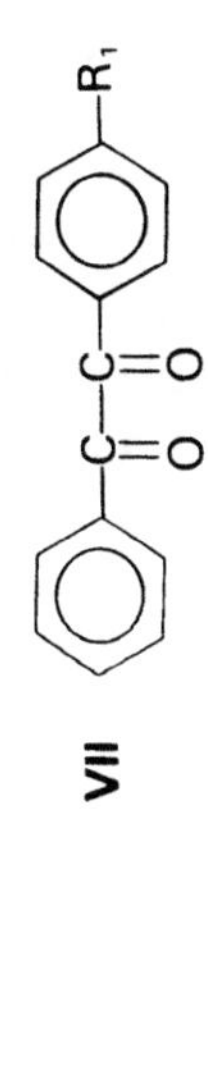

XI
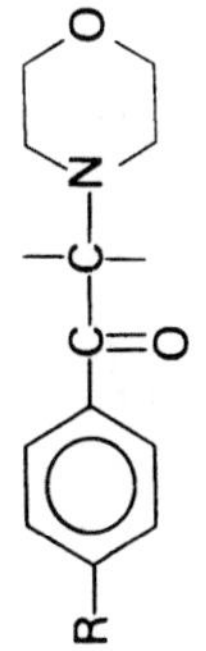

XII
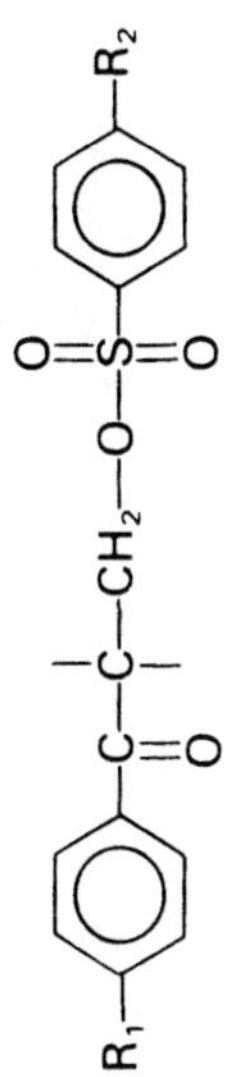

XIII
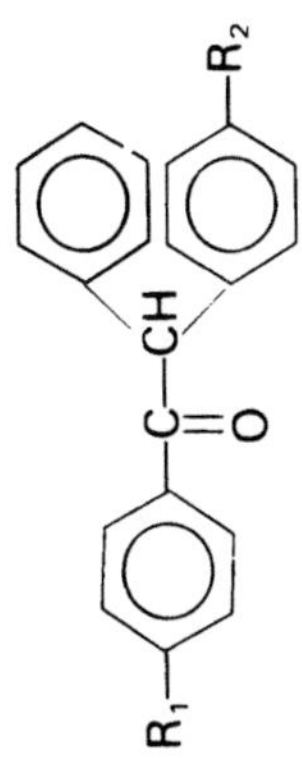

XIV
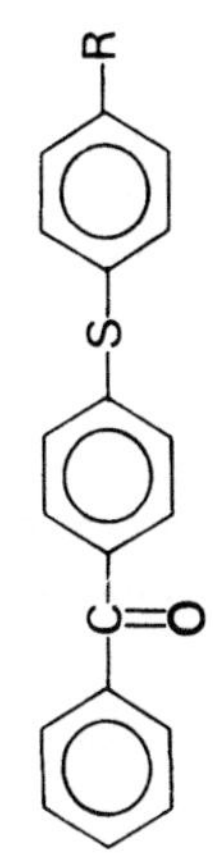

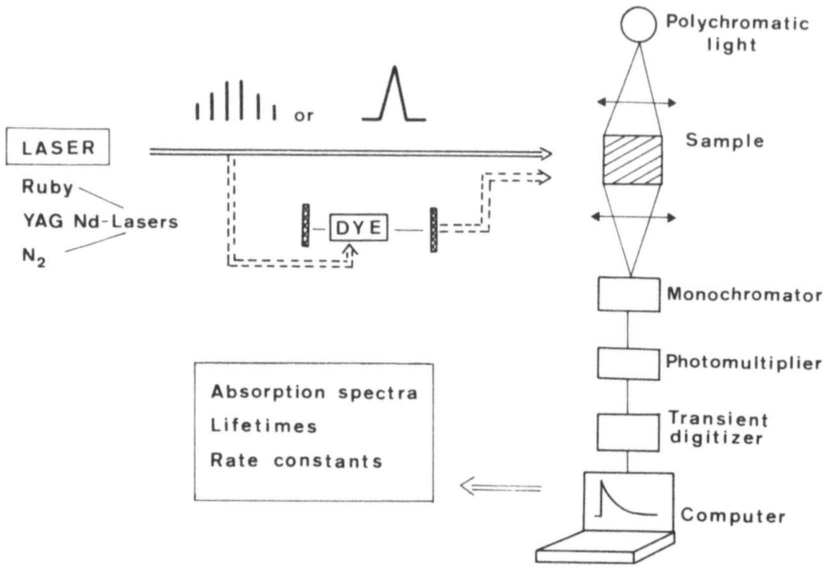

FIG. 4. Typical nanosecond absorption laser spectroscopy device.[22]

multiplier is read by a transient digitizing system and stored. A computer controls the experiment, gathers the data, processes the information, produces suitable files on storage devices and provides copies of graphics and other useful information. Several synchronizers activate the different units (shutters, pulser, lasers) with suitable delays. The risetime of the detection system is *c.* 2 ns, when terminated into 50 Ω, and the actual 'window' in which measurements can be carried out extends up to 200 μs.

In some cases, when the wavelengths of the excitation pulses which are directly available from the Nd/YAG lasers are not adapted to the requirements of a specific experiment, a laser-pumped dye laser is used. The pulse corresponding to the second, third or fourth harmonic of the Nd/YAG laser is used as the pumping source of an adequately chosen dye solution and placed in a laser cavity. The secondary laser emission obtained from the dye serves as a new excitation source of the molecules studied; its time profile is similar to that of the pump laser whereas the spectral characteristics are completely different and dependent upon the laser dye excited: 4 mJ at $\lambda = 430$ nm (with Coumarin 420) or 25 mJ at $\lambda = 600$ nm (with Rhodamine 6G).

If a much higher time resolution is needed, a picosecond spectroscopy apparatus should be used: this excitation source consists generally of a mode locked laser which delivers short pulses of light (up to a few picoseconds). Sophisticated optical and electronic analysing devices make it possible to work down to the true picosecond time range (e.g. Ref. 25). The price of these devices increases with the degree of sophistication; fortunately, the majority of the processes relevant to polymer photochemistry occur above the nanosecond range. However, picosecond spectroscopy appears to be a powerful technique, allowing the study of specific fast processes, e.g. photoscission in the singlet state, generation of short-lived triplet states with fast intersystem crossing [our results support the view that the radical generation in benzoin ethers occurs below 100 ps (Ref. 4 and references therein)], and charge-transfer complex formation and deactivation.

Figure 5 shows a typical oscillogram (recorded at a given analysis wavelength) exhibiting the relaxation of two excited species, the benzophenone triplet (^{3}BP) and a ketyl radical (K·). The kinetic treatment giving the rate constants of the processes is exemplified by:

$$^3\mathrm{BP} + \mathrm{RH} \rightarrow \mathrm{R}\cdot + \mathrm{K}\cdot \quad k_r$$

$$\mathrm{K}\cdot + \mathrm{K}\cdot \rightarrow \mathrm{K{-}K} \quad k_c$$

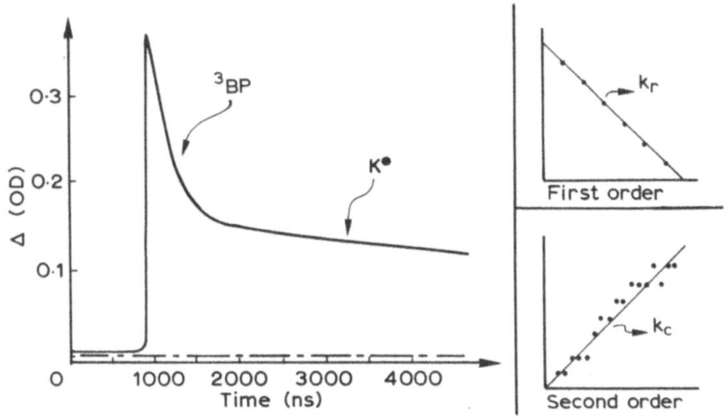

FIG. 5. Typical overall transient absorption of ^{3}BP and K· at $\lambda = 540$ nm. Second-order and first-order kinetic treatment at long and short times of analysis, respectively, yields the rate constants of disappearance of the ketyl radical and the triplet state.

The transient absorption spectra are recorded by plotting the optical density of the solution against the wavelength at a given time after the excitation (e.g. at the maximum optical density). For example, Fig. 6 [26] shows the absorptions of three transients: ^{3}BP, K· and K·$^-$ (resulting from an acid–base equilibrium):

$$\mathrm{K\cdot + B \rightleftharpoons K\cdot^- + \overset{+}{B}H}$$

The progressive modification of the overall absorption is apparent from the absorption spectra recorded at different times (Fig. 7), and provides evidence for the superimposition of several absorptions.[26]

The modification of the deactivation rate constant (or the reciprocal lifetime, τ, in the case of a first-order process) as a function of the concentration of an additive Q (e.g. monomer, electron- or hydrogen-donor), yields the rate constant of the bimolecular quenching k_q, according to the usual Stern–Volmer plot:

$$\frac{1}{\tau} = \frac{1}{\tau_0} + k_q[\mathrm{Q}]$$

Such a plot is shown by Fig. 8, for the quenching of ionic benzo-

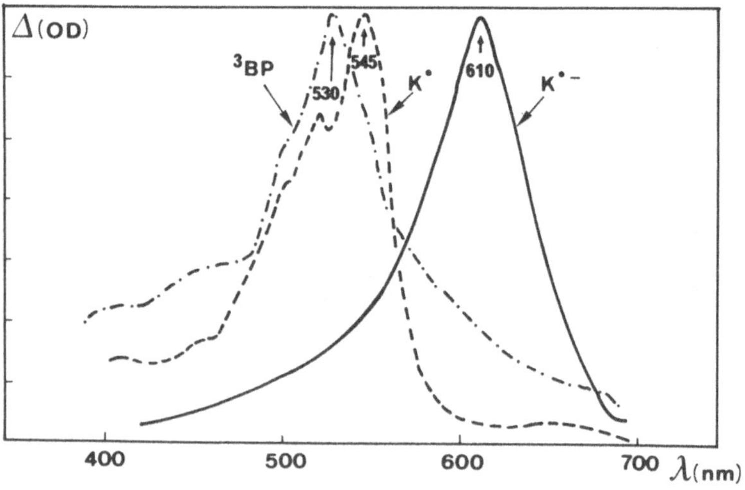

FIG. 6. Transient absorption spectra of the benzophenone triplet (^{3}BP) and the ketyl structures, the radical (K·) and the radical anion (K·$^-$) in NaLS micellar solution.[26]

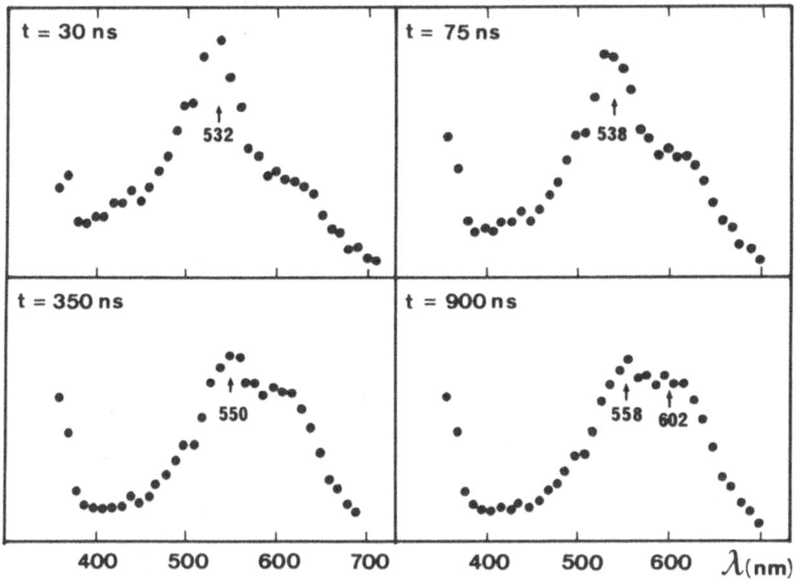

FIG. 7. Modification of the transient absorption for benzophenone in a NaLS micelle.[26]

phenones **VI** ($BP^-:R_1 = CH_2SO_3^-Na^+$; $BP^+:R_1 = CH_2N^+(CH_3)_3Cl^-$) by acrylamide in water.[27]

A thorough investigation of the different processes involved[27,28] leads to a better insight into the overall evolution of the excited state (e.g. Scheme 4 for BP^- and BP^+). As recently mentioned, the initiation process usually occurs after an electron (CT_1 complex) and proton transfer between the triplet state and an amine: BP^+ exemplifies this typical behaviour. However, in order to explain the experimental dependence of R_p on [M] and [AH] in the case of BP^-, another initiation route has to be considered.[27] Quenching of the triplet state by monomer results, indeed, in the generation of a CT_2 complex which can undergo proton transfer and/or attachment of the two radical ions (B), deactivation through a back electron transfer, formation of a triplex (CT_3) with amine and subsequent generation of a free amine-derived radical. Ketyl and ketyl anion radicals are also suspected of playing a decisive role.

Such diagrams have been established in the past for phenylacetophenones,[29–32] hydroxyalkylketones,[33] thioxanthones[34] and

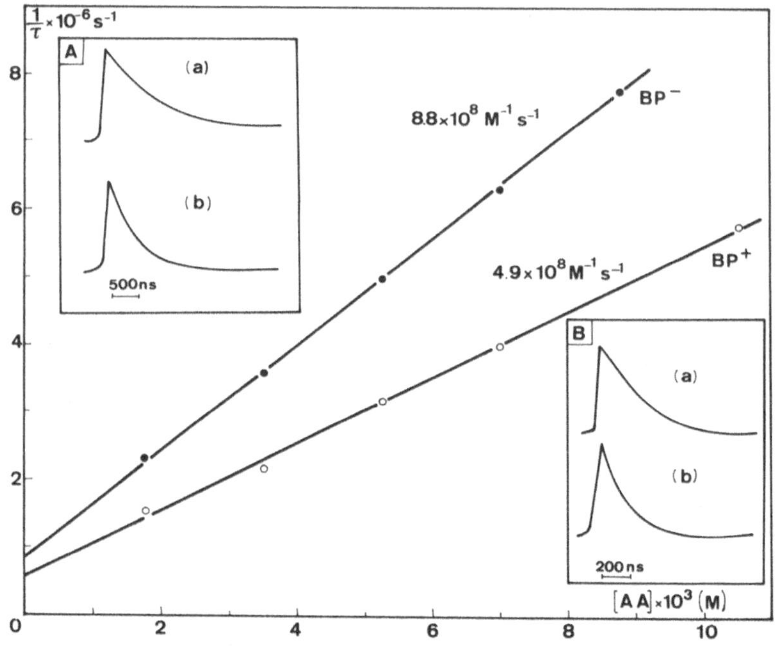

FIG. 8. Stern–Volmer plot of the quenching of the water-soluble benzophenone triplet state by acrylamide in aqueous solution.[27] A: TT absorption of BP^- in the absence (a) and in the presence (b) of $1{\cdot}8\times10^{-3}$ M acrylamide. B: TT absorption of BP^+ in the presence of acrylamide; (a) 5×10^{-3} M, (b) 10^{-2} M.

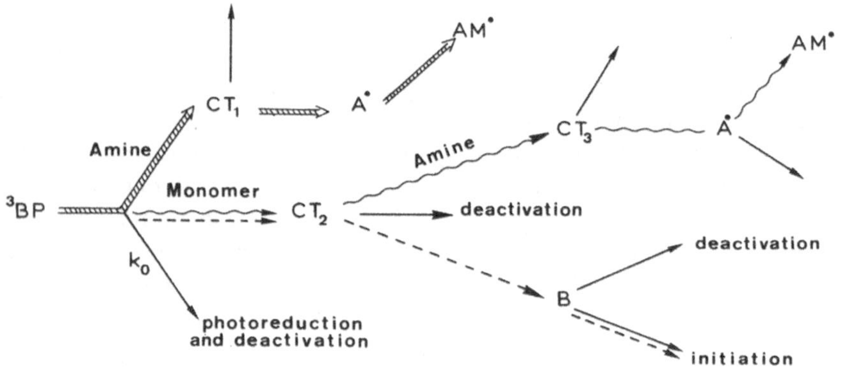

SCHEME 4. Overall diagram of the primary processes involved in water-soluble benzophenones during the initiation step of acrylamide polymerization.[27]

benzophenones[35,36] and more recently for benzoylphosphine oxides,[37] α-morpholinoacetophenones,[22] dialkoxy- and α,α-diphenyl-acetophenones,[39] benzoyl oxime ester,[40] benzoyl-aryl sulphides (Fouassier, J. P., unpublished), α-sulphonylacetophenones,[38] ionic thioxanthones[41] and benzils.[41,42] They provide a good insight into the dynamics of the excited states.

3.1.2 Examples of applications

(a) Quantum yields of the primary processes: At the present time, many rate constants corresponding to cleavage or/and electron transfer and monomer quenching have been measured (see Table 2 for a non-exhaustive list). The related quantum yields ϕ_C and ϕ_{CT} can be computed. By putting these values together with those of the rates of polymerization obtained in well-defined conditions, one is able to calculate the quantum yield of radical attack on the monomer unit, ϕ_{RM} or $\phi_{RM}\phi_R$:

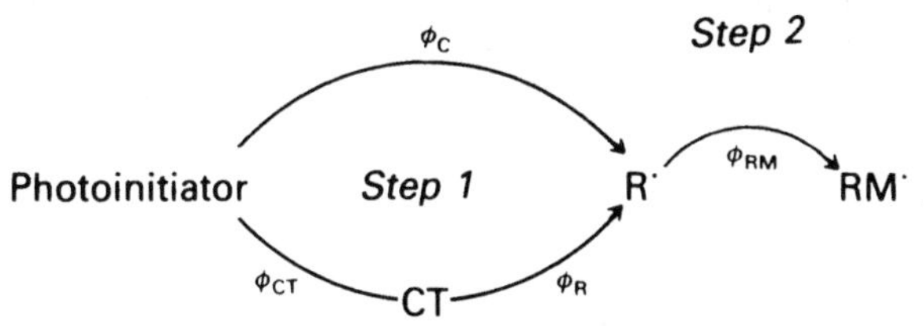

The following main conclusions are drawn.

(1) Photoinitiators working through fast cleavage generally prevent monomer and O_2 quenching. This may also hold true to some extent for systems exhibiting electron transfer (e.g. **V**, where monomer quenching is not efficient).
(2) Structural and substituent effects may strongly affect the rate constants of the processes involved.
(3) The efficiency is partly governed by the primary processes (step 1) leading to the R· radical generation. However, both the reactivity of the radical (step 2) towards the monomer (bond-opening reaction or hydrogen-abstraction process) and other subsequent reactions—proton transfer, back electron transfer, radical ion formation, chemical radical reaction of the excited fragments, acid–base equilibrium, etc.—can still strongly affect the overall efficiency of the initiation process. Table 3 shows the

TABLE 2

Rate Constants,[a] Quantum Yields[b] and Relative Rates of Polymerisation (R_p)

Experiments are performed in toluene or water according to the photoiniaters used

Structure	*Substituent*	*Rate constants* $10^{-9}k_{\alpha}(s^{-1})$	$10^{-9}k_e$ $(M^{-1}s^{-1})$	$10^{-6}k_q$ $(M^{-1}s^{-1})$
I	R_1 = alkyl; $R_2 = H$	>10	—	~200
	$R_1 = CH_3$; $R_2 = OCH_3$			
II	e.g. $R_1 = R_2 = OC_2H_5$	>1		~100
III		~1		~300
IV	$R_1 = H$	~0·03		8
	$R_1 = CH_3$			
V	$R_4 = R_2 = R_3 = H$; $R_1 = Cl$		3	2
	$R_4 = R_2 = R_3 = H$; $R_1 = OCH_2COOH$		0·065	<0·1
	$R_2 = R_3 = CH_3$; $R_4 = H$		0·02	<0·1
	$R_1 = OCH_2CH(OH){-}CH_2{-}N(CH_3)_3Cl$			
	$R_4 = CH_3$; $R_1 = H = R_3$		0·7	<0·01
	$R_2 = COOEt$			
VI	$R_1 = CH_2SO_3Na$		1·2	880
	$R_1 = CH_2N(CH_3)_3Cl$		1·6	490
XIV	$R_1 = CH_3$	7×10^{-4}[d]	1·6	0·7
VIII	$R_1 = OC_2H_5$	>1		—
IX	$R_1 = R_2 = R_3 = CH_3$	>1		
X	$R_1 = CH_3$; $R_2 = H$	0·04[e]		150
XI	$R_1 = SCH_3$	0·1		2
XII		<1		—
XIII	$R_2 = H$; $R_1 = CH_3$	12×10^{-3}	0·9	700
VII	$R_1 = CH_2SO_3Na$		0·8	<0·1

[a] Rate constants of cleavage (k_{α} or k_{β}), electron transfer (k_e), monomer quenching (k_q).
[b] Corresponding quantum yields of cleavage (ϕ_{α}), electron transfer (ϕ_{CT}), radical generatior (ϕ_R) and monomer radical formation (ϕ_{RM}).
[c] With amine.

TABLE 2—*contd*

Ref.	*Calculated parameters*					*Relative* R_p [42]		*Proposed main initiating radicals*
	ϕ_α	ϕ_{CT}	*Rel.* ϕ_{RM}	*Rel.* $\phi_R\phi_{RM}$		*MMA/ toluene*	*Acryl- amide/ water MDEA*	
				MMA	*Acryl- amide*			
3	~1		1	1		41		$C_6H_5\dot{C}O$
39	>0·2		1	>0·2		31		$C_6H_5\dot{C}O$
[f]	0·3		3			42		$C_6H_5\dot{C}O$, 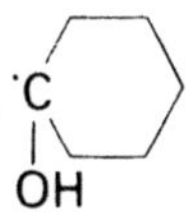
33	~0·2		5	1		41		$C_6H_5\dot{C}O$ $\dot{C}OH(CH_3)_2$
34		>0·9		0·2		18[c]		
[f]								
42		>0·9			4·4		2	
42		>0·9			8·7		2·8	
41								Amine-derived radical (34–27)
[f]		>0·9		0·2		18[c]		
27		0·12			14		1·3	
27		0·25			64		4	
[f]	$\phi_{S\cdot} = 0{\cdot}2$		0·25	0·12		9		⁓S·
[f]	—		—	—		27		$C_6H_5\dot{C}O$
40								?
37			>1	>0·4		31		$C_6H_5\dot{C}O$ ·P⩸O $C_6H_5\dot{C}O\dot{C}<$
[f]	$\phi_\beta = 0{\cdot}05$		0·3	0·015		6		⁓SO_2
22	0·8		0·7	0·56		27		$CH_3SC_6H_5\dot{C}O$
[f]						32		$C_6H_5\dot{C}O$
39		0·5				5[c]		Amine-derived radical (39, 42)
42		>0·9			0·01		0·1	

[d] $K_{cs\cdot}$

[e] K_β

[f] Fouassier, J. P. unpublished.

TABLE 3
Rate Constants of the Reaction R· + M and Relative Values of ϕ_{RM}

R·	$10^{-5}\,k\ (s^{-1})$	*Refs*	*Rel.* ϕ_{RM}[a]
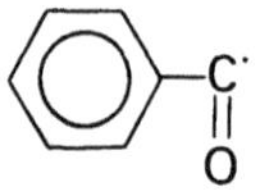	1	3, 32	1
$CH_3\cdot$			0·2
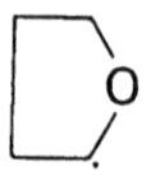			4·5
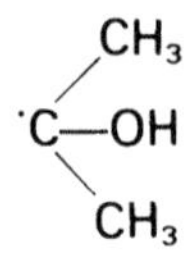			5·5
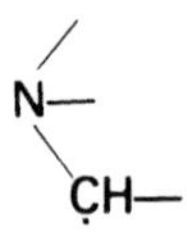			0·2
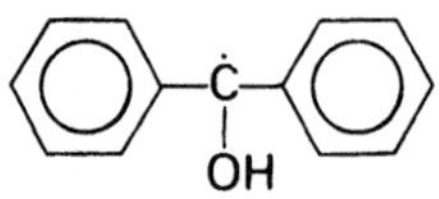	0·1	35	
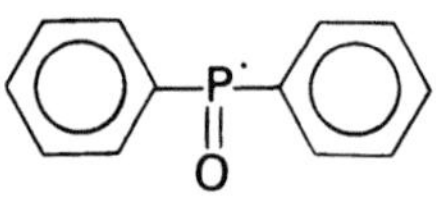	500	37	
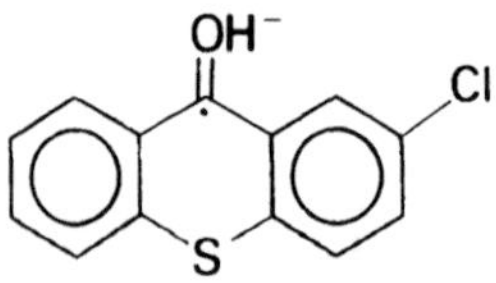	$<10^{-3}$	33	

[a] Ref. 4.

rate constants of the monomer quenching of initiating radicals and the estimated relative efficiency of these species (ϕ_{RM}).

(b) Investigation of structural effects: Quantification of the reactivity of a series of compounds as a function of the substituents is generally not so easy, because of a large variety of effects. In some cases, but

not generally speaking, electronic effects can explain the modification of a rate constant, predominantly, as a function of the substitution. For example, the α-cleavage process of deoxybenzoin[43] or α,α-diphenylacetophenone[39] derivatives can be expressed in terms of the stabilization of the transition state, since the electron-donating group on the benzoyl moiety destabilizes this state and decreases the rate constant of the process (Fig. 9).

A second example is found when studying the quenching of a series of benzophenones by MMA in micellar solution (Fig. 10): the rate constants correlate well with the Hammett constant for electronic effects.[26]

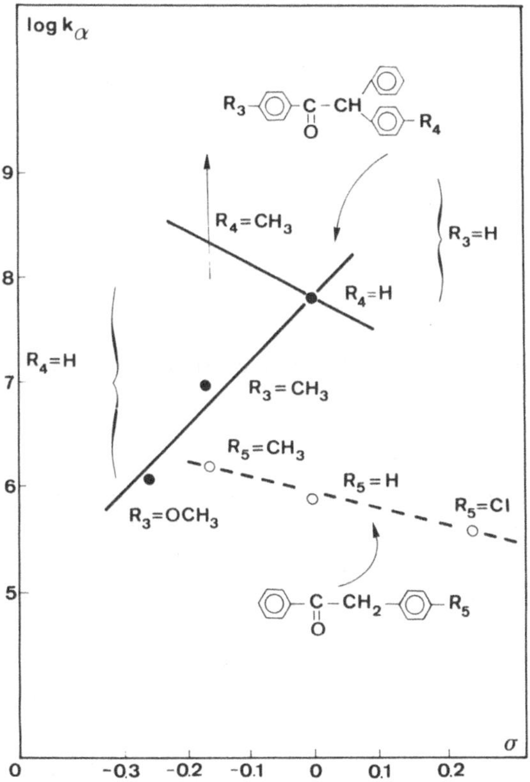

FIG. 9. Plot of the α-cleavage rate constant vs Hammett constant for α,α-diphenylacetophenone[39] and deoxybenzoins.[43]

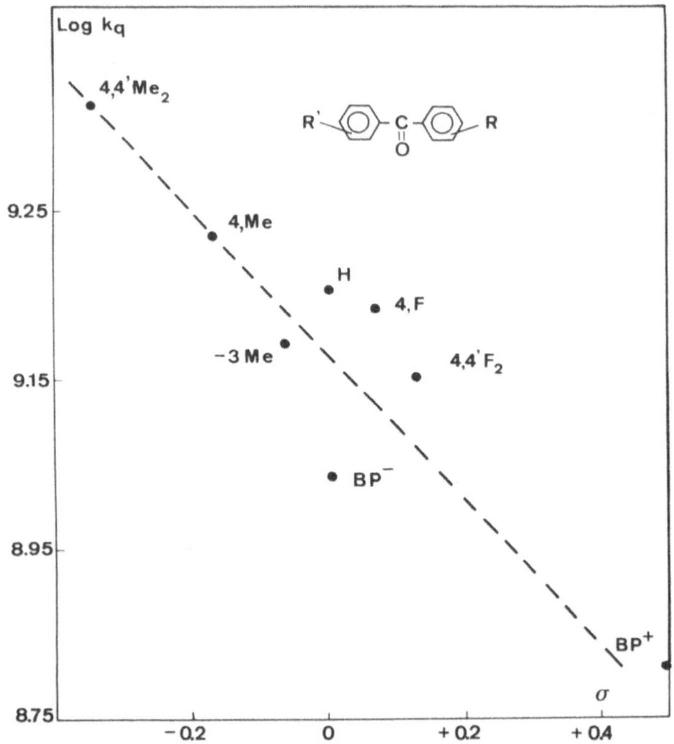

FIG. 10. Quenching rate constant vs σ constants for benzophenone derivatives in a monomer (MMA, 0·5 M)–NaLS (0·5 M) micelle.[26]

A last example has reference to the quenching of a given compound by a number of amines and monomers. Laser spectroscopy leads directly to an estimation of the process efficiency. Apart from a few specific cases, quenching generally occurs through an electron transfer process, in which the ketone plays the role of an acceptor. As implied by earlier studies,[44] satisfactory relations seem to be found from the usual plot of rate constant versus ionization potential (Fig. 11) although a correction for steric effects still improves the correlation.[45] The parameters governing the overall process of electron and proton transfer leading to the formation of free radicals (ionization potential of the donor, reduction potential of the acceptor, solvent interaction, electronic and steric effects, etc.) have so far not been fully elucidated.

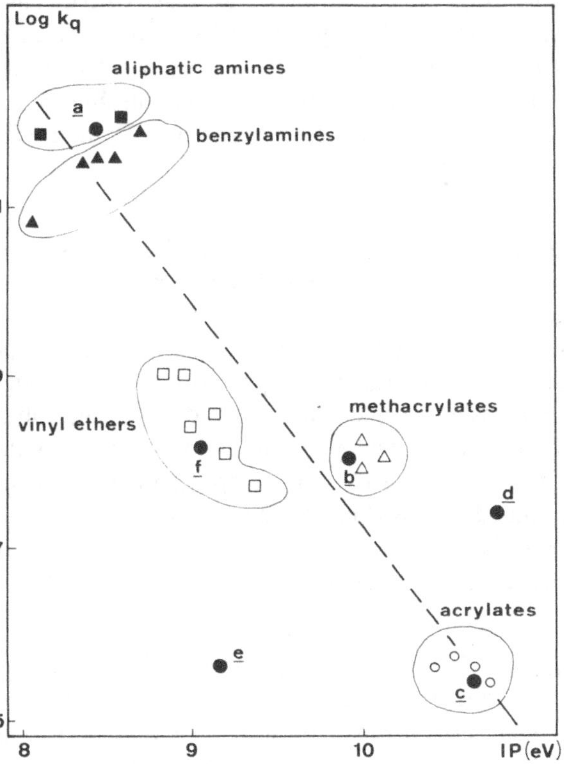

FIG. 11. Correlation between the quenching rate constants of the benzophenone triplet state and the ionization potential of a series of amines and monomers. Styrene *a* acts mainly through energy transfer; methyl methacrylate *b*, methylacrylate *c*, acrylonitrile *d*, vinyl acetate *e* and vinyl ether *f* through electron transfer.[45]

(c) Role of the water solubility: In the photopolymerization of acrylamide (AA) in water (under monochromatic excitation at $\lambda =$ 366 nm), BP^+, BP^-, ATX (**V**; $R_2 = R_3 = R_4 = H$, $R_1 = OCH_2COOH$) and QTX (**V**; $R_2 = R_3 = CH_3$, $R_4 = H$, $R_1 = OCH_2CHOHN(CH_3)_3Cl$) act as efficient photoinitiators. In some instances, polymerization quantum yields reach 800 (number of monomer units polymerized per absorbed photon). Addition of amine is necessary (e.g. MDEA: methyl diethanolamine), except in the case of BP^- which is able to initiate polymerizations even in the absence of amine (with a rate about three times lower). Amine and monomer interactions are

displayed in Table 2. The triplet states of benzophenone and benzil derivatives are only slightly quenched by monomers; nevertheless their ability as photoinitiators[46] is very important. The laser spectroscopy investigation suggests that step 2 is unfavourable in ATX and QTX and limits the overall efficiency, whereas in BP^- and BP^+, step 1 is the driving force of the process.[42]

(d) Synergistic effects: Synergism in photosensitizer/photoinitiator combinations (e.g. Ref. 47) has been recognized for many years as being able to increase the overall efficiency of a UV-curable formulation and was investigated mainly from a practical point of view. Broadly speaking, synergism is expected to occur through electron or/ and energy transfer mechanisms. A steady-state photolysis study of the interaction between benzophenone and Michler's ketone was published a few years ago.[48] Laser spectroscopy investigations were generally considered as difficult because of the overlapping of absorption spectra. Development of laser-pumped dye laser spectroscopy[22] recently made such studies possible (see Section 3.2).

(e) Role of the environment: The radical photopolymerization reaction in microheterogeneous media (micelle–microemulsion–emulsion) generally exhibits both a high rate and a high degree of polymerization. Fundamental questions arise from the specific role played by the photoinitiator[49,50] and related to its photophysical, photochemical and chemical behaviour in the aggregate (processes involved; reactivity of the excited states;[51] interaction with the environment;[52] localization,[26] etc). Other points of interest include the water-solubility character of the photoinitiator, the exit efficiency of the radicals from the micelle, the micropolarity and microbasicity effects, the charge of both the initiator and the micelle, the acid–base equilibrium, the local concentration effects and so on.

For example, the behaviour of positively and negatively charged benzophenones has been depicted through a laser spectroscopy investigation. Upon UV excitation of the n–π^* transition, the generation of the triplet state is very fast (presumably as fast as in homogeneous solution) and it actually follows the picosecond laser excitation. Moreover, neither excitation intensity (which increases the number of excited molecules) nor concentration effect (which varies the occupancy) is detected. This observation suggests that, owing to

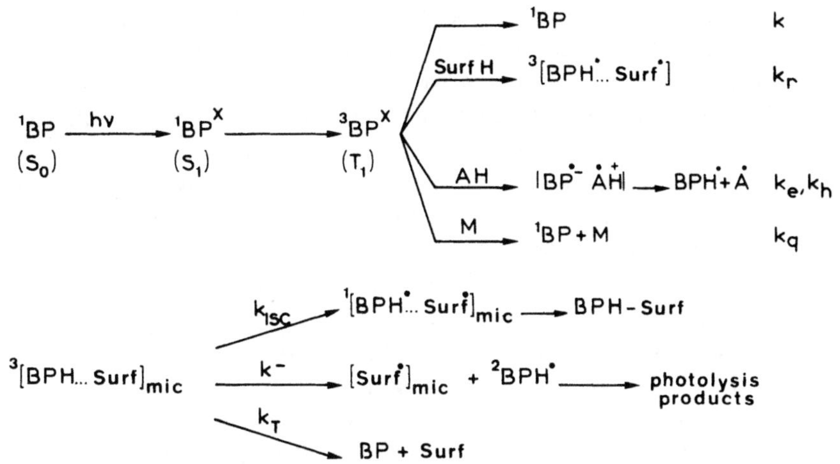

SCHEME 5. Excited-state processes in micelle polymerization.

compartmentalization of the reactants, the bimolecular reactions which take place in homogeneous solution[53] are avoided in micelles.

The photoinitiator in its triplet state decays according to four relaxation processes (Scheme 5): deactivation through a non-radiative process (k), formation of a charge transfer complex with an amine (k_e) yielding a radical pair through proton transfer (k_h), generation of a radical pair through hydrogen abstraction from the surfactant (k_r)[52,54] and quenching by a monomer (k_q). The radical pairs are formed in their triplet state; they convert to singlet radical pairs, through intersystem crossing k_{isc} (yielding photolysis products) and to free radicals k_{esc} after decorrelation and exit.

Recent research work[55] showed that the exit of a radical depends on the nature of the solute (e.g. BP or BP^-), surfactant (e.g. SDS or DTAC), or counter-radical R· (e.g. surfactant or 1,4 cyclohexadiene); the yield of exit was observed to increase when a cyclohexadienyl counter-radical replaces the surfactant-derived radical and hence the conclusion that the 'mobility' of the partners is a predominant factor for the escape of the radical. Higher yields of exit are obtained in the presence of a magnetic field: this magnetic field effect (MFE), first recognized in Ref. 49, results from the Zeeman splitting of the triplet sublevel: the external magnetic field H_0 decreases the intersystem crossing between the triplet and the singlet radical pair and thereby indirectly favours the yield of escape.

As usual, a strong decrease of the triplet state lifetime is observed in the presence of amines, but the relaxation process becomes complex in longer time scales: it appears to be governed by the formation of radical pairs, the exit of the radicals and the acid–base equilibrium between the excited species.

Just as in homogeneous solution, the quenching by the monomer (MMA) results only in a decreased triplet state lifetime; neither long-lived residual absorption nor static quenching is observed. The quenching by the monomer is a dynamic process. In addition, the Stern–Volmer plot remains altogether valid up to the high concentrations used in photopolymerization experiments.

In other cases, the mechanism of interaction between the photoinitiator excited states and the amine/monomer system may become more complicated. For example, singlet quenching occurs in positively charged thioxanthone by amine addition[41,42,56] which decreases, at least, the yield in triplet molecules. Whilst triplet quenching seems also to occur through a static mechanism in the presence of styrene, possible simultaneous singlet quenching cannot be excluded (Fouassier, J. P., unpublished).

Typical data referring to the bimolecular quenching rate constants (by MDEA and MMA), the triplet state lifetimes and the rates of polymerization are shown in Table 4. At first glance, this table shows

TABLE 4
Characteristics of the Micelle[a] Photopolymerization Reaction in Presence of Water and Oil-Soluble Photoinitiators[b]

		$10^{-9}k_e$ ($M^{-1}s^{-1}$)	$10^{-6}k_q$ ($M^{-1}s^{-1}$)	ϕ_{CT}	*Rel. R_p* With amine[b]	*Rel. R_p* Without amine
VI	BP^+	0·4	0·7	0·4	250	~0
	BP^-	1·5	1	0·65	45	15
	BP	1·5	1·5	0·40	50	~0
V	ATX	0·04	1	0·8	105	~0
	CTX	0·34	1·4	>0·95	70	~0
I	DMPA					90
XIV	SQS	0·3	<0·1	>0·9	50	~0

[a] [NaLS] = 0·5 M; [MMA] = 0·5 M.
[b] [MDEA] = 0·05 M.

that there is no direct correlation between the primary process efficiency (ϕ_{CT}) and the rate of polymerization. A thorough laser spectroscopy investigation[51,53,57] suggests that, besides the usual factors [the respective localization of the reactants in the micellar assembly, the monomer quenching and the electron transfer (which are governed by substituent electronic effects), the proton transfer, the exit of the radical pair, the water-soluble character of the molecules used as photoinitiators, and the ionic or dipolar interactions of charged molecules in ionic micelles], a chemical parameter governing the reaction efficiency seems to play a significant role: the reactivity of the ketyl species, i.e. the neutral radical and the anion radical (which act as terminating agents of the growing polymer chains), whose concentration depend on the acid–base equilibrium between the excited species, and whose efficiency in the termination reactions is strongly related to their chemical nature.

(f) Cationic initiation processes: In a similar manner, an investigation of the primary processes in cationic photopolymerization could be very helpful in elucidating the mechanism of the initiation. Recent research work was carried out on onium salts. Progress has been achieved concerning the mechanism of the spectral sensitization from the excited states of the donor.[58,59]

(g) Processes in sensitizers: Visible photosensitizers have always been considered as attractive systems either for direct initiation through visible light (conventional or laser beams) or to increase the energy dose received by a substrate subjected to polychromatic excitation. Recent work on dye sensitization was carried out by laser spectroscopy.[60,61]

(h) Polymeric or copolymerizable photoinitiators: These compounds generally increase the compatibility of the photoinitiator in the coating and, better still, can prevent it from migrating outside the polymer film (if copolymerizable). Such systems, which are shown in Table 5, should lead to further developments. Some of them were studied by laser spectroscopy: **XV**[62,63] and **XVI**, (Fouassier, J. P., unpublished). In addition to the processes usually observed in the monomeric structure, hydrogen-abstraction reactions, having little effect on the intrinsic reactivity, were detected between the ketonic structure and the alkyl chain.

TABLE 5
Polymeric or Copolymerizable Photoinitiators

Compd	*Ref*	*Structure*
XV	63	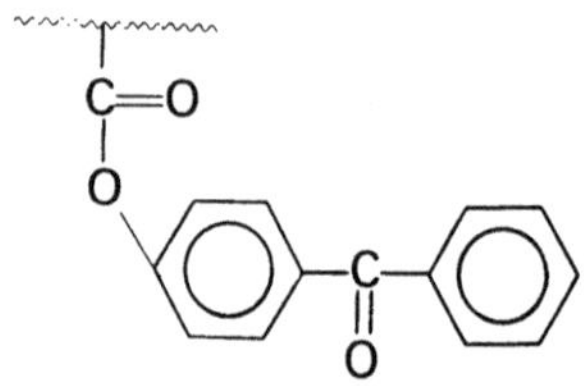
XVI	19	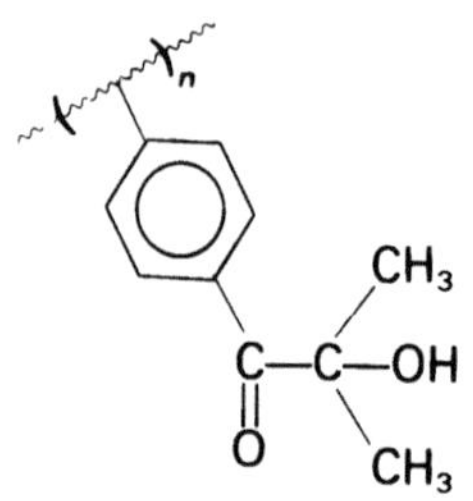
XVII	18	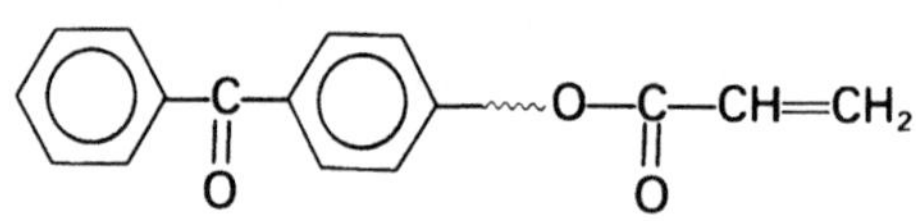
XVIII	18	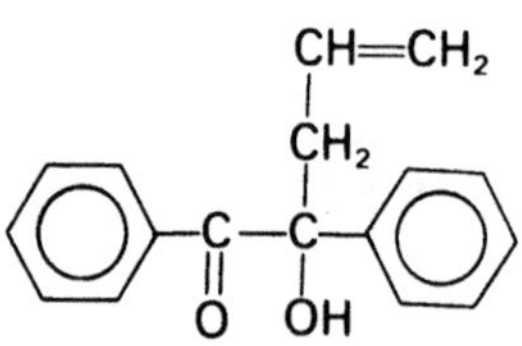

3.2 Laser-Pumped Dye Laser Spectroscopy

As stated above, in order to extend the feasibility of the usual laser spectroscopy technique to other spectral ranges of excitation, one can resort to the emission of a laser-pumped dye laser (LPDL). In most cases, because of the rather broad electronic absorption spectra of the emitting species, a LPDL device is not required. However, one may be faced with specific problems demanding a very selective excitation in a mixture of different molecules with closely related absorption spectra. An appropriate application of such a technique[22] concerns its use for the investigation of the excitation exchange (energy and electron transfer) between two molecules used as photoinitiators of polymerization. Indeed, it was recently shown that mixing ETX (in the absence of any H or electron donor) with TPMK, extends its photosensitivity towards the visible wavelengths.[47]

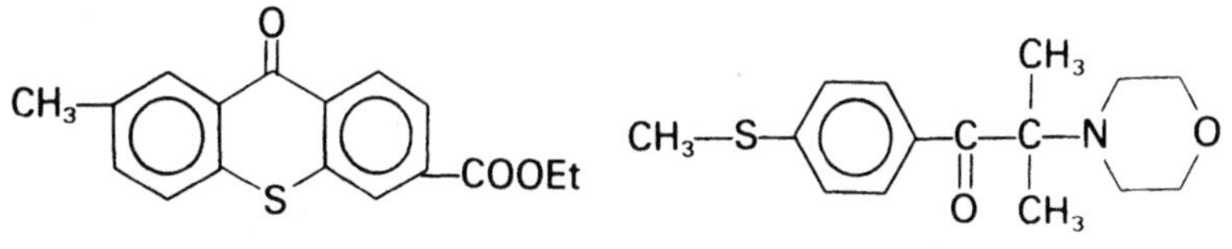

The UV absorption spectra of ETX and TPMK shown in Fig. 12 reveal why a very selective excitation must be used to excite ETX while leaving TPMK unaffected, in order to investigate the excitation transfer from ETX to TPMK.

It is apparent[22] (Scheme 6) that ETX and TPMK exhibit the usual behaviour of thioxanthone derivatives and aryl alkyl or aryl aryl ketones. The triplet state of ETX is long-lived in de-aerated solution. Fluorescence and triplet absorption are quenched when MDEA is added to solutions of ETX. New long-lived absorptions, assigned to the thioxanthone-derived ketyl radical K· and radical anion $K^{-}\cdot$, are

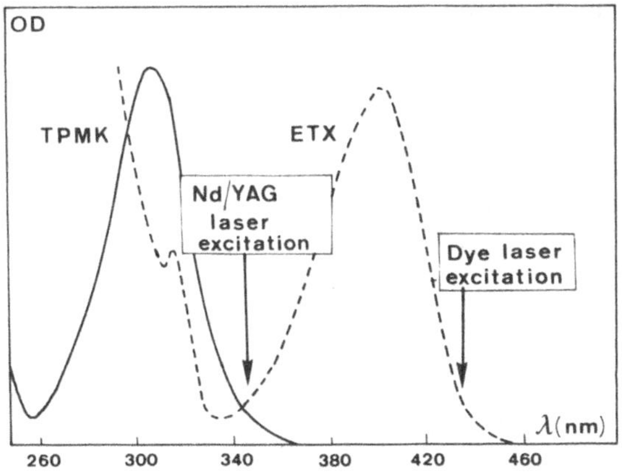

FIG. 12. UV absorption spectra of ETX and TPMK.

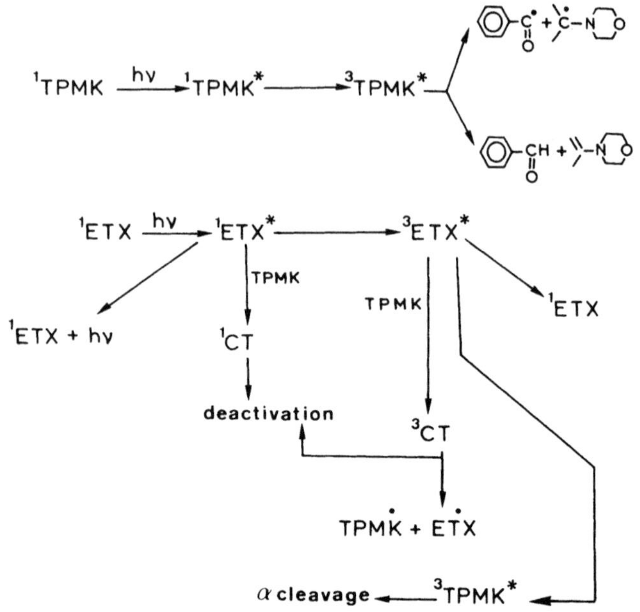

SCHEME 6. Excited-state processes of ETX, TPMK and ETX/TPMK.[22]

detected. Monomer quenching is rather inefficient. In the case of TPMK, fast cleavage occurs in a short-lived triplet state, generating a methylthiobenzoyl–morpholinoisopropyl radical pair, which undergoes disproportionation or separation as shown in Ref. 47 through CIDNP studies. The benzoyl radical transient absorption is detected (Fouassier, J. P., unpublished).

The interaction between the excited states of ETX and TPMK as well as between morpholine and thioanisole—used as models of the TPMK moieties—were investigated through LPDL spectroscopy. It was shown that the thioether group and the morpholino moiety of TPMK are responsible for the interaction with the S_1 and T_1 states of ETX, respectively. A thorough investigation of the rate constant modification (Table 6) of the excitation exchange process in the triplet state as a function of solvent and the nature of the thioxanthone (e.g. isopropylthioxanthone, ITX), suggests that the energy transfer process competes with the electron transfer; the relative efficiencies of these processes are dependent on the stabilization of the lowest excited state by solvation and on the relative positions of the triplet energy levels of

TABLE 6
Rate Constants[22] of the Processes Involved in the Triplet States of TPMK (Cleavage) and ITX–ETX (Electron Transfer, k_e; Monomer (MMA) Quenching, k_q; Excitation Transfer, k)

	$10^{-9}k_\alpha$ [a] (s^{-1})	$10^{-9}k_e$ [a] ($M^{-1}s^{-1}$)			$10^{-6}k_q$ [b] ($M^{-1}s^{-1}$)	$10^{-8}k$ (excitation exchange) ($M^{-1}s^{-1}$)	
		MDEA	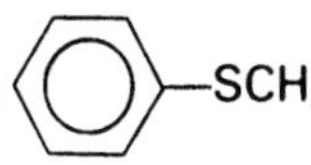	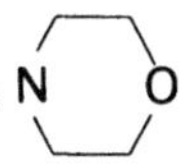	*MMA*	*Toluene*	*Methanol*
TPMK	0·10	—	≤10^{-2}	0·4	2	—	—
ETX	—	0·7	<10^{-4}	0·3	0·02	0·12	1·1
ITX	—	—	—	0·1	3	0·6	0·55

[a] In methanol.
[b] In toluene.

TPMK and ETX or ITX. Whilst in toluene the excitation transfer would occur mainly through electron transfer for ETX and energy transfer for ITX, in methanol energy transfer would predominate for both ITX and ETX.[22]

3.3 Multiphotonic Spectroscopy

The above-mentioned considerations derived from laser spectroscopy and the models elaborated for the initiation step of the photopolymerization reaction are primarily valid under the following experimental conditions: oxygen-free monomer solutions; low light absorption conditions; continuous quasimonochromatic light; and low-intensity irradiation.

When using conventional industrial irradiation devices in order to polymerize commercial formulations, the reaction develops rapidly in air with a higher intensity of irradiation (by two orders of magnitude). Under short-exposure laser excitation, the instantaneous power density can reach a few megawatts per square centimetre. An understanding of the reactions occurring in these formulations would require the investigation of the consequence of high-intensity excitations (e.g. multiphotonic effects, bimolecular reactions) and the typical influence of the film matrix on the reactivity of the excited states.

Multiphotonic spectroscopy involves the simultaneous use of at least two photons, one produced by the pump laser and the other by a

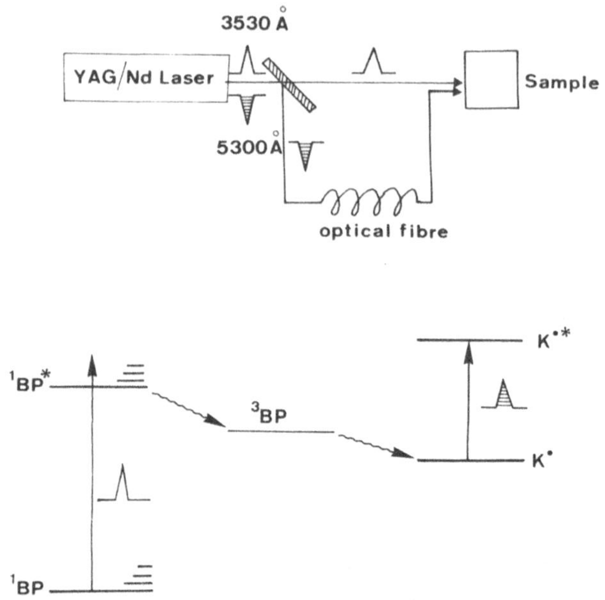

FIG. 13. Typical device for two-photon time-resolved laser spectroscopy. Excited-state processes involved.

secondary pulsed light source (a second laser or a flash or the same pump laser, but with another output wavelength). For example, the device illustrated in Fig. 13 was built (Lougnot, D. J., unpublished): the YAG laser is used to excite directly the sample (e.g. BP) with the 353 nm beam. After having been delayed in an optical fibre, the green part of the laser light (530 nm) excites the K· radical of BP formed by an electron transfer between ^{3}BP and an amine. The relaxation of the excited K· radical thus generated can be explored by emission measurements:

$$^1\mathrm{BP} \xrightarrow{h\nu_1} {}^1\mathrm{BP}^* \longrightarrow {}^3\mathrm{BP} \longrightarrow \mathrm{K}\cdot \xrightarrow{h\nu_2} \mathrm{K}\cdot^* \longrightarrow \mathrm{K}\cdot + h\nu'$$

Another type of device involves the synchronously delayed irradiation of the sample with two lasers: this has provided evidence[64] for the photofragmentation of benzil (B) according to a biphotonic process:

$$^1\mathrm{B} \xrightarrow{h\nu_1} {}^1\mathrm{B}^* \longrightarrow {}^3\mathrm{B} \xrightarrow{h\nu_2} {}^3\mathrm{B}^* \longrightarrow \phi\dot{\mathrm{C}}\mathrm{O}$$

The new and original investigation of photophysical and photochemical processes originating from upper excited states seems very promising; it opens a new field of research in the understanding of specific initiation mechanisms occurring under high-energy beams delivering polychromatic (usual mercury lamps) or monochromatic (laser) light.[41]

3.4 Holographic Spectroscopy

Holography-based devices seem to hold great promise for the investigation of photochemical reactions in organic materials[65] with a view to understanding the mechanisms involved and developing applications in the manufacture of optical elements. We are at present using such a technique for the study of the chemical and the photochemical reactivity of photoinitiators and photosensitizers incorporated in an oligomer matrix polymerizing under laser light beams.[66]

A typical experimental set-up is shown in Fig. 14. At the moment, the hologram is produced with a He–Ne laser ($\lambda = 633$ nm). The writing light is split into two beams of equal intensity. The interference pattern results from the superposition of two coherent plane waves.

The hologram is read with an Ar^+ laser. The intensity of the diffracted beam and, thereby, the diffraction efficiency ($\eta = I_D/I_R$) is measured with a powermeter connected to a recorder. As its wavelength (488 nm) is selected in such a way that this light is little absorbed by the sample, the hologram growth can be recorded as a

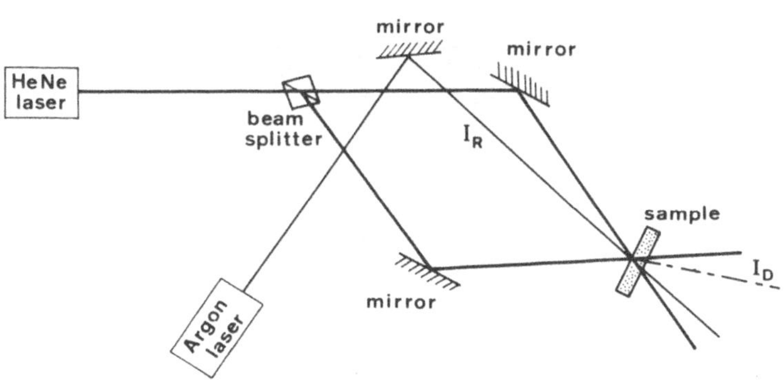

FIG. 14. Typical holographic set-up for recording and reading information in a polymer matrix.[66]

TABLE 7
Relative Efficiency of the Amine-Derived Radical for the Photopolymerization of Acrylamide Monomers in Film in Presence of a Methylene Blue/Amine Combination as Photoinitiator[66]

	$N(C_2H_4OH)_3$	$NH(C_2H_4OH)_2$	$NH_2(C_2H_4OH)$	$(CH_3)_2N$– 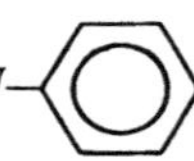 –$CO_2C_2H_4OC_4H_9$	$(CH_3)_2NCH_2$– 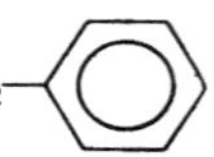
Relative efficiency	1	0·52	0·94	0·01	~0

function of time, $\eta = f(t)$, without destruction of the index modulation.

A mathematical treatment, developed elsewhere,[66] for the phase hologram theory applied to a photochemical reaction exhibiting a square-root light intensity dependence, gives access to the relative efficiency of the initiating amine-derived radicals towards the opening of the monomer double bond (Table 7). Other parameters corresponding to the formation of the polymer network seem to be attained by this method, which considerably extends the possibility of investigation based on usual means (IR spectroscopy, GC of extractable residues, dilatometry, refractometry, DSC analysis, etc). Moreover, because of the availability of laser sources, one easily imagines that the writing beam can also be operating in the UV and visible part of the spectrum (e.g. with an Ar^+ laser).

4 LASER-INDUCED POLYMER CHEMISTRY

The use of lasers for inducing photochemical events in monomer film or polymer matrix is a tempting subject not only because of the magic character of the laser but also because of practical and fundamental specific advantages, e.g. high energy concentration onto a small surface, high spatial resolution, very short exposure times allowing a scanning of the film surface by the laser spot, easy focusing, spectral selectivity and narrow bandwidth of the emission. This procedure of laser-induced chemical reactions seems to be a very promising method and applications are currently being explored or/and seriously envisaged in the fields[67,68] of microelectronics, microlithography, recording and storage of information, manufacture of holographic optical elements, integrated optics, etc.

The basic features bear a strong resemblance to those of light-induced reactions under conventional lamps or flashlamps. The chemical and photochemical processes correspond to photopolymerization, photocrosslinking and photomodification (change of solubility, colour, refractive index, adhesion etc.). Typical behaviour is observed in excimer laser microlithography where the photodecomposition becomes ablative.[69] When the laser-induced reaction really corresponds to a usual photochemical initiation event, one has to sensitize the curable formulation either by using a system which absorbs the laser light directly (by incorporating photoinitiators) or is sensitive to the UV, visible or near-IR part of the spectrum, according to the type of laser used. Because of the high power densities available, large amounts of excited species are produced, which presumably have a great influence on the subsequent development of the reactions involved: as described above, holographic spectroscopy appears to be a versatile technique allowing the investigation of these reactions.

Taking as a basis the rather sparse data reported in the literature, the general experimental behaviour of a UV or visible laser-induced polymerization reaction in a monomer film containing a photosensitive agent may be described as follows.

The medium becomes rapidly crosslinked and completely insoluble. The reaction is greatly accelerated (and the curing speeded up) either by increasing the source power density or by focusing the laser beam. The rate of polymerization is almost linear with the incident light intensity, because of the unimolecular termination process occurring presumably through radical occlusion in the polymerizing matrix where the mobility decreases as a function of time.[70]

The sensitivity remains comparable[60,71] whatever the laser source (e.g. 15 W continuous-wave Ar^+ laser, mode-locked ruby laser, excimer laser, nitrogen laser, modulated Nd laser, Cu vapour laser) even if the energy is delivered within a few tens of nanoseconds and with very high power density. Moreover, it was found that photopolymerization can be induced by only a single shot of pulsed laser, which provides evidence for an ultrafast reaction process.

The characteristics of the photosensitizer can affect the overall efficiency to a considerable extent. This point is discussed in Refs 60, 70, 72 and 73. It stands to reason that all these types of performance can really be improved in more complex laser-curable formulations; industrial research laboratories could, probably, provide such data (e.g. in a recent patent, a sensitivity of 0·1 mJ cm^{-2} under visible light was claimed[74]). This research in laser-induced chemical processing arouses a great deal of interest in various areas of basic science such as physics, optics, organic and polymer chemistry, and photochemistry, but the industry is likely to remain the driving force.

5 CONCLUSION

Evidence has been given that lasers can prove helpful in one specific aspect of polymer photochemistry. Accordingly, the development of laser-based technologies using photosensitive polymer substrates may be expected: the answer lies with the industrial companies. As regards laser spectroscopy, the reader has probably been convinced of the interest of investigating in real time the processes involved in the excited states of photoinitiators, in order to define reactivity/efficiency relationships. Further experiments on new available systems will be continued, (i) to provide a deeper insight into the reaction mechanisms and (ii) to promote the design of new efficient compounds. However, as mentioned above, it should be borne in mind that the dynamics of the excited states in only one of the key factors in the definition of the most suitable photoinitiator in a given light-curable formulation for a wide range of practical applications.

REFERENCES

1. Pappas, S. P. (ed.), *UV Curing: Science and Technology*. Stamford Technology Marketing, Norwalk, CT, 1978.
2. Ledwith, A., *Pure Appl. Chem.*, **49** (1977) 431.
3. Schnabel, W., *Photogr. Sci. Engng.*, **23** (1979) 154.
4. Fouassier, J. P., *J. Chim. Phys.*, **80** (1983) 339.
5. Crivello, J. V., *Adv. Polym. Sci.*, **62** (1984) 1.
6. Pappas, S. P. & Jilek, J. H., *Photogr. Sci. Engng*, **23** (1979) 140.
7. Green, G. E., Stark, B. P. & Zahir, S. A., *J. Macromol. Sci., Rev. Macromol. Chem.*, **C21**(2) (1981) 187.
8. Fouassier, J. P., *Double Liaison Chim. Peint. No.* 356 (1985) 173/19–187/33.
9. LiBassi, G., *Pitt. Vernici*, **5** (1983) 23.
10. Schnabel, W., *J. Radiat. Curing*, **13** (1986) 1.
11. Pappas, S. P., *Prog. Org. Coat.*, **13** (1985) 35.
12. Hageman, H. J., *Prog. Org. Coat.*, **13** (1985) 123.
13. Christensen, J. E., Jacobine, A. F. & Scanio, C. J. V., *J. Radiat. Curing*, **8**(3) (1981) 5.
14. Davis, M. J., Gawne, G., Green, P. N. & Green, W. A., *Proc. Radcure Baltimore*, Technical paper FC 86-802, 1986.
15. Gaur, H. A., Groenenboom, C. J., Hageman, H. J., Hakwoort, G. T. M., Osterhoff, P., Overeem, T., Polman, R. I. & v.d. Werf, S., *Makromol. Chem.*, **185** (1984) 1795.
16. Berner, G., Rist, G., Rutsch, W. & Kirchmayer, R., *Proc. Radcure Basel*, Technical paper FC 85-446, 1985.
17. Fratelli Lamberti, European Patent 192967 (1986).
18. Baumer, W., Koehler, M. & Langemach, J., *Proc. Radcure Baltimore*, Technical paper FC 86-443, 1986.
19. Li Bassi, C., Nicora, C., Cadora, L. & Carlini, C., European Patent Application EP 161463 (1985).

20. Laufer, M. & Dreeskamp, H., *J. Magn. Res.*, **60** (1984) 357.
21. Phillips, D., Moore, J. N., & Hester, R. E., *Proc. XI IUPAC Photochemistry Lisbon,* 1986.
22. Fouassier, J. P., Lougnot, D. J., Payerne, A. & Wieder, F., *Chem. Phys. Lett.*, **135** (1987) 30.
23. Faure, J., Fouassier, J. P., Lougnot, D. J. & Salvin, R., *Nouv. J. Chim.* **1**(1) (1977) 15.
24. Scaiano, J. C., *J. Am. Chem. Soc.*, **102** (1980) 7747.
25. Rentzepis, P. M., *Nato ASI Ser. C,* **139** (1984) 373.
26. Jacques, P., Lougnot, D. J., Fouassier, J. P. & Scaiano, J. C., *Chem. Phys. Lett.*, **129**(2) (1986) 205.
27. Fouassier, J. P., Lougnot, D. J., Zuchowicz, I., Green, P. N., Timpe, H. J., Kronfeld, K. P. & Muller, U., *J. Photochem.*, **36**(3) (1987) 347.
28. Lougnot, D. J., Jacques, P., Fouassier, J. P., Casal, H. L., Kim-Thuan, N. & Scaiano, J. C., *Can. J. Chem.* **63**(11) (1985) 3001.
29. Fouassier, J. P. & Merlin, A., *Can. J. Chem.*, **57**(21) (1979) 2812.
30. Fouassier, J. P. & Merlin, A., *J. Photochem.*, **12** (1980) 17.
31. Amirzadeh, G., Kuhlmann, R. & Schnabel, W., *J. Photochem.*, **10** (1979) 133.
32. Kuhlmann, R., & Schnabel, W., *Polymer,* **18** (1977) 1163.
33. Eichler, J., Herz, C. P., Naito, I. & Schnabel, W., *J. Photochem.*, **12** (1980) 225.
34. Amirzadeh, G. & Schnabel, W., *Makromol. Chem.*, **182** (1982) 2821.
35. Kuhlmann, R. & Schnabel, W., *Polymer,* **17** (1976) 419.
36. Merlin, A., Lougnot, D. J. & Fouassier, J. P., *Polym. Bull.*, **3** (1980) 1.
37. Sumiyoshi, T. & Schnabel, W., *J. Photochem.*, **30** (1985) 63.
38. Fouassier, J. P., Lougnot, D. J. & Scaino, J. C., *Chem. Phys. Lett.* (in press).
39. Fouassier, J. P. & Lougnot, D. J., *J. Chem. Soc., Faraday Trans.*, **83** (1987) 2935.
40. Schnabel, W., In *Applications of Lasers in Polymer Science and Technology,* ed. J. P. Fouassier, and J. F. Rabek. CRC Press, Boca Raton (in press).
41. Fouassier, J. P. & Lougnot, D. J., *J. Polym. Sci., Polym. Chem.* **26** (1988) 1021.
42. Fouassier, J. P. & Lougnot, D. J., *Proc. Radcure Baltimore,* Technical Paper FC 86-41, 1986.
43. Lewis, F. D., Hoyle, C. H., Magyar, J. G., Heine, H. G. & Hartmann, W., *J. Org. Chem.*, **40** (1975) 4.
44. Davidson, R. S., *Adv. Phys. Org. Chem.*, **19** (1983) 223.
45. Jacques, P., Payerne, A. & Fouassier, J. P., *Proc. XI IUPAC Photochemistry Lisbon,* 1986, p. 85.
46. Green, P. N., *Polymer Paint Resins,* **175** (1985) 246.
47. Rutsch, W., Berner, G., Kirchmayr, R., Hüsler, R., Rist, G. & Bubler, N., *Organic Coatings Science and Technology 8,* ed. G. Parfitt and A. Patsis. M. Dekker, New York, 1986, p. 175.
48. McGinniss, V. D., Kuo, C. & Allopo, A. G., *Macromolecules,* **11** (1978) 393.

49. Turro, N. J., Chow, M. J., Chung, C. J. & Tung, C. H., *J. Am. Chem. Soc.*, **102** (1980) 7390.
50. Merlin, A. & Fouassier, J. P., *Polymer*, **21** (1980) 1363.
51. Fouassier, J. P. & Lougnot, D. J., *Polym. Photochem.*, **3** (1983) 79.
52. (a) Jacques, P., Lougnot, D. J. & Fouassier, J. P., *Surfactants in Solution —2*, ed. K. L. Mittal, Vol. 2. Plenum Press, New York, 1984, p. 1177; (b) Lougnot, D. J., Floret-David, D., Jacques, P. & Fouassier, J. P., *J. Chim. Phys.* **82**(5) (1985) 505.
53. Fouassier, J. P. & Lougnot, D. J., *J. Appl. Polym. Sci.*, **32**(8) (1986) 6209.
54. Scaiano, J. C. & Abuin, E. B., *Chem. Phys. Lett.*, **81** (1981) 209.
55. Scaiano, J. C. & Lougnot, D. J., *J. Phys. Chem.*, **88** (1984) 3379.
56. Allen, N. S., Catalma, F., Green, P. N. & Green, W. A., *Eur. Polym. J.*, **22**(5) (1986) 347.
57. Fouassier, J. P. & Lougnot, D. J., *J. Appl. Polym. Sci.*, **34** (1987) 477.
58. Pappas, S. P., Pappas, B. C., Gatechair, L. R. & Schnabel, W., *Polym. Photochem.*, **5** (1984) 1.
59. Crivello, J. V. & Lee, J. L., *Polymer J.*, **17**(1) (1985) 73.
60. Fouassier, J. P. & Chesneau, E., *Angew. Makromol. Chem.*, **135**(2166) (1985) 41.
61. Bottcher, A., Fisher, M., Denk, O., Wang, W. K. & Schnabel, W., *Proc. XI IUPAC Photochemistry Lisbon*, 1986, p. 110.
62. Carlini, C., Gardelli, F., Donati, D. & Gurzoni, F., *Polymer*, **24** (1983) 599.
63. Flamigni, L., Barigelletti, F., Bortolus, P. & Carlini, C., *Eur. Polym. J.*, **20**(2) (1984) 171.
64. Scaiano, J. C. & Johnston, L. J., *Pure and Applied Chemistry*, **58** (1986) 1273.
65. Brauchle, C. & Burland, D. M., *Angew. Chemie, Int. Ed. Engl.*, **22** (1983) 582.
66. Carre, C., Lougnot, D. J. & Fouassier, J. P., *Macromolecules*, **22** (1989) 791.
67. Jain, K., *Lasers and Applic.*, **9** (1983) 49.
68. Osgood, R. M. & Deutsch, T. F., *Science*, **227** (1985) 709.
69. Jellinek, H. H. G. & Srinvasan, R., *J. Phys. Chem.*, **88** (1984) 3048.
70. Decker, C. In *Materials for Microlithography*, ed. L. F. Thompson, C. G. Willson & J. M. J. Frechet, ACS Symp. Series, **266** (1984).
71. Fouassier, J. P., Lougnot, D. J. & Pilot, T., *J. Polym. Sci., Polym. Chem. Ed.*, **23** (1985) 569.
72. Sadhir, R. K., Smith, J. D. B. & Castle, P. M., *J. Polym. Sci., Polym. Chem. Ed.*, **23** (1985) 411.
73. Iwata, K., Hegiwara, T. & Matzuzawa, H., *J. Polym. Sci., Polym. Lett. Ed.*, **22** (1984) 215.
74. Yamakoa, T., Koseki, K. & Goto, Y., Belgian Patent 897694 (1984).

Chapter 8

POST-CURED STABILITY OF RADIATION-CURED RESINS

N. J. WHITE,* N. S. ALLEN, P. J. ROBINSON

Faculty of Science and Engineering, Manchester Polytechnic, Manchester, UK

&

R. CLANCY

Harcros Chemicals Limited, Manchester, UK

LIST OF ABBREVIATIONS

BZ	Benzophenone
Di-PETA mono, DPMHP	Dipentaerythritol monohydroxypenta-acrylate
EB	Electron beam
ESCA	Electron spectroscopy for chemical analysis
HALS	Hindered amine light stabiliser
HDDA, HDODA, 1,6-HDODA	1,6-Hexanediol diacrylate
IOTG	Iso octyl thioglycolate
ITX	2-Isopropyl thioxanthone
MDEA, MDEOA	Methyldiethanolamine
MDI	Methylene diisocyanate
MPL	Maximum practical level

* *Present address*: Banbury Laboratories, Alcan International Limited, Southam Road, Banbury, Oxfordshire OX16 7SP, UK.

PETA PETA-tetra	Pentaerythritol tetra-acrylate
PETA-tri hydroxy-PENTA	Pentaerythritol triacrylate
TEGDA	Triethylene glycol diacrylate
TMPEOTA	Trimethylolpropane ethylene oxide triacrylate
TMPPOTA TMPPTA	Trimethylolpropane propxylate triacrylate
TMPTA	Trimethylolpropane triacrylate
TPGDA	Tripropylene glycol diacrylate

1 INTRODUCTION

Since its introduction in the early 1970s, radiation curing[1–4] has become a diverse and well-developed technology. At its inception the technology offered energy savings over conventional thermal curing technologies with a reduction in the requirement for solvents. With the diversification of such applications these advantages are almost taken for granted and end-users are now looking for particular performance characteristics or improvements. One area of post-cure performance which is currently of interest is the post-cure stability of radiation-cured coatings, with the main areas being thermal stability and photostability. Various applications require thermal stability; solder masks, abrasives, wire/cable coatings, solar collectors or reflectors, adhesives and optical fibre coatings. Applications requiring photostability include no-wax flooring, automotive varnishes, solar collectors or reflectors and various varnishing applications.

Polymer degradation has been widely studied for many years; the degradation processes of a number of polymers are well understood and various stabilization methods are available.[5–7] There has been little work undertaken in the area of radiation-cured coating stability. The aim of this chapter is to review the present status of research in this field, with the emphasis on thermal and photostability.

2 THERMAL STABILITY OF RADIATION-CURED COATINGS

The thermal stabilities of polyacrylate materials such as poly(methyl methacrylate) and poly(ethyl acrylate) have been extensively studied.

The degradation of methacrylates by heat is well known and is essentially a depolymerization process producing mainly monomer in >90% yield:

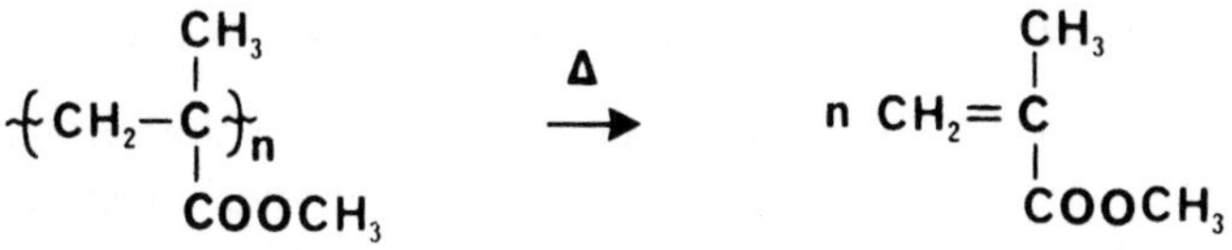

(1)

The thermal degradation of acrylates is a more complex process and is believed to involve a chain auto-oxidation mechanism as illustrated in Scheme 1.

The majority of radiation-curable resins/oligomers are acrylate and not methacrylate in nature. It is envisaged therefore that because of this, and the fact that complex three-dimensional networks are produced on curing, the thermal degradation processes will be more complex in radiation-cured coatings.

Armbruster & Molina have carried out a detailed study of the thermal stability of benzoin ethyl ether UV-cured and electron-beam (EB)-cured coatings.[8] The effects of multifunctional monomer diluent, oligomer and curing conditions were investigated. The thermal stabi-

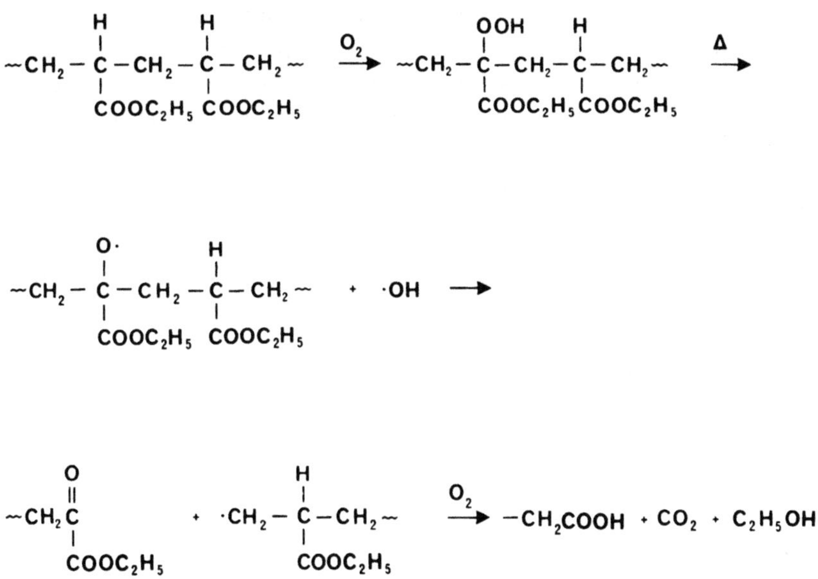

SCHEME 1

lity of prepared coatings was characterised by the use of TGA under both air and nitrogen. The effect of multifunctional monomer structure was found to follow the trend of decreasing thermal stability, thus:

$$\text{TMPTA} = \text{PETA} > \text{HDODA} \gg \text{TMPPOTA} = \text{TMPEOTA} > \text{TPGDA} = \text{TEGDA}$$

There were two factors which appeared to govern the observed order of thermal stability. The first relates to monomer functionality and crosslink density. For a given series, e.g. alkyl or ether type, the higher functional monomers are more stable than the difunctional monomers. The crosslink density is likely to increase with increased functionality leading to lower oxygen permeability and improved thermal stability. The second factor is the backbone structure of the monomer. Those multifunctional monomers that contain polyether backbones are more prone to thermal degradation than their alkyl counterparts; compare for example TMPTA with TMPPOTA. It is well known that the CH bonds α to ether linkages in polymer backbones are susceptible to thermo-oxidative degradation by the mechanism of Scheme 2.

The photoinitiator concentration was also found to have an effect on the thermo-oxidative stability; on reduction from 5% to 2·5 wt % the stability of TMPTA was reduced. This effect is probably due to the lower crosslink density, in the low photoinitiator sample, due to the film being undercured. The fact that the multifunctional monomers containing a polyether backbone are more susceptible to thermo-oxidation than their alkyl counterparts was shown by thermal experiments carried out under a nitrogen atmosphere. Under these conditions the polyether-based multifunctional monomers exhibited a dramatic increase in thermal stability: for example, TEGDA exhibited a weight loss reduction from 5% to 7·4% at 350°C, under nitrogen. Interestingly the aliphatic alkyl multifunctional monomers such as TMPTA exhibit little nitrogen effect.

This finding indicates that both thermolysis and thermo-oxidative mechanisms of degradation occur in the aliphatic alkyl multifunctional monomers, whereas the polyether-based materials degrade mainly by thermo-oxidation. The electron-beam-cured coatings tested indicated that EB-cured multifunctional acrylates are more susceptible to thermal degradation than their UV-cured counterparts. This was explained in terms of the relative crosslink densities of the two systems with the UV-cured system having a higher crosslink density than the EB-cured system.

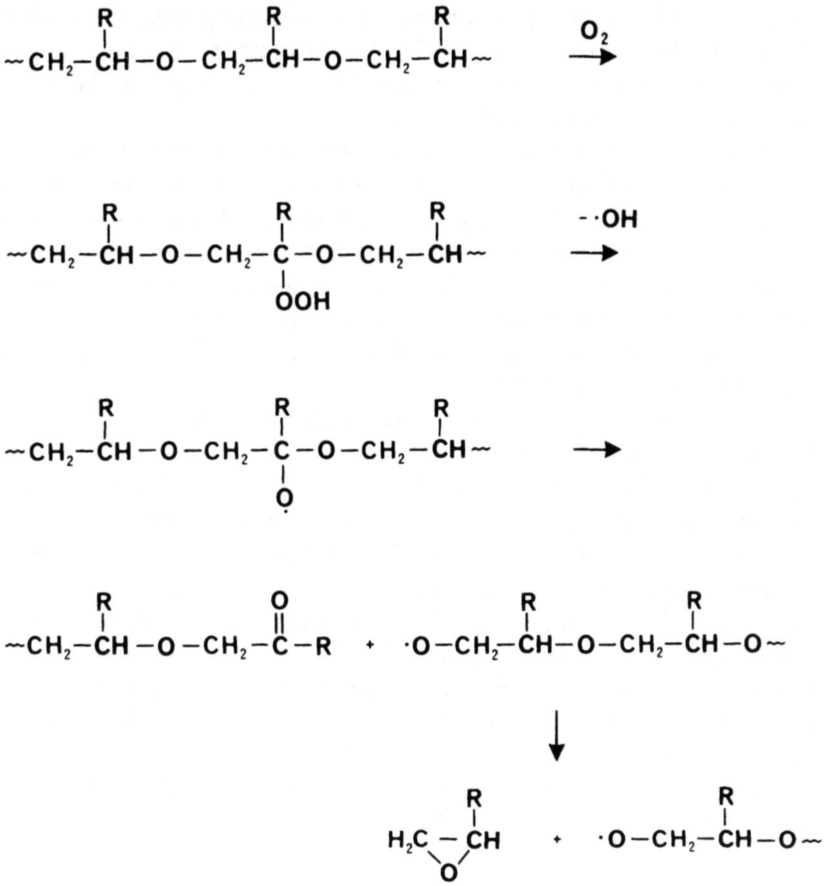

SCHEME 2

From the above multifunctional monomer study, Armbruster & Molina concluded that TMPTA was the least sensitive to thermal degradation and proceeded to evaluate this multifunctional monomer with a range of acrylated oligomers including acrylated epoxies and acrylated urethanes. It was found that the thermal stability of the acrylated epoxy materials tested was improved when thermal degradation was carried out under a nitrogen atmosphere, indicating that a thermo-oxidative degradation mechanism occurs under air. The acrylated urethanes tested were found to be generally inferior to the acrylated epoxies. Only a small increase in thermal stability was observed under a nitrogen atmosphere, suggesting that thermolysis plays a significant part in the mechanism of urethane acrylate thermal degradation. Aliphatic urethane acrylates were found to be more stable than the aromatic urethane acrylates.

Thalacker & Boettcher have carried out an interesting investigation of the thermal stability of EB-cured systems which complements the work by Armbruster & Molina outlined above.[9] These workers have also used TGA measurements as a method for assessing the thermal stability of radiation-cured coatings. Similar multifunctional monomers and acrylated oligomers were studied. The TGA thermal degradation profiles of 3MRad EB-cured multifunctional monomer samples are illustrated in Fig. 1 and Table 1.

These results compare well with those of Armbruster & Molina for UV-cured systems.[8] The difunctional acrylate monomers such as 1,6-HDODA, TPGDA and TEGDA are more susceptible to thermo-oxidative degradation than the multifunctional acrylate monomers such as TMPTA, PETA-tri and PETA-tetra. Also the polyether-containing multifunctional acrylate monomers such as TMPPTA are more susceptible to thermo-oxidative degradation than their aliphatic multifunctional analogues such as TMPTA. These workers postulated that monomers of high functionality produce three-dimensional polymer networks of higher crosslink densities than those produced by the difunctional monomers. The effect of EB dose on thermal stability was also investigated; the results are illustrated in Fig. 2 and Table 2 for the thermal degradation of 1,6-HDODA. Here it can be seen that on

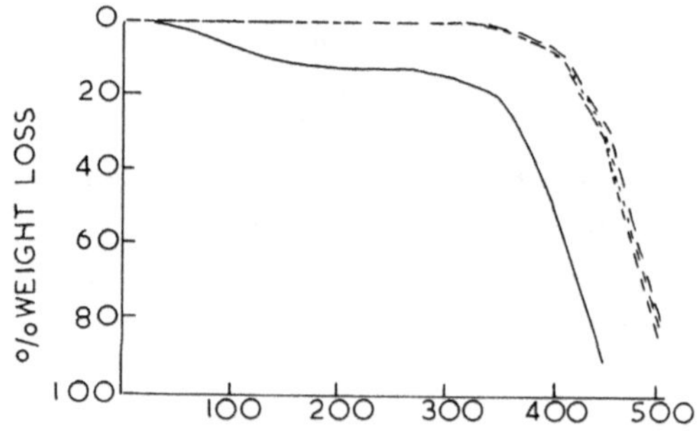

FIG. 1. Effect of functionality on thermal stability at a dose of 3MRad. ——, 16-HDODA; – – –, PETA-tetra; - - - -, TMPTA; · · · ·, DPMHP. (Reprinted courtesy of the Society of Manufacturing Engineers. Copyright 1985; from *Radcure Europe 1985 Proceedings*.)

TABLE 1
Monomer Type and Functionality

Monomer	*Trade designation*	*Weight loss (%)*								
		100°C	*150°C*	*200°C*	*250°C*	*300°C*	*350°C*	*400°C*	*450°C*	*500°C*
HDODA	SR238	7	12	14	14	15	22	50	92	
TEGDA	SR268	0	0	0	3	12	30	67	84	
TPGDA	Celanese	0·5	0·7	1	7	17	37	78	88	91
TMPTA	SR351	0	0	0	0	1	2	10	30	85
TMPPTA	P4072	0	0	0·5	1	6	23	60	72	88
PETA-tri	SR444	0	0	0	0·5	1	3·5	14	38	81
PETA-tetra	SR295	0	0	0	0	0	2	10	33	80
Di-PETA mono Hydroxy-penta	SR399	0	0	0	0·5	1	3	8	35	85

Reprinted courtesy of the Society of Manufacturing Engineers. Copyright 1985; from the *Radcure Europe 1985 Proceedings*.

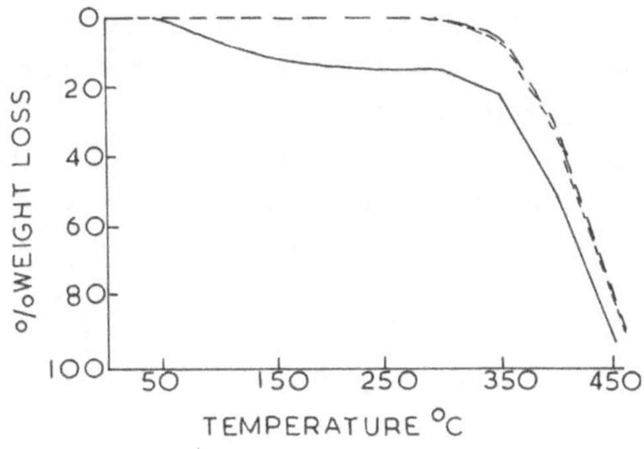

FIG. 2. Thermal stability of 1,6-HDODA, EB-cured at 3, 5 and 10 MRad. ——, 3 MRad; - - - -, 5 MRad; – – –, 10 MRad. (Reprinted courtesy of the Society of Manufacturing Engineers. Copyright 1985; from *Radcure Europe 1985 Proceedings*.)

TABLE 2
HDODA Dose Versus Percentage Weight Loss

Dose (MRad)	*Weight loss (%)*							
	100°C	*150°C*	*200°C*	*250°C*	*300°C*	*350°C*	*400°C*	*450°C*
3	7	12	14	14	15	22	50	92
5	0	0	0	0	1	7	33	82
10	0	0	0·5	0·5	1	5	31	82

Reprinted courtesy of the Society of Manufacturing Engineers. Copyright 1985; from the *Radcure Europe 1985 Proceedings*.

TABLE 3
Oxygen Permeability

Material	*EB cure dose (MRad)*	*Permeability ($cm^3 m^{-1} day^{-1}$)*
HDODA	3	135
HDODA	5	124
HDODA	10	110
Celrad® 3500	3	65
Celrad® 3500	5	59
Celrad® 3500	10	45

increasing the dose used to cure the coating, the thermal stability is improved, due to the increased crosslink density. Evidence for this crosslink density effect was provided by solid-state ^{13}C-NMR and oxygen permeability measurements.

The oxygen permeability data shown in Table 3 indicate that with increased dose the permeability is reduced, the likely reason being a higher crosslink density with increased dose. The solid-state ^{13}C-NMR spectra illustrated in Fig. 3(A) are located in the carbonyl region for resonance. The acrylate carbonyl occurs slightly upfield (165 ppm) from the unconjugated carbonyl associated with the reacted acrylate double bond (176 ppm). These two peaks can be used to quantify the extent of cure and by implication the crosslink density. These plotted data are illustrated in Fig. 3(B). It can be seen that at higher doses the extent of cure approaches 100%. The oxygen permeability data in Table 3 also show the difference between the permeability of a difunctional acrylate monomer and oligomer; for a given dose the oligomer has a much lower oxygen permeability; this correlates well with the relative thermal stability of the two materials as tested by TGA. A comparison of EB- and UV-cured coatings with thermally cured conventional coatings showed that the radiation-cured coatings have superior stability.

O'Hara has carried out a study of the effect of relatively low temperatures (100°C) on UV- and EB-cured clear and pigmented coatings.[10] In this study various film properties were assessed, including film colour, gloss and loss of volatile material. The effect of 30 min

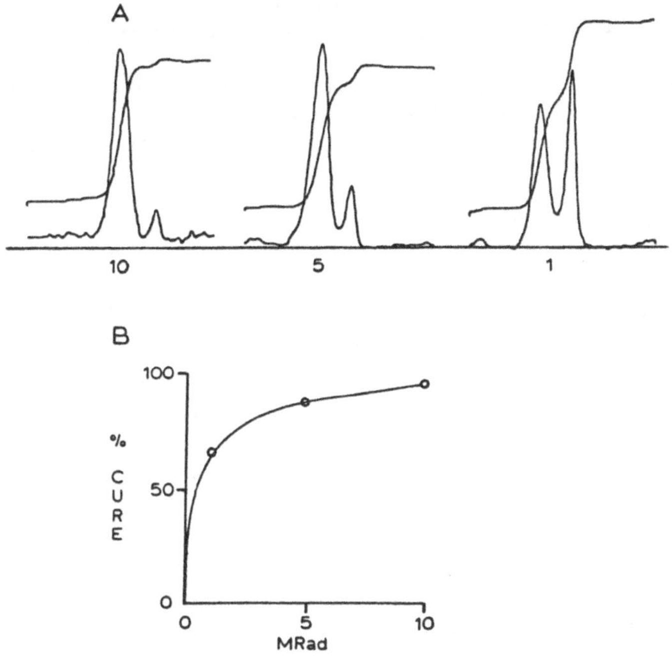

FIG. 3. (A) Solid-state ^{13}C-NMR spectra of carbonyl region for acrylate cured by 10, 5 and 1 MRad dose. (B) Percentage cure versus MRad dose. (Reprinted courtesy of the Society of Manufacturing Engineers. Copyright 1985; from *Radcure Europe 1985 Proceedings.*)

of heat ageing at 100°C on the colour and gloss of an epoxy acrylate-based clear varnish is illustrated by the data in Table 4.

As one would expect, the EB-cured lacquer is superior to the UV-cured lacquer with regard to colour stability; however, it is surprising to see that, for the system studied, the UV-cured lacquers initially have a superior colour to their EB-cured counterparts. The results do not indicate any trends in colour performance relating to initiator content or type, and cure conditions, e.g. line speed for UV or dose for EB. However, there appears to be a trend between gloss changes on heat ageing and dose. In all cases the degree of gloss is reduced on heat ageing. The gloss difference of UV-cured coatings before and after heat ageing is generally worse that that of EB-cured coatings before and after ageing. O'Hara suggested that the poor performance of the UV-cured coatings, in relation to gloss, may be

TABLE 4
Effect of Heat Ageing on Epoxy Acrylate Clear Lacquers

Initiator	*Level (%)*	*Cure rate ($M\ min^{-1}$ per lamp)*	*Yellowing index*		*20°Gloss*	
			Before	*After*	*Before*	*After*
UV-cured						
BZ/amine	3/1	5	−4·3	−0·9	64·2	37·7
BZ/amine	3/1	2	−2·6	2·0	66·2	47·2
BZ/amine	5/2	5	−3·3	0·8	62·4	33·2
BZ/amine	5/2	2	−2·3	1·9	60·3	41·6
Irgacure® 184	1	10	−5·4	2·4	58·5	41·5
Irgacure® 184	1	5	−2·5	1·2	55·9	46·6
Irgacure® 184	2	20	−4·7	−1·3	54·4	44·6
Irgacure® 184	2	10	−3·8	−0·7	62·2	52·4
EB-cured		*Dose (MRad)*				
None		2	−2·6	−1·9	71·2	48·9
None		4	−2·9	−2·4	67·1	51·1
None		6	−2·2	−1·8	64·8	51·3

System: Synocure 3101/TPGDA.
Substrate: Chromated aluminium coated with white acrylic/isocyanate basecoat.
Test: 30 min at 100°C.

TABLE 5
Effect of Heat Ageing Pigmented Epoxy Acrylate Films

Initiator	*Level (%)*	*Cure rate ($min\ min^{-1}$ per lamp)*	*Yellowing index*		*20° gloss*	
			Before	*After*	*Before*	*After*
UV-cured						
ITX/BMS/BEA[a]	1/1/4	5	7·6	4·8	77·1	36·5
ITX/BMS/BEA	1/1/4	2	12·1	7·5	68·7	44·5
EB-cured		*Dose (MRad)*				
None		4	−5·2	−5·3	66·4	39·6
None		8	−3·1	−3·1	61·7	57·4

[a] Quantacure® BMS and BEA

due to the inhibition of photopolymerisation by O_2 at the coating surface; this would produce different extents of polymerisation at the surface compared with the coating 'bulk'. This effect is greatly reduced in the case of the EB-cured lacquers; these are cured under a blanket of nitrogen, i.e. <200 ppm O_2. Data relating to the heat ageing of TiO_2-pigmented systems are illustrated in Table 5.

The EB-cured pigmented coating gave similar results to the EB-cured clear lacquer. The UV-cured pigmented coating was initially more coloured than the UV-cured clear coating. On heat ageing, however, the pigmented system exhibited a reduction in colour whereas the UV-cured clear lacquers showed increased colour. In the case of the UV-cured clear coatings the photoinitiators benzophenone (**I**) with *N*-methyldiethanolamine (**II**) and 1-benzoylcyclohexanol (**III**) were used, whereas for the pigmented coatings a mixture of 2-isopropylthioxanthone (**IV**), 4-benzoyl-4′-methyldiphenyl sulphide (**V**) and 2-*n*-butoxyethyl 4-(dimethylamino)benzoate (**VI**) was used.

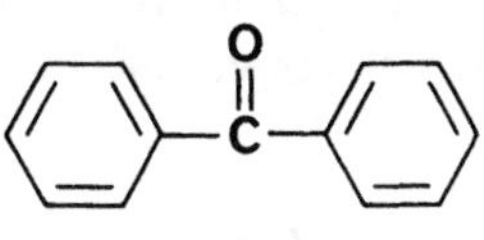
I

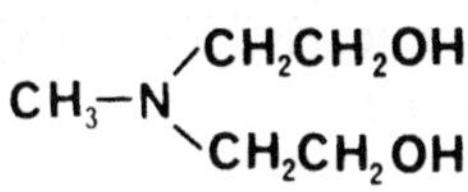

II

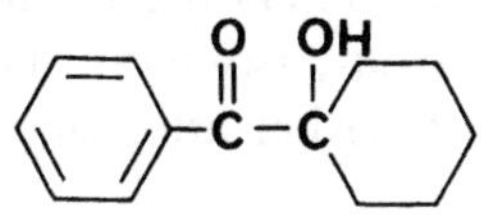

III

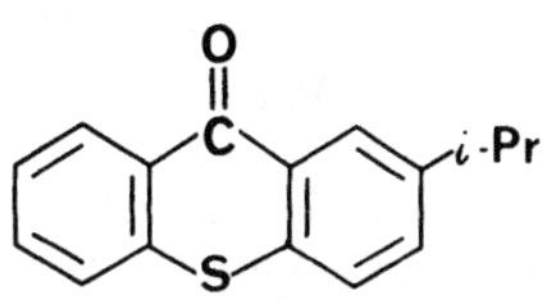

IV

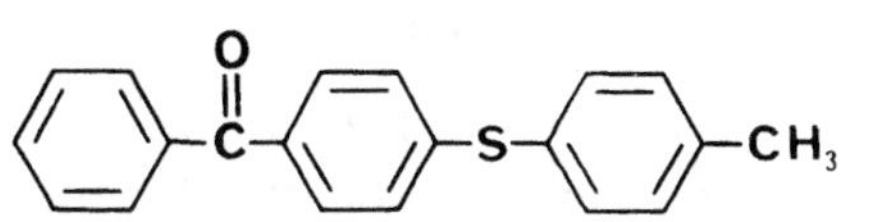

V

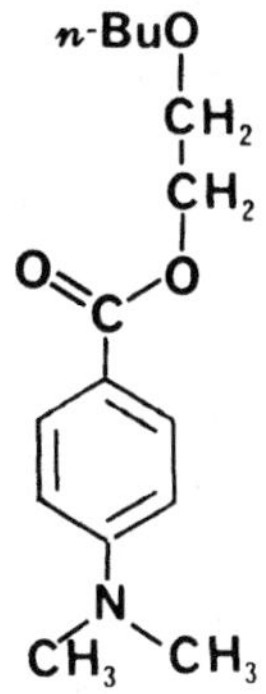

VI

The initial strong colour in the UV-cured pigmented system is probably due to chromophoric impurities produced from photoinitiator by-products from the 'three-compound' mixture used. These chromophoric impurities are thermally labile. Hult & Ranby have investigated this phenomenon, which is observed with the isopropylthioxanthone (**IV**)–*N*-methyldiethanolamine (**II**) system used in their studies on the photodegradation of pigment UV-cured coatings[11] (See Section 3.1.1).

Relatively little work has been published on the stabilisation of radiation-cured coatings against thermal degradation. During their study on thermal degradation of radiation-cured multifunctional monomers with acrylate oligomers, Armbruster & Molina briefly investigated stabilisation to thermal degradation.[8] They adopted two approaches to the stabilisation of radiation-cured coatings. The first involved the use of ethylthioethyl methacrylate (ETEMA; **VII**) which copolymerises with the multifunctional monomers and/or acrylate oligomers used. The second approach was to include a chain transfer agent, iso-octyl thioglycolate (IOTG; **VIII**), which is converted to a thioether after chain transfer.

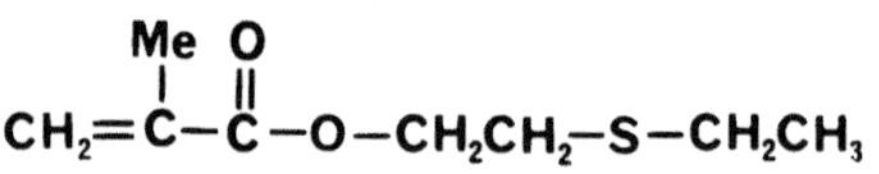

VII

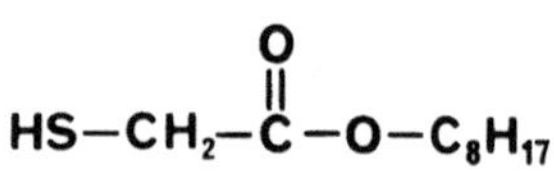

VIII

Thioethers are believed to act as thermal stabilisers by scavenging oxygen, thus oxidising the thioether to sulphoxide and sulphone. They found that these compounds stabilised multifunctional polyether acrylate monomers against thermo-oxidation, which further indicates that these multifunctional monomers are thermally degraded by an oxidative process. These stabilisers exhibited a marginal improvement with epoxy acrylate oligomers and a marked improvement with a urethane acrylate oligomer. This suggests that, since the urethane is degraded predominantly by a thermolysis process, these stabilisers are also acting as radical scavengers.

Heyward *et al.* have investigated a number of antioxidants as thermo-oxidative stabilisers for UV-cured coatings.[12] They studied the effect of thermal stabilisers on the curing process when using 2,2-dimethoxy-2-phenylacetophenone as photoinitiator, the solubility of the stabilisers in an acrylate-based coating formulation and the effect of these stabilisers on the onset of oxidation as measured by DSC and oxygen absorption of the system. Three types of antioxidant stabiliser were studied: phenolic antioxidants, hindered-amine light stabilisers and thio- or phosphate-based peroxide decomposers. A number of these were quickly excluded from further investigation on solubility grounds. Some of the stabilisers decreased the photoinitiator conversion whereas others appeared to sensitise photoinitiator decomposition. The degradation results showed that the hindered-amine light stabiliser, bis(2,2,6,6-tetramethyl-4-piperidinyl) sebacate (**IX**) and the hindered phosphonate, bis(-*n*-octodecyl-3,5-di-*tert*-butyl-4-hydroxybenzyl) phosphonate (**X**) improve the thermal stability of UV-cured epoxy acrylate systems, without any significant effect on the curing process.

Although there is little published work on the thermal stability of radiation-cured coatings the studies outlined above have already provided an insight into the mechanisms of thermal degradation and possible routes of stabilisation.

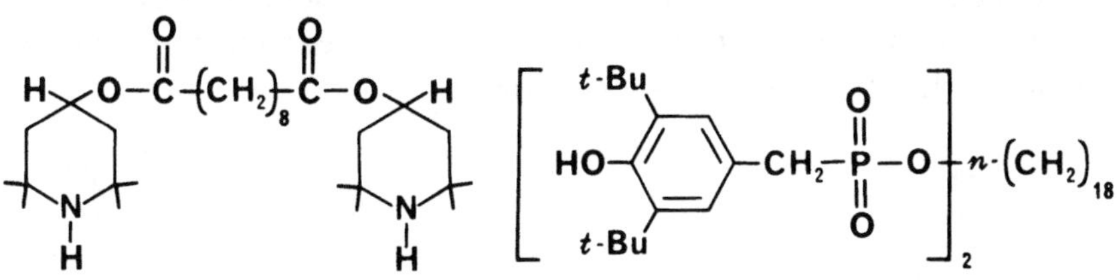

IX X

3 PHOTOSTABILITY OF RADIATION-CURED COATINGS

The photostability of radiation-cured coatings is becoming increasingly important. A number of applications require the UV- or EB-cured coating, either clear or pigmented, to provide stable colour and gloss performance under ambient or artificial lighting conditions. Other applications require the physical and mechanical properties of the films to be maintained under these conditions. A number of studies have been carried out to investigate photoyellowing phenomena in radiation-cured coatings and the photo-oxidation/photodegradation processes which occur in these systems.[13] Progress has also been made in the stabilization of radiation-cured coatings to improve their post-cure photostability.[13,14]

3.1 Photoyellowing Phenomena

Photoyellowing in radiation-cured coatings is dependent on both photoinitiator and resin structure. It can be apparent immediately after and as a consequence of the curing process or only after prolonged photoageing. The phenomena can be separated into photoinitiator-related and acrylate-resin-related photoyellowing processes.

3.1.1 Photoinitiator Effects

Ohngemach *et al.* have investigated a number of photoinitiators for their initiating efficiency in the photocuring of acrylate coatings.[15] During this investigation the photoyellowing performance of some of these photoinitiators was evaluated in a Xenotest 150 accelerated weatherometer. The photoinitiators studied included 2-isopropyl thioxanthone (**IV**), benzophenone (**I**), dibenzosuberon (**XI**), benzil dimethyl ketal (**XII**), benzoin isopropyl ether (**XIII**), α,α-dimethyl-α-hydroxyacetophenone (**XIV**), α,α-dimethyl-α-hydroxy-4-*t*-butyl-acetophenone (**XV**) and α,α-diethoxyacetophenone (**XVI**). Photoinitiators **I**, **IV** and **XI** were investigated in conjunction with *N*-methyldiethanolamine (**II**) as a coinitiator.

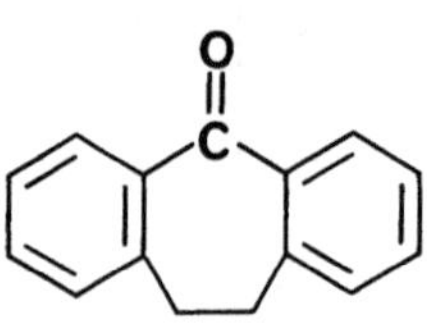

XI

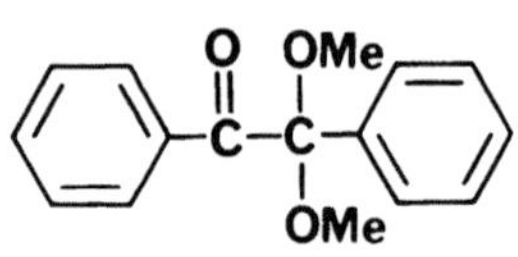

XII

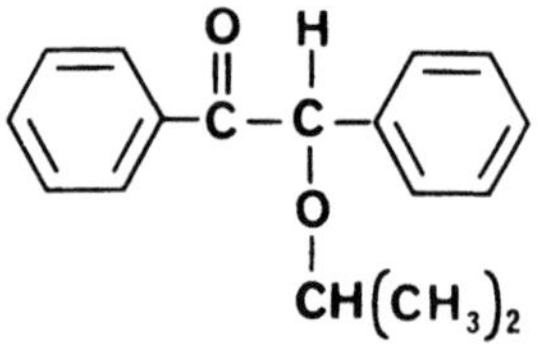

XIII

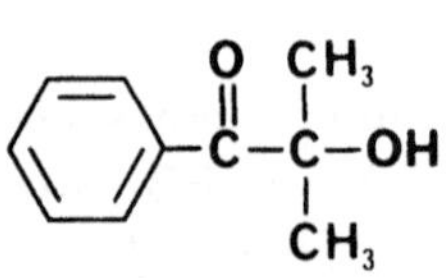

XIV

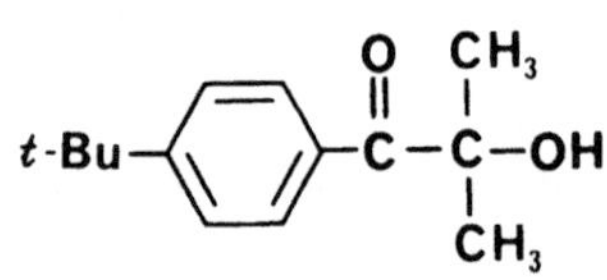

XV

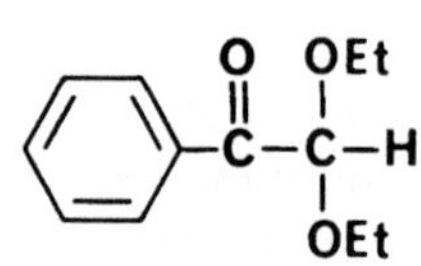

XVI

The following order of decreasing tendency to photoyellow was observed:

$$\text{IV/II} \gg \text{XI/II} > \text{I/II} > \text{XII} = \text{XIII} > \text{XIV} = \text{XV/XVI}$$

The coinitiator systems studied, which consisted of a hydrogen-abstracting type of photoinitiator in combination with the tertiary amine (**II**), exhibited the greatest tendency towards photoyellowing. The benzoin- and benzil-based photoinitiators **XIII** and **XII**, which both contain two aromatic moieties, were the second most susceptible group to photoinduced yellowing. Finally, the acetophone-based photoinitiators **XIV**, **XV**, and **XVI** produced coatings with the greatest resistance to photoyellowing. No indication was given of the nature of the discolouration products produced by the photoinitiators in this study.

Schmid has carried out a similar study to that of Ohngemach *et al.*, investigating a range of photoinitiators with various acrylate oligomers, both urethane and epoxy, with TMPTA as the multifunctional monomer.[16] The cured coatings were exposed for 24 hours to two Westinghouse FS-40 fluorescent sunlamps, with the yellowness index being measured before and after exposure. Like Ohngemach, Schmid evaluated photoinitiators **I**, **XIV**, **XVI** and **XII**; he also evaluated **III** and *p*-phenoxy-2,2-dichloroacetophenone (**XVII**). All of these photoinitiators were evaluated in the presence of one of three amines; *N*-methyldiethanolamine (**II**), diethylamine (**XVIII**), and hexamethyleneimine (**XIX**). the results are shown in Tables 6 and 7.

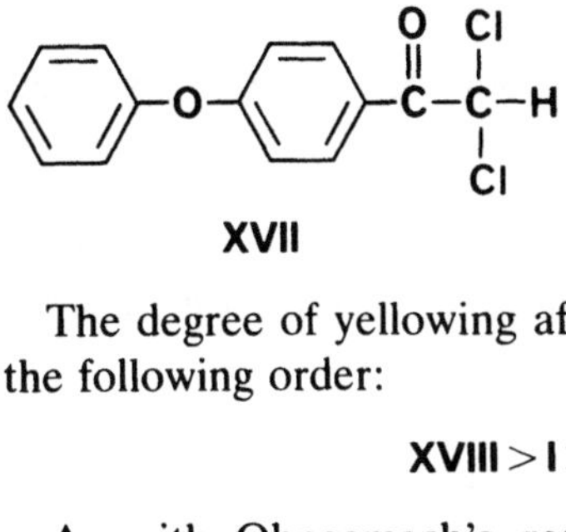

XVII

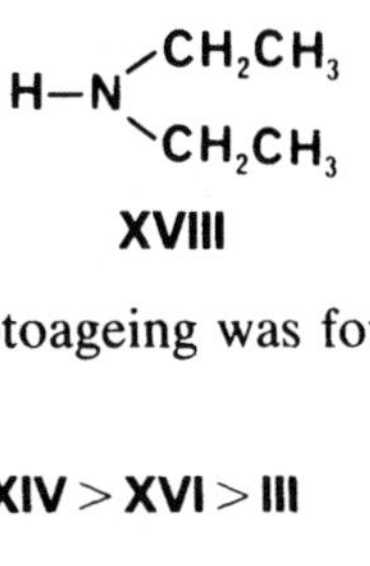

XVIII

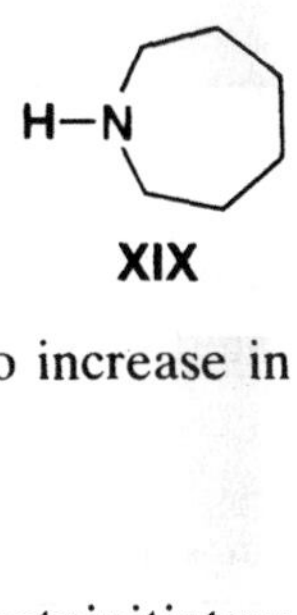

XIX

The degree of yellowing after photoageing was found to increase in the following order:

$$\mathbf{XVIII} > \mathbf{I} > \mathbf{XII} > \mathbf{XIV} > \mathbf{XVI} > \mathbf{III}$$

As with Ohngemach's results it appears that the photoinitiators containing two aromatic moieties (**XVII**, **I** and **XII**) are more susceptible to photoyellowing than those that contain one aromatic group. Schmid suggested that this was due to recombination side reactions such as:

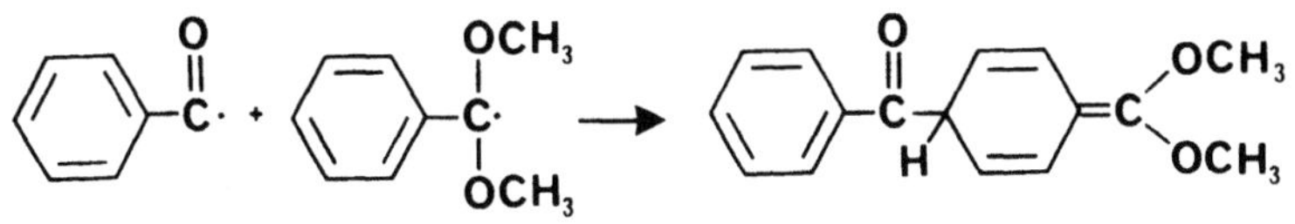

(2)

Filipescu & Minn have postulated similar products for the photoreduc-

TABLE 6
Yellowing Index Values for TMPTA–Aliphatic Oligomer Systems

TMPTA–Aliphatic epoxy acrylate						*TMPTA–Aliphatic urethane acrylate*					
MDEOA		*DEA*		*HMI*		*MDEOA*		*DEA*		*HMI*	
I-184	2·88	I-184	3·04	D-1173	3·89	I-184	2·34	DEAP	2·59	DEAP	3·20
D-1173	3·14	BP	3·46	DEAP	3·92	DEAP	2·61	I–184	3·19	D-1173	3·22
I-651	3·24	DEAP	3·60	I-651	3·94	D-1173	2·84	D-1173	3·53	I-184	3·55
DEAP	3·34	I-651	4·05	I-184	4·24	I-651	2·93	S-1000	3·90	I-651	4·72
S-1000	3·90	S-1000	4·60	S-1000	Gel	BP	3·33	I-651	4·08	S-1000	Gel

Reprinted with permission from *Journal of Radiation Curing,* **11,** No. 2, April 1984. Published by Technology Marketing Corp., Norwalk, CT, USA.

BP	Benzophenone
D-1173	Darocure 1173
DEAP	2,2-Diethoxyacetophenone
I-184	Irgacure® 184
I-651	Irgacure® 651
S-1000	Sandoray 1000
MDEOA	Methyldiethanolamine
DEA	Diethylamine
HMI	Hexamethyleneimine

tion of **I** by isopropyl alcohol, which produces unstable coloured non-radical intermediates.[17]

It is believed that the absence of a secondary aromatic group in the least yellowing photoinitiators prevents the formation of these conjugated structures. It was found that the degree of yellowing associated

TABLE 7
Yellowing Index Values for TMPTA–Aromatic Oligomer Systems

TMPTA–Aromatic Epoxy Acrylate						*TMPTA-Aromatic Urethane Acrylate*					
MDEOA		*DEA*		*HMI*		*MDEOA*		*DEA*		*HMI*	
1-184	3·08	1-184	4·12	1-651	5·22	1-184	4·37	1-184	4·48	D-1173	5·35
DEAP	4·45	DEAP	4·53	DEAP	5·57	DEAP	4·49	DEAP	5·05	1-184	5·72
1-651	4·68	1-651	4·65	D-1173	5·88	S-1000	5·03	D-1173	5·56	DEAP	5·92
D-1173	4·79	D-1173	4·98	1-184	6·59	BP	5·23	1-651	5·70	1-651	6·08
BP	4·85	S-1000	5·76	S-1000	6·67	1-651	5·44	BP	5·95	S-1000	7·04
S-1000	5·81	BP	6·15	BP	7·30	D-1173	6·33	S-1000	6·48	BP	7·40

Reprinted with permission from *Journal of Radiation Curing,* **11,** No. 2, April 1984. Published by Technology Marketing Corp., Norwalk, CT, USA.
Abbreviations: see footnote to Table 6.

with amine type used decreased in the following order:

hexamethyleneimine > diethylamine > methyldiethanolamine

It must be noted that it is likely that the hexamethyleneimine and diethylamine secondary amines would have reacted with the acrylate unsaturation on either the TMPTA or acrylate oligomers by an exothermic Michael type addition:

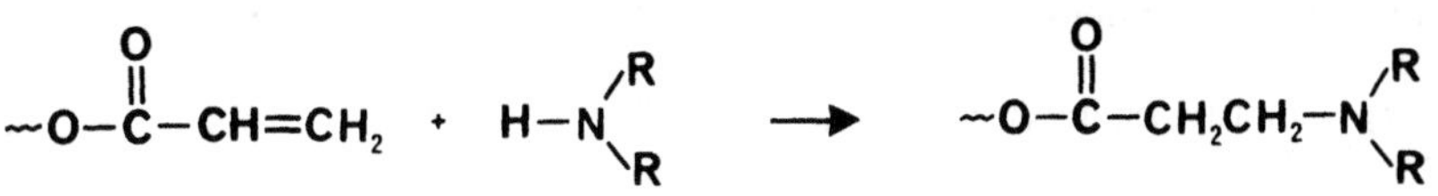

(3)

This reaction effectively binds the secondary amines into the acrylate system, whereas the tertiary amine **II** is not bound into the system and can be lost during ageing by volatilisation. This may account for the reason why cured coatings containing tertiary amine **II** are more resistant to photoyellowing. The role of bound amine in photoyellowing processes will be discussed further in Section 3.12.

The findings of both Schmid and Ohngemach have been confirmed by Gatechair in a study carried out using Florida weathering conditions.[13] The following photoinitiators in addition to those already mentioned above were used; 3,4-dimethyl-α,α-dimethyl-α-hydroxyacetophenone (**XX**), 4-chloro-α,α-dimethyl-α-hydroxyacetophenone (**XXI**), and 4,4′-bis dichloro methyl benzophenone (**XXII**). The yellowing data obtained from this study are shown in Table 8. Considering the findings of Schmid, Gatechair concluded that the

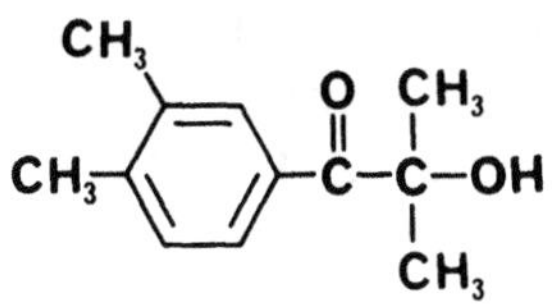

XX

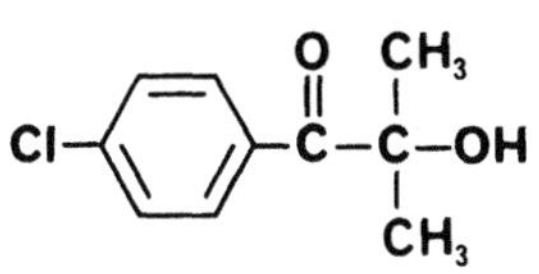

XXI

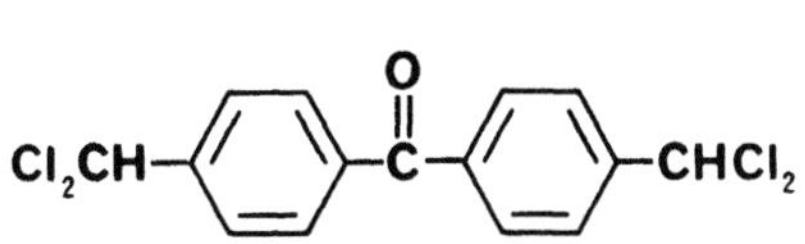

XXII

TABLE 8
Photoinitiator Comparison: Yellowness Index of UV-Cured Plaques During Florida Weathering
Model aliphatic urethane acrylate formulation. Plaques 20 mils thick, exposed at 45° facing South.

Photoinitiator comparison (4·75%)	*Initial*	*3 months*	*6 months*	*9 months*	*12 months*
Diethoxyacetophenone	1·9	3·0	4·7	5·4	4·7
α-Hydroxycyclohexyl phenyl ketone	1·5	3·9	5·6	6·2	5·8
α-Hydroxy-α-methyl-phenylpropanone (HMPP)	1·6	4·5	5·6	6·4	X[a]
3,4-Dimethyl-HMPP	1·6	3·8	5·5	6·3	5·3
4-Chloro-HMPP	1·9	25·2	20·8	18·5	X[a]
Methyl glyoxalate	2·2	3·8	5·8	5·6	X[a]
α,α-Dimethoxy-α-phenylacetophenone (**XII**)	1·8	16·6	13·9	13·3	11·8
4,4′-Bis(dichloromethyl)benzophenone	2·1	83·9	84·5	83·5	86·6

Reprinted with permission from *UV Curing: Science and Technology*, Vol. II, ed. S. P. Pappas. Published by Technology Marketing Corp., Norwalk, CT, 1985.
[a] X, Plaque failed via cracking; testing not possible.

presence of chlorine in the photoinitiator molecule contributed considerably to photoyellowing; however no mechanism was proposed to account for this.

Hult and Ranby have studied the photoyellowing phenomena associated with the use of 2-isopropylthioxanthone (**IV**) in conjunction with *N*-methyldiethanolamine (**II**), as the initiating system for TiO_2-pigmented coatings.[18] They utilised an Atlas UVCON light exposure unit for ageing tests; this is a laboratory instrument for exposing materials to fluorescent ultraviolet light. During their study they observed an initial strong yellowing during the curing process which disappeared within a few minutes in the dark (Fig. 4). A broad absorption band at 400–450 nm was thought to be responsible for this yellow colour (Fig. 5).

When the film was aged on the UVCON light exposure unit this yellowing relaxation process was reversed and the films became yellow again. The same phenomenon was observed in irradiated films which did not contain amine coinitiator. It was suggested that this photoyellowing was due to the oxidation of intermediate ketyl radicals (**XXIII**) during the curing process to produce quinone (**XXIV**) and semiquinone (**XXV** and **XXVI**) intermediates:

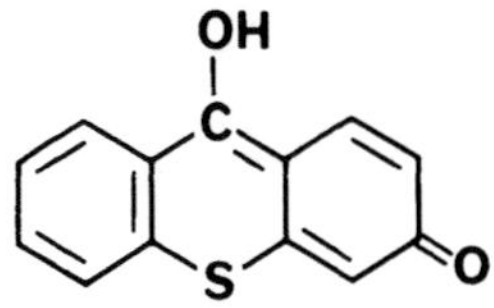

XXIV

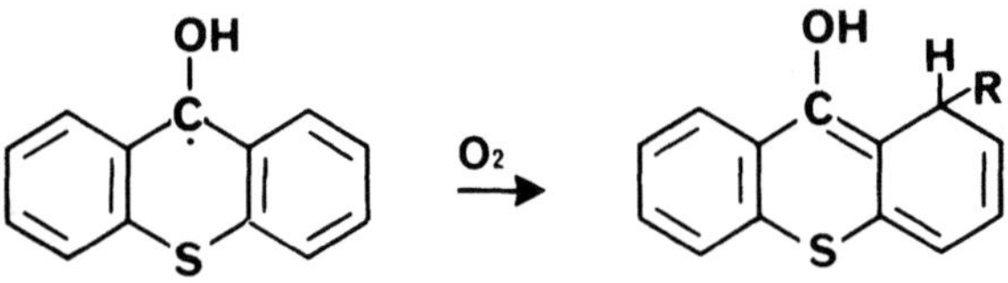

XXIII **XXV** (4)

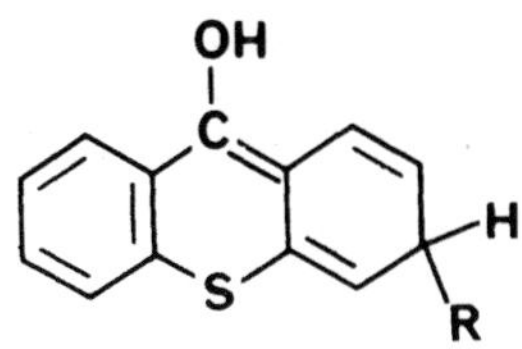

XXVI

It was found that the concentration of **IV** affects the rate of yellowing during the first 24 hours of ageing in the UVCON apparatus. After this initial period the rate of yellowing becomes independent of the photoinitiator concentration. In a separate study these workers found that photoinitiator **XII** was inferior to **III** and **XXVII** for photoyellowing in epoxy acrylate coatings.[19]

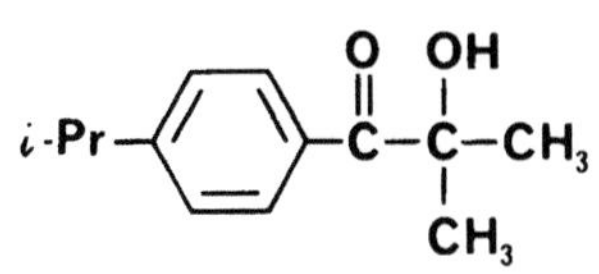

XXVII

O'Hara observed a similar phenomenon with the use of 2-isopropylthioxanthone (**IV**) as a photoinitiator for TiO_2-pigmented coatings.[10] He found that **IV** produced more highly coloured coatings than **I**, which in turn produced more highly coloured coatings than **III**.

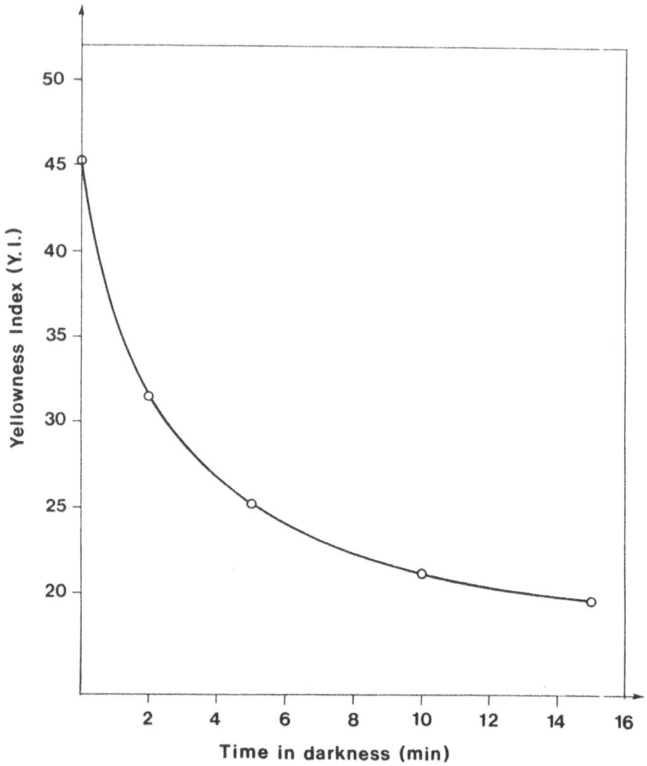

FIG. 4. Initial yellowing decay of a 2-isopropyl thioxanthone/amine photo-cured coating during dark storage.[18]

The choice of photoinitiator obviously plays a major role in the photoyellowing of a UV-cured coating. By judicious choice it should be able to produce a cured coating with acceptable yellowing properties. Hydrogen-abstracting photoinitiators such as **I** and **IV** in conjunction with amine coinitiators produce the most highly coloured coatings. The best photoinitiators for acceptable photoyellowing are those with one aromatic group, such as **III**, **XV** and **XXVII**.

3.1.2 Acrylate Binder Effects

Radiation-curing formulations prepared for coating applications usually consist of one or more acrylated oligomers or prepolymers and at least one multifunctional acrylate monomer. The acrylated oligomers are usually epoxy or urethane materials and can be either

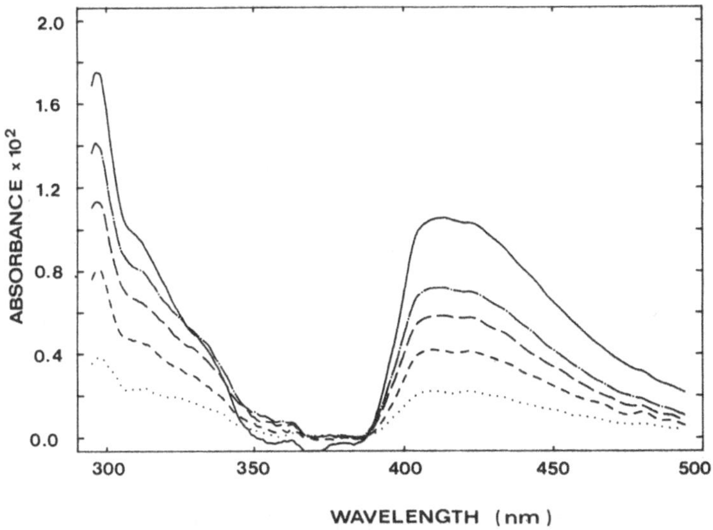

FIG. 5. UV–Vis light absorption spectra of an ITX/amine photocured coating. Difference spectra between the recordings at zero time (directly after irradiation) and after storage in darkness for 1 min (· · · · · ·), 2 min (- - - - - -), 3 min (– – –), 4 min (— · —), or 20 min (——).[18]

aliphatic or aromatic in nature. Each part of the formulation is likely to produce its own contribution to the photoyellowing phenomenon, above that due to photoinitiator residues, which are observed in UV- and EB-cured coatings on photoageing.

Gismondi has investigated several commercially available acrylate monomers for their resistance to UV-induced yellowing.[20] He found that those containing aromatic functionality or nitrogen were prone to photoyellowing, whereas binary mixtures of TEGDA and TMPTA were the most resistant to yellowing. In his studies on thermal and photochemical degradation O'Hara found that epoxy acrylate oligomers were considerably inferior to urethane acrylates (Synocure 3101 is a bisphenol-A-based epoxy acrylate and Synocure 3134 is an aliphatic urethane acrylate)[10] (Fig. 6).

Schmid found that aromatic urethane- and aromatic epoxy-based oligomers were more prone to photoyellowing than their aliphatic analogues.[16] The aliphatic urethane oligomer was found to be the most resistant to photoyellowing. The photoyellowing associated with aromatic-based oligomers is believed to be due to the formation of

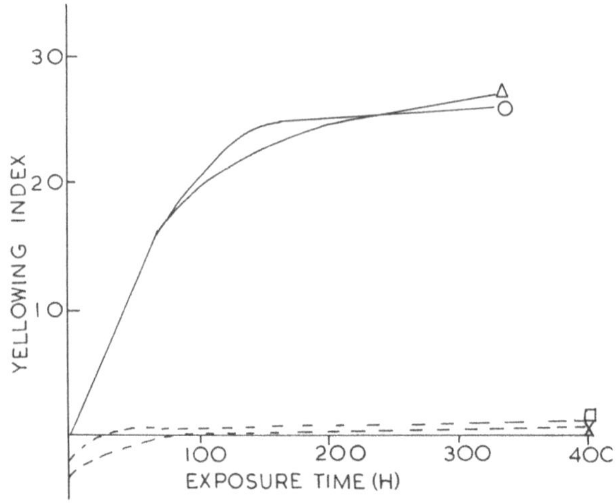

FIG. 6. Effect of QUV weathering on colour retention of epoxy and urethane acrylate films: (△) Synocure® 3101 UV Irgacure® 184 2%; (○) Synocure 3101 EB 4MRad; (□) Synocure 3134 UV Irgacure 184 2%; (×) Synocure 3134 EB 4MRad. Substrate: white basecoated Al panels. (Reprinted courtesy of the Society of Manufacturing Engineers. Copyright 1985; from *Radcure Europe 1985 Proceedings*.) Synocure® is a registered trade name of Cray Valley Products; Irgacure® is a registered trade name of Ciba–Geigy.

quinoid structures; Allen and McKellar have demonstrated that aromatic urethanes based on MDI oxidise to a highly coloured quinone-imide product (Scheme 3).[21] Toluene di-isocyanate-based oligomers can undergo α-cleavage reactions which may be followed by a photo-Fries type rearrangement with the production of highly coloured quinoid structures (Scheme 4).[13] A large number of epoxy acrylates are based on bisphenol A; a typical example is **XXVIII**.

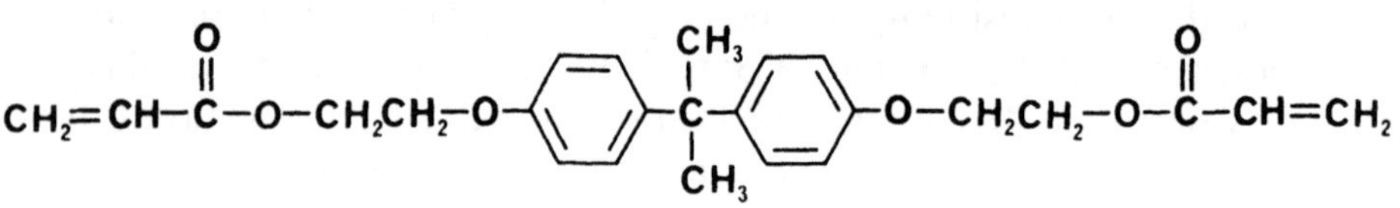

XXVIII

Pappas *et al.* have studied the photochemistry of bisphenolic polymers and model compounds.[22] It was found that the aromatic chromophores induced photocleavage and photo-oxidation reactions; the production of quinoid products may account for the yellowing of

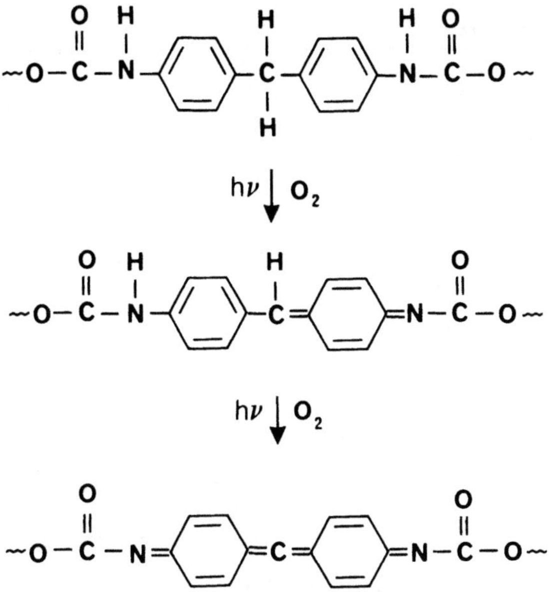

SCHEME 3

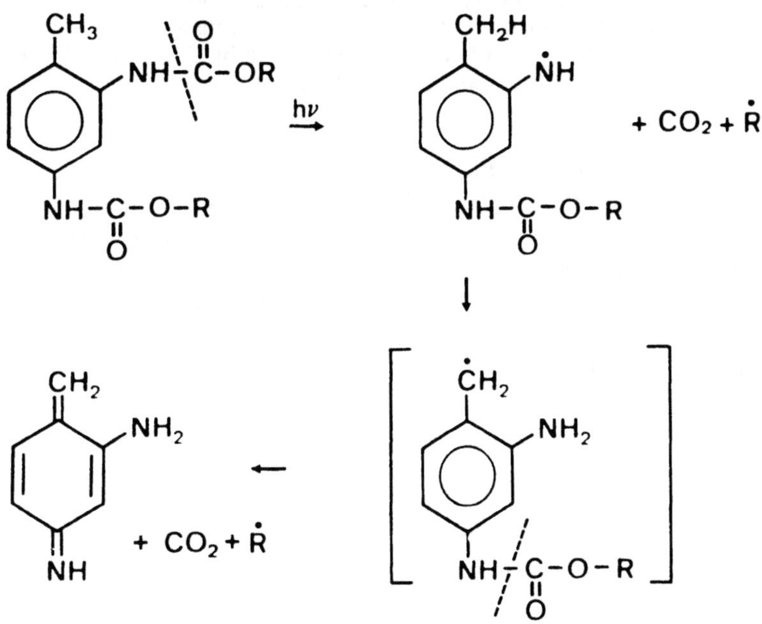

SCHEME 4

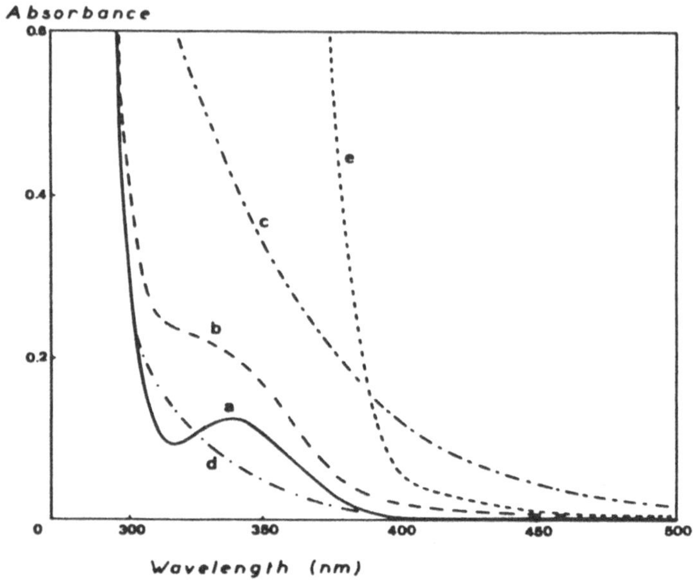

FIG. 7. UV absorption spectra of a 20 μm epoxy acrylate coating: (a) before UV curing; (b) after UV curing for 0·4 s; (c) after photo-oxidation at >250 nm for 150 h; (d) after photo-oxidation at >300 nm for 150 hours; (e) after UV curing with 1% Tinuvin 900®. [Reprinted with permission from *Journal of Radiation Curing,* **11** No. 2 (April 1984). Published by Technology Marketing Corporation, Norwalk, CT, USA.]

these materials under light exposure. Decker and Bendaikha have studied the photodegradation of UV-cured epoxy acrylate networks using IR and O_2 uptake measurements.[23,24] With irradiation at $\lambda >$ 250 nm the coatings developed a slight discolouration which remained constant after irradiation for 5 h (Figs 7 and 8). It was believed that this initial yellowing was due to residual photoinitiator photolysis products and not to the photodegradation or photo-oxidation of the polymer network. When wavelengths greater than 300 nm were used this initial yellowing was photobleached. It appears that this photo-yellowing phenomenon is not an oxidative process; in fact the yellow chromophore is photobleached in the presence of O_2. After the disappearance of this initial photoyellowing phenomenon Decker observed that these systems were stable to light exposure in their weatherometer even after 2000 h. Interestingly, after only 200 h of exposure there is a dramatic reduction in certain functional groups

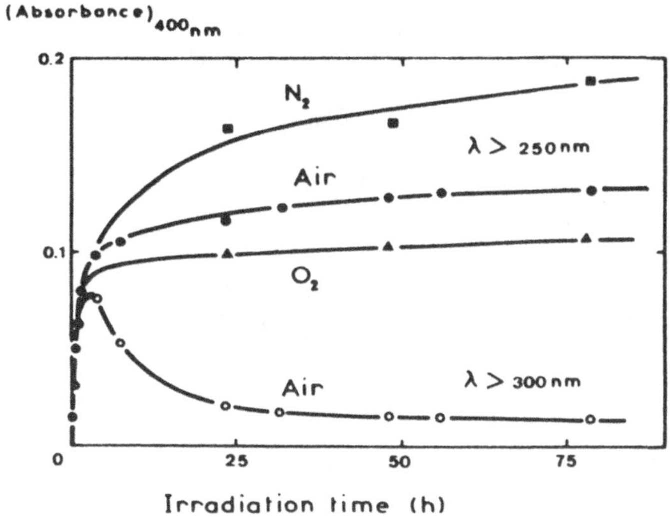

FIG. 8. Discolouration of epoxy acrylate coatings UV irradiated in the presence of air, N_2 or pure O_2. [Reprinted with permission from *Journal of Radiation Curing*, **11**, No. 1 (April 1984). Published by Technology Marketing Corporation, Norwalk, CT, USA.]

present in the cured epoxy acrylate network. The most notable deterioration is in the phenyl ring absorptions as measured by infrared spectroscopy; these absorptions are associated with the bisphenol A groups of the epoxy acrylate network and Scheme 5 accounts for this disappearance. It would be expected, however, that the production of the quinoid structure shown in the scheme would result in yellowing of the cured coating.

Allen *et al.* have carried out a study of the photodegradation and photoyellowing occurring in both UV- and EB-cured coatings.[25–27] During their studies they have observed two photoyellowing phenomena associated with multifunctional diluent monomer and oligomer structures.[25–27] During preliminary studies on multifunctional monomers both UV-cured and EB-cured, severe photoyellowing was observed with amine diacrylate monomers such as **XXIX**. This photoyellowing was associated with the growth of a band, in the second-derivative UV/Vis absorption spectrum, centred at 280 nm (Fig. 9) and the growth of two new bands in the infrared spectrum situated at 1605 and 1680 cm^{-1} (Fig. 10). The photoyellowing and

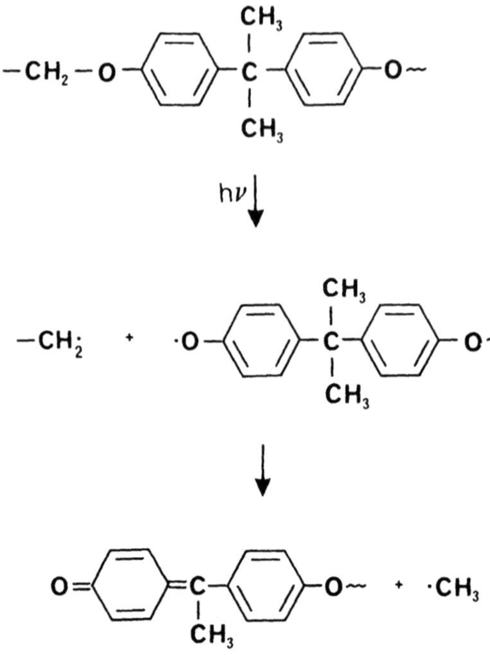

SCHEME 5

these bands increased in intensity over the first 50 h of exposure in a Microscal fadeometer and then remained constant; this was followed by photobleaching on prolonged ageing (Fig. 11). Coatings based on the trifunctional monomer used to produce **XXIX** did not photoyellow

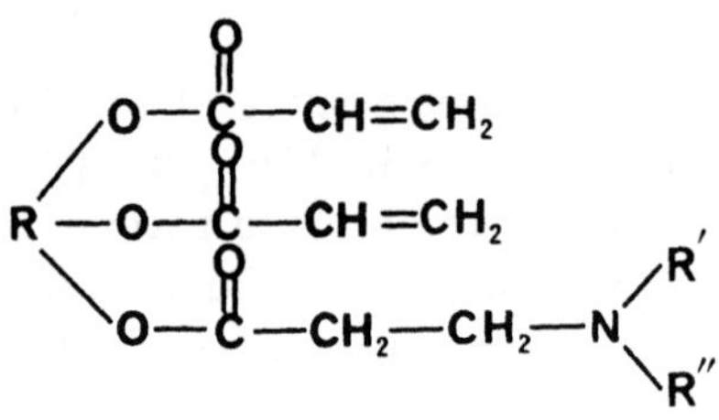

XXIX

and did not exhibit these spectral changes. It was found that the curing process and type of photoinitiator has a major effect on the photoyellowing phenomena as illustrated in Figs 12 and 13.

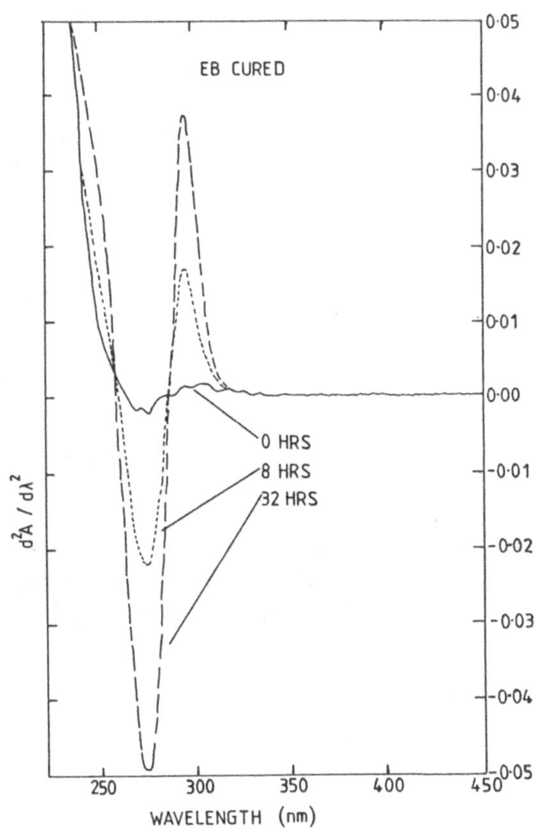

FIG. 9. Variations in the second-order derivative spectra ($d^2A/d\lambda^2$) of EB-cured **XXIX** during irradiation in a Microscal unit. (Reprinted with permission from Ref. 25. Copyright 1984, Pergamon Journals Ltd.)

It can be seen that the rate of formation is greater for systems which have been cured with benzophenone as opposed to photofragmenting initiators. However, the greatest amount of yellow product is produced by the EB-cured system. This could be due to the residual carbonyl initiators or their photoproducts in the UV-cured systems sensitizing the photodecomposition of the chromophore responsible for yellowing.

Another interesting effect illustrated in both Fig. 12 and Fig. 13 is that the photoexposure under N_2 resulted in only a minor amount of photoyellowing and product formation as measured spectroscopically.

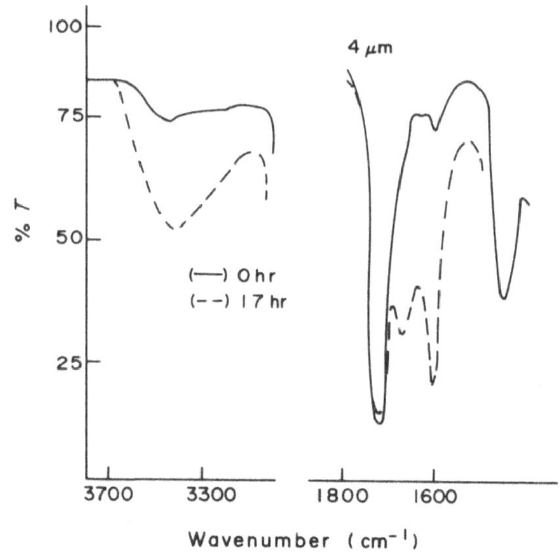

FIG. 10. Infrared spectra of EB-cured **XXIX**, (- - -) before and (——) after exposure in a Microscal unit. (Reprinted with permission from Ref. 19. Copyright 1984, Pergamon Journals Ltd.)

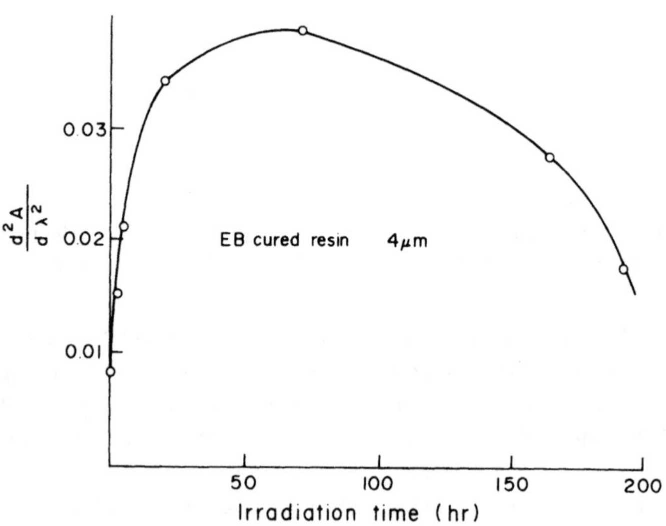

FIG. 11. Development and decomposition of the absorption product at 280 nm during irradiation in a Microscal unit of an EB-cured film (4 μm thick) of resin **XXIX**. (Reprinted with permission from Ref. 25. Copyright 1984, Pergamon Journals Ltd.)

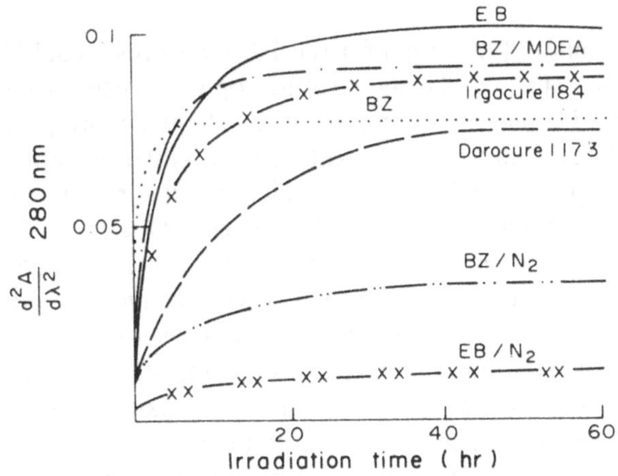

FIG. 12. Increase in intensity of UV absorption band at 280 nm during irradiation in a Microscal unit for resin **XXIX** (12 μm thick) cured by electron beam (——) and by UV with benzophenone (· · · · · ·), benzophenone/MDEA (– · – · –), Darocure® 1173 (- - - -) and Irgacure® 184 (— × —) exposed in air and for resin **XXIX** cured by electron beam (— × × —) and by UV with benzophenone (— · · —) and exposed under N_2. (Reprinted with permission from Ref. 25. Copyright 1984, Pergamon Journals Ltd.) Darocure® is a trademark of E. Merck.

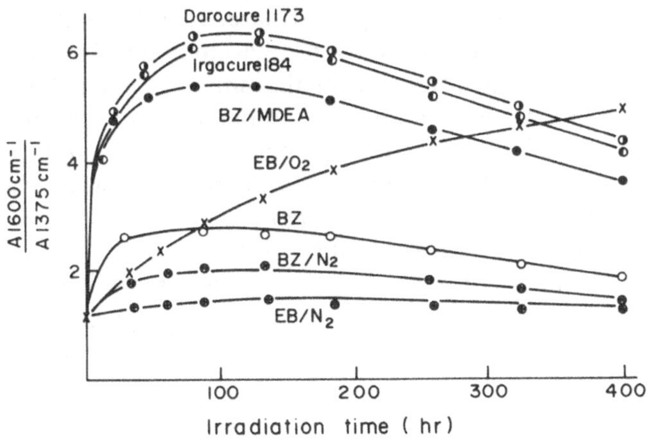

FIG. 13. Production of IR absorption band at 1605 cm^{-1} during irradiation in the Microscal unit of resin **XXIX** cured by electron-beam (×) exposed in air and (⊕) in nitrogen and cured by UV with (○) benzophenone; (●) benzophenone/MDEA; (◐) Irgacure® 184; (◑) Darocure® 1173 exposed in air and (⊗) benzophenone exposed in nitrogen. (Reprinted with permission from Ref. 25. Copyright 1984, Pergamon Journals Ltd.)

The small amount observed for the EB-cured resin could be attributed to exposure to air during spectroscopic measurements. The benzophenone UV-cured coating exhibited slightly more product than the EB-cured resin under N_2, indicating that oxidation products from the UV curing process, which is carried out in air, are possibly responsible for this slight increase. Overall, the results clearly show that this photoyellowing phenomenon is an oxidative one and does not occur in multifunctional triacrylate monomers without amine modification; it is a process requiring the photo-oxidation of the amine group. The nature of the chromophore responsible for the yellowing is unclear; however, some indication of its structure can be gleaned from the spectroscopic evidence. The simultaneous formation of absorption bands at $1680\,cm^{-1}$ and $1605\,cm^{-1}$ is indicative of a carbonylic species which is conjugated to unsaturation. A conjugated species of this type would also account for the strong absorbance observed between 250 and 300 nm in the UV/Vis spectrum of the amine-modified acrylate monomer. It is known that simple conjugated enones and enals absorb strongly at 230 nm with a weak n–π^* band situated at about 290 nm. Conjugated dienones and dienals absorb strongly at 280 nm due to a strong π–π^* transition; the n–π^* transition of a dienone or dienal is too weak to be observed by second-order derivative UV/Vis spectroscopy at the concentrations expected here. It appears therefore from spectroscopic data that the coloured chromophore is likely to be a dienone or dienal associated with the amine functionality; a simple dienone, β-ionone, was chosen for comparison purposes only and the spectra are shown in Fig. 14 together with those produced with a typical irradiated EB-cured film. The spectra of the model show a very close match to those of the irradiated EB-cured film, confirming the production of a dienone or dienal species.

In a later study Allen *et al.* reported on a second photoyellowing phenomenon which was associated with bisphenol-A-containing coatings.[27] During this study coatings formulated with non-amine-containing multifunctional diluents and aliphatic urethane oligomers exhibited good photoyellowing properties. However, when the aliphatic urethane oligomer was replaced by an aromatic bisphenol-A-based epoxy acrylate, photoyellowing was observed on extended ageing. This photoyellowing phenomenon is undoubtedly associated with the bisphenol A structure in the polymer backbone. On initial ageing the residual photoinitiator is removed photochemically within the first 6 h of irradiation, as can be seen by the loss of photoinitiator absorptions in the UV/Vis spectra situated at 250 nm; the absorption remaining at

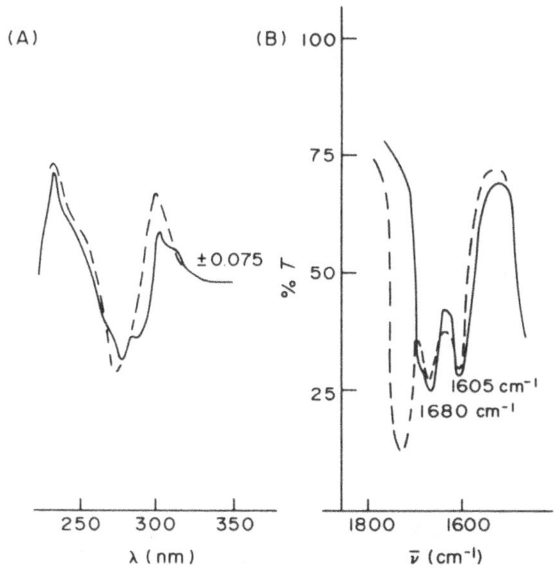

FIG. 14. (A) Second-order derivative UV absorption spectra of an EB-cured resin **XXIX** film (12 μm thick) irradiated for 2·5 h in a Microscal unit (- - -) and β-ionone (10^{-4} M) in *n*-hexane (——). (B) Infrared spectra in the region 1800–1500 cm^{-1} of an EB-cured resin **XXIX** film (4 μm thick) irradiated for 85 h in a Microscal unit (- - -) and β-ionone film between salt flats (——). (Reprinted with permission from Ref. 25. Copyright 1984, Pergamon Journals Ltd.)

270 nm is due to the bisphenol A structure present in the cured coating (Fig. 15). This absorption remains constant for 1000 h of exposure in a Microscal exposure unit and the polymer films remain colourless. On extended ageing the films become yellow and the absorption at 270 nm begins to decrease and broaden with an increase in absorption of the tail towards longer wavelengths and into the visible part of the spectrum. This observation clearly implicates the bisphenol A unit in the formation of a yellow chromophore during photo-oxidative ageing. This can be attributed to highly conjugated quinone-type chromophores such as **XXX**.

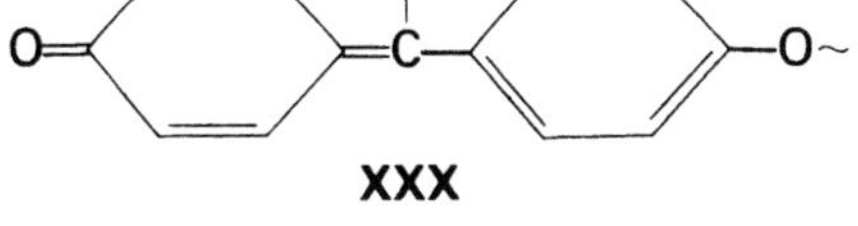

XXX

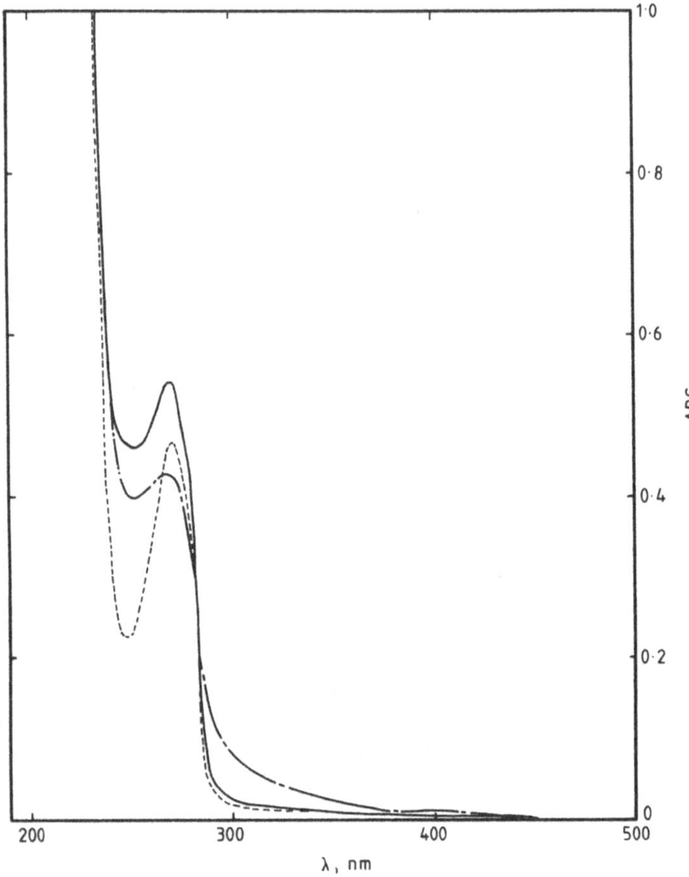

FIG. 15. Normal UV/Vis absorption spectrum of **XXXII** and **XXXI** UV-cured with 5 wt % benzophenone before (——) and after 6 h (– – –) and 1100 h (– · – ·) of irradiation in a Microscal unit. (Reprinted courtesy of the Society of Manufacturing Engineers. Copyright 1985; from the *Radcure Europe 1985 Proceedings.*)

It is clear from the above discussion that the choice of both multifunctional monomer and acrylated oligomer is important when considering photoyellowing performance. Oligomers which contain a pre-quinoid aromatic structure and multifunctional monomers modified with amino groups are particularly susceptible to photoyellowing. This process appears to be photo-oxidative in nature with hydroperoxide species being the key intermediates. Further investigation is necessary to elucidate the full mechanism for the formation of dienone and/or dienal species in amine-modified acrylate systems.

3.2 Photo-oxidation of Radiation-cured Coatings

There are relatively few data available on the photo-oxidation of radiation-cured coatings. Most of the early interest in the photodegradation of radiation-cured coatings has centred around their photoyellowing performance. A number of studies have concentrated on the physical performance of coatings during photodegradation or photo-oxidation with little mechanistic investigation; coatings were exposed until film failure occurred and empirical relationships were obtained between structure and photodegradation resistance. Whilst this allows the best use of currently available materials, a thorough investigation of the photochemical basis for the photodegradation will lead to a better understanding of what structural conditions improve the photodegradation performance and lead to new acrylate systems with greatly improved photostability.

Workers at Celanese have found that a bisphenol-A-based epoxy acrylate and two aliphatic urethane acrylates were superior to an aromatic urethane acrylate in their photodegradation performance as determined by film failure during Atlas Weatherometer ageing.[13,28] These oligomers were studied in conjunction with a multifunctional diluent, trimethylolpropane triacrylate, at about 30% concentration and the benzophenone dimethylethanolamine photoinitiator system. During the same study the effect of multifunctional acrylate types was studied; the data collected indicated that for a given functionality and therefore crosslink density, i.e. hexanediol diacrylate versus tetraethyleneglycol diacrylate, the polyether-containing monomers were more susceptible to photodegradation than the polyether free monomer. In the absence of ether linkages it appears that the higher the functionality and therefore the higher the crosslink density, the more stable the UV-cured coating to Atlas Weatherometer ageing. It appears that the polyether-containing monomers are more susceptible to photo-oxidation due to the ease of oxidation of the α-CH bonds to the ether linkage in the molecule (Scheme 6).

During the QUV® ageing of EB-cured and UV-cured coatings, O'Hara found that the EB-cured epoxy acrylate was less stable than a UV-cured counterpart as indicated by film failure.[10] The same relationship was found, with respect to curing conditions, with an

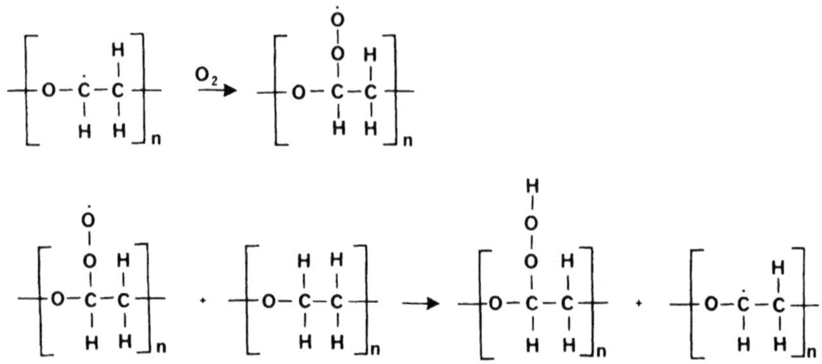

SCHEME 6

aliphatic urethane acrylate. However, surprisingly the aliphatic urethane systems were significantly less stable than the bisphenol-A-based epoxy acrylate systems. During this study the epoxy-based systems were formulated with the polyether containing diacrylate; tripropylene glycol diacrylate whereas the urethane acrylate systems were formulated with a trifunctional polyether diluent monomer, glycerol propoxytriacrylate. It is possible that the higher concentration of diluent used in the urethane system is the reason for its greater instability compared with the bisphenol-A-based epoxy acrylate system.

In a study carried out by Gismondi, the photodegradation of a wide range of monomers was investigated in conjunction with two polyester-based acrylate oligomers.[20] The effect of photodegradation (100 h in a lightbox) on the physical properties of the coatings was monitored by use of an Instron. It was found surprisingly that the multifunctional acrylates containing polyether segments performed well; it was suggested that the multifunctional nature of the monomers and the high crosslink density masked any chain scission effect due to the polyether structure.

Hult & Ranby have carried out detailed studies into the photodegradation pathways which occur in UV-cured coatings.[11,18,19,29] During an initial study they investigated the secondary reactions occurring in organic coatings photocured with 2-isopropylthioxanthone (**IV**).[11]

Irradiation of a PMMA film containing **IV**, as a model for a UV-cured coating, resulted in the rapid photoreduction of **IV** due to hydrogen abstraction from the polymer by photoexcited **IV**. By the use of spin-trapping techniques and ESR these workers were able to show that photoexcited **IV** could decompose hydroperoxides and peroxides by an energy transfer process. Similar radical species were identified during the irradiation of PMMA films containing peroxide species and **IV** for which the mechanism of Scheme 7 was proposed. These alkoxy radicals are highly reactive and may easily abstract a hydrogen atom from the polymer backbone, furthering the degradation of the polymer. Thus, residual 2-isopropylthioxanthone (**IV**) in the cured coating can sensitise the photodegradation of this coating via direct hydrogen abstraction or by the photosensitised decomposition of peroxides and/or hydroperoxides in the coating after curing.

These workers have also investigated the photodegradation of epoxy acrylate and urethane acrylate coatings formulated with hexanediol

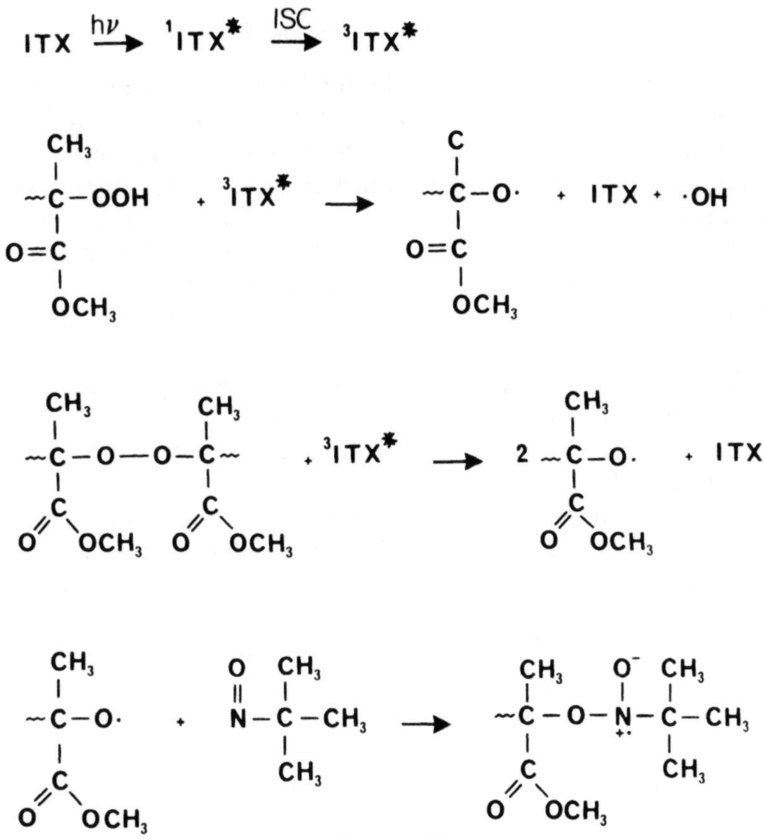

SCHEME 7

diacrylate and tripropyleneglycol diacrylate which have been photocured with a 2-isopropylthioxanthone/methyldiethanolamine photoinitiator mixture. The photodegradation was followed quantitatively by using ESCA and qualitatively by using both reflection and transmission infrared spectroscopy. They found that the photocuring of an epoxy acrylate system in the presence of oxygen leads to a rapid surface oxidation (of the top 10 nm).

The ESCA spectra showed an increase in the formation of carbon–oxygen single bonds during curing which may be due to the formation of hydroxyl, hydroperoxide, peroxide or ether groups. The spectra indicated that 30% of the total surface photo-oxidation, after 96 h of photodegradation, occurred during the curing process. During photodegradation, ESCA spectroscopy indicated the formation of

$$>C{-}O,\ >C{=}O \text{ and } O{-}\overset{|}{C}{=}O$$

groups. Approximately 75% of the total oxidation occurs in the first 24 h (top 10 nm). A comparison of reflection and transmission IR spectroscopy of a 30 μm film confirms that the majority of this photo-oxidation occurs at the coating surface. Corresponding photo-oxidation studies were also carried out with a photostable urethane acrylate resin. With this system about 40% of the total oxidation after 96 h of photodegradation occurred during the photocuring process. The overall oxidation, however, was much less than that observed with the epoxy system.

These workers have also investigated the use of photofragmenting initiators **III**, **XII** and **XXVII** and their effects on the photodegradation of UV-cured coatings by studying polystyrene as a model in solution and film.

Addition of these photoinitiators to polystyrene solutions in benzene resulted in a rapid decrease in molecular weight on irradiation; similar effects were observed in polystyrene films; also, the polymer became partially crosslinked. This degradation was caused by hydrogen abstraction by radicals which originated from the added photoinitiator. The order of degradation was found to be:

$$\textbf{XII} > \textbf{III} > \textbf{XXVII}$$

Initiator **XII** has the strongest absorption between 300 and 400 nm and also produces two highly reactive radicals on photoexcitation, whilst **XXVII** and **III** produce only one highly reactive radical.

The significant difference between **XXVII** and **III** may be that **XXVII** has an isopropyl substituent which will stabilise the benzoyl radical produced and make it less reactive. Furthermore, initiator **XII** can form acetophenone via in-cage or near-cage recombination of the benzoyl and methyl radicals.[30] The acetophenone produced can sensitise the photodegradation through hydrogen-atom abstraction from the polymer.

Decker & Bendaikha have carried out a mechanistic and kinetic study of the photodegradation of a UV-cured epoxy acrylate coating.[23,24] The epoxy diacrylate used was derived from the glycidyl ether of bisphenol A and was studied in conjunction with tripropylene glycol as reactive diluent and **XII** as photoinitiator. The photodegradation was carried out using a medium-pressure mercury lamp with and without a Pyrex filter. The photodegradation was monitored by GLC analysis of volatiles produced, oxygen uptake measurements and infrared spectroscopy. No significant changes in the pendulum hardness could be noticed during the photo-oxidation of the coating, even after extending the exposure time up to 1000 h; this was explained as being due to the high crosslink density of the UV-cured coatings. It was observed that on photoageing with radiation at wavelengths >250 nm the epoxy acrylate coatings suffer important chemical modifications after only 140 h (Fig. 16).

Clearly under these irradiation conditions the phenyl ring of the bisphenol A unit is particularly photosensitive. The IR spectra show the loss of various functional groups with a slight initial growth of the OH band and broadening of the carbonyl band at 1735 cm^{-1}. Photolysis was also carried out in the absence of oxygen and with filtered light of wavelengths >300 nm. The quantum yields of functional group disappearance are shown in Table 9.

Under N_2 exposure the films were much more resistant to UV degradation with wavelengths >250 nm. The disappearance of various functional groups was three- to ten-fold less pronounced than under air exposure. The bisphenol A unit was the least affected by atmospheric conditions, indicating that its UV absorption results in photolysis. The alkyl groups were hardly changed under N_2, indicating that hydroperoxidation is the most important photodegradation mechanism under air exposure. Decker & Bendaikha concluded that 80% of the functional group loss under air was due to oxidation reactions.

Under irradiation at wavelengths >300 nm the bisphenol A unit was again the most labile part of the coating; however, it appeared that the

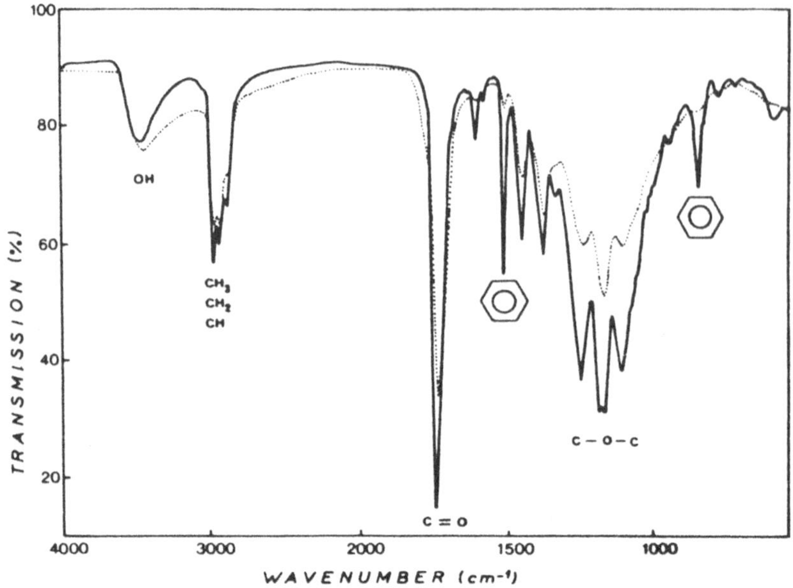

FIG. 16. Infrared spectra of epoxy acrylate coatings before (——) and after (· · · · · ·) photo-oxidation at >250 nm for 140 h. (Reprinted with permission from *Journal of Radiation Curing*, **11**, No. 2, April 1984. Published by Technology Marketing Corp., Norwalk, CT, USA.)

photons absorbed by the n–π^* absorption at wavelengths >300 nm were about half as efficient as those absorbed at wavelengths <300 nm by a π–π^* transition. The ester and ether groups also appeared to be more resistant under photolysis at wavelengths >300 nm. The —CH_3, —CH_2 and —OH groups were more sensitive to irradiation at wavelengths >300 nm due to an efficient chain peroxidation process.

Allen *et al.* have carried our a series of studies into the photostability of both UV- and electron-beam-cured acrylate films.[25–27] In an initial study they compared the performance of two multifunctional diluents **XXXI** and **XXIX**, where R = poly(ether polyol), which were

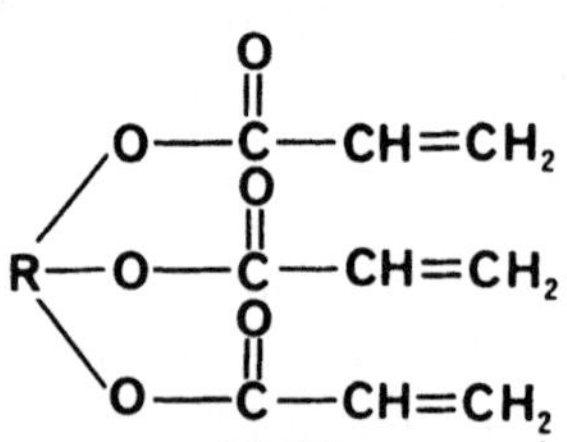

XXXI

TABLE 9
Quantum Yield of Disappearance of Functional Groups in the Photolysis of Epoxy-Acrylate Coatings

Functional group	*IR absorption wavenumber (cm^{-1})*	*Initial concentration ($mol\ litre^{-1}$)*	*Quantum yield (10^{-4} mol einstein^{-1})*		
			>250 nm[a]		*>300 nm*[b]
			N_2	*O_2*	
Ester	1735	7	0·8	3	0·8
Ether	1180	7	0·9	3	0·7
Phenyl	1510 830	1·7	4·5	11	6
Methyl	2980	9·7	0·7	4	8
CH_2	2940	17·7	0·2	2	5
CH	2880	16·8	0·2	2	6
OH	3470	1·7	0	−2[c]	−3·2[c]

Reprinted with permission from *Journal of Radiation Curing*, **11**, No. 2, April 1984. Published by Technology Marketing Corp., Norwalk, CT, USA.

[a] Photon flux = $3{\cdot}8 \times 10^{-8}$ einstein s^{-1} cm^{-2}.

[b] Photon flux = 2×10^{-8} einstein s^{-1} cm^{-2}.

[c] Quantum yield of OH groups accumulation.

UV-cured with **I**, **III**, **XV** and a combination of **I** and **II**. Films were also EB-cured with an ESI electrocurtain curing apparatus at a dose of 2MRad. Photo-oxidation was induced by ageing in a Microscal apparatus utilising a 500 W high-pressure Hg/tungsten lamp and monitored by the use of transmission IR spectroscopy and hydroperoxide titration. The main changes observed in the infrared spectra associated with oxidation (photoyellowing phenomena have been covered in Section 3.1.2) were the growth of a hydroxyl band centred at 3450 cm^{-1} and broadening of the carbonyl band centred at 1720 cm^{-1} (Fig. 17). The rate of formation of this hydroxyl band with irradiation time was monitored as a hydroxyl index (Figs 18 and 19).[25]

Three main features were evident from the results. Firstly, multifunctional monomer **XXIX** is generally more photostable than multifunctional monomer **XXXI**. Secondly, the EB-cured resins are more photostable than those cured by a UV-initiated process. Thirdly, the photostabilities of the UV-cured resins are highly dependent on the structure of the carbonyl initiator used. The latter is more evident with **XXXI** than with **XXIX**.

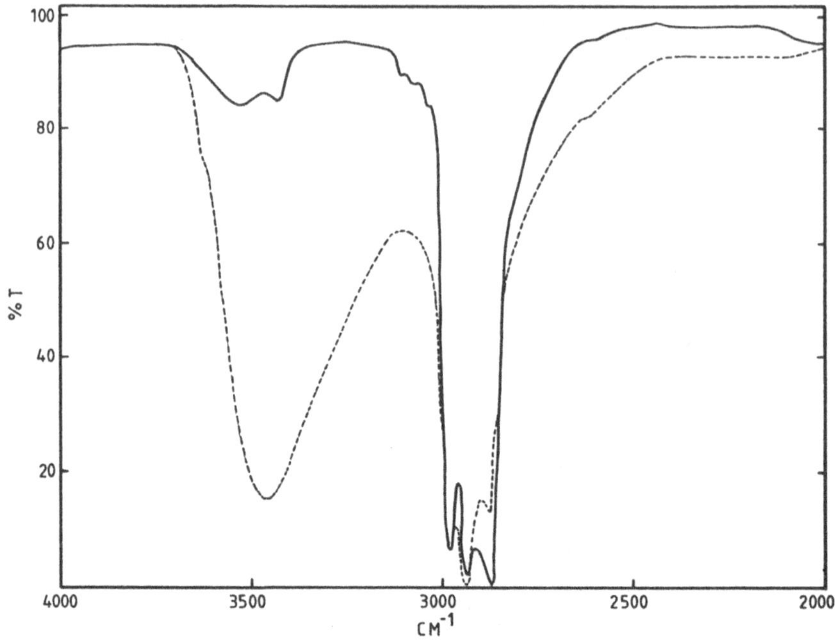

FIG. 17. Infrared spectrum of resin **XXXI** UV-cured with 5 wt % benzophenone before (——) and after (– – –) 172 h of irradiation in a Microscal unit. (Reprinted courtesy of the Society of Manufacturing Engineers. Copyright 1985; from the *Radcure Europe 1985 Proceedings*.)

The higher photostability of the EB-cured coatings is not unexpected as there is no residual photoactive photoinitiator as in the case of the UV-cured systems. In the UV-cured systems hydrogen-abstracting photoinitiators such as **I**, and **I** with **II** are more damaging that **III** and **XV**. The generally greater stability of multifunctional monomer **XXIX** compared with **XXXI** is probably due to the presence of the terminal diethylamino group present in the polymer after curing. It is well known that alkylated amines can operate as labile hydrogen-donating sites for photoactive carbonyls, e.g. according to Scheme 8.

The radical site on the α-carbon atom will then be preferentially attacked by oxygen, thus protecting the polymer. Essentially, the amine will act as a scavenger for oxygen and, in fact, this is one way in which amines operate with aromatic carbonyls to accelerate the curing process.[31]

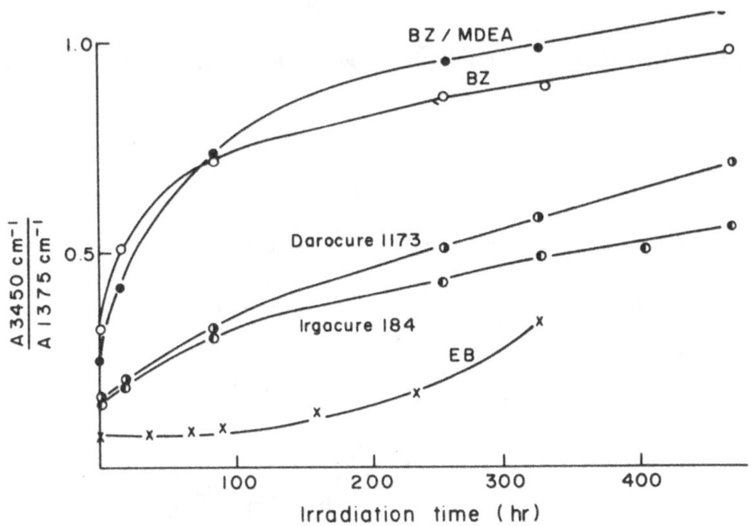

FIG. 18. Rate of hydroxyl production at 3450 cm^{-1} during irradiation in a Microscal unit of resin **XXXI** after curing by (×) electron beam, and by UV with (○) benzophenone; (●) benzophenone/MDEA; (◑) Darocure® 1173; (◐) Irgacure® 184. (Reprinted with permission from Ref. 25. Copyright 1984, Pergamon Journals Ltd.)

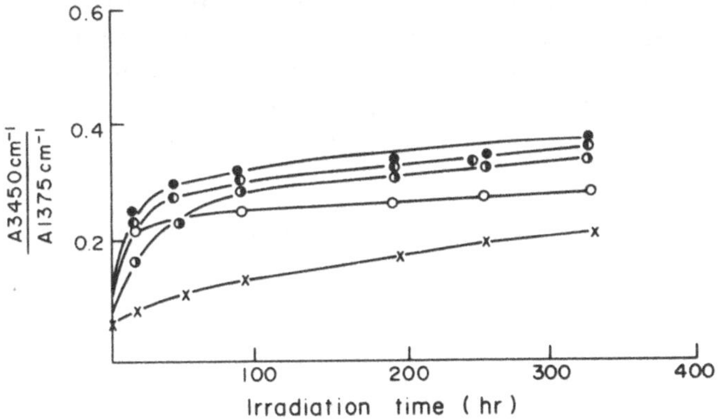

FIG. 19. Rate of hydroxyl production at 3450 cm^{-1} during irradiation in a Microscal unit of resin **XXIX** after curing by (×) electron beam and by UV with (○) benzophenone; (●) benzophenone/MDEA; (◑) Darocure® 1173; (◐) Irgacure® 184. (Reprinted with permission from Ref. 25. Copyright 1984, Pergamon Journals Ltd.)

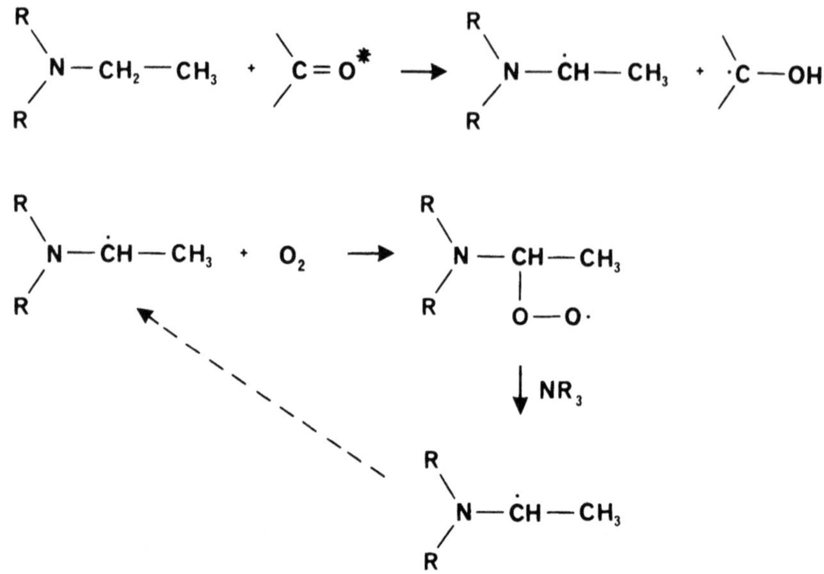

SCHEME 8

In a follow-up study, Allen *et al.* investigated the role of hydroperoxides in the photo-oxidative degradation of these two multifunctional monomers, by the use of hydroperoxide titration, SO_2 treatment and monochromatic irradiation at 365 nm.[26] Under polychromatic light carbonyl absorption is important whereas under 365 nm irradiation hydroperoxide photolysis predominates. Treatment with SO_2 after curing destroys hydroperoxides formed during the curing process:

$$ROOH + SO_2 \longrightarrow R-O-\overset{\overset{\displaystyle O}{\|}}{\underset{\underset{\displaystyle O}{\|}}{S}}-OH \longrightarrow \text{Further Products} \qquad (5)$$

The most significant features of the results of hydroperoxide measurement are shown in Figs 20 and 21. In Fig. 20 it can be seen that multifunctional monomers **XXIX** and **XXXI** showed markedly different hydroperoxidation behaviour on photoexposure. Resin **XXXI** is rapidly attacked by the residual photoinitiator **I**, which disappears in the first 3-4 h of irradiation, with the rapid production of hydroperoxides as a result of oxidation. Multifunctional acrylate monomer **XXIX**,

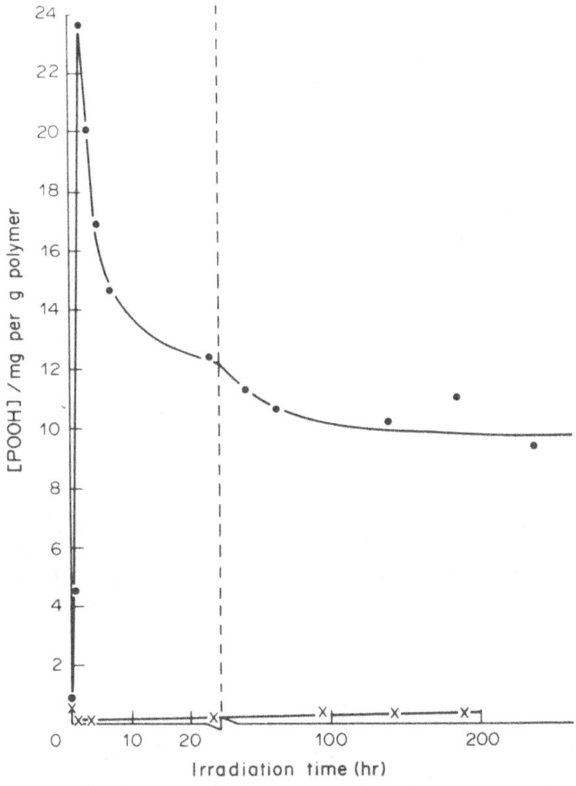

FIG. 20. Hydroperoxide formation in UV-cured (5 wt % benzophenone) resin **XXXI** (●) and resin **XXIX** (×) during post-cure polychromatic irradiation in a Microscal unit. Hydroperoxide is expressed as equivalent mass of cumene hydroperoxide (POOH) per unit mass of polymer. (Reprinted with permission from Ref. 26. Copyright 1985, Pergamon Journals Ltd.)

however, is not rapidly oxidised, as evidenced by the low level of hydroperoxides produced. Figure 21 illustrates that the addition of methyldiethanolamine **II** reduces the level of hydroperoxide formation in the UV-cured multifunctional acrylate monomer **XXXI** and that the hydroperoxide build-up in EB-cured multifunctional monomer **XXXI** follows very closely the hydroxyl index profile for this sample. Sulphur dioxide treatment does not markedly affect the photo-oxidation of monomer **XXXI** either UV- or EB-cured (Fig. 22). Also irradiation at 365 nm markedly affects the EB-cured film but not the UV-cured film

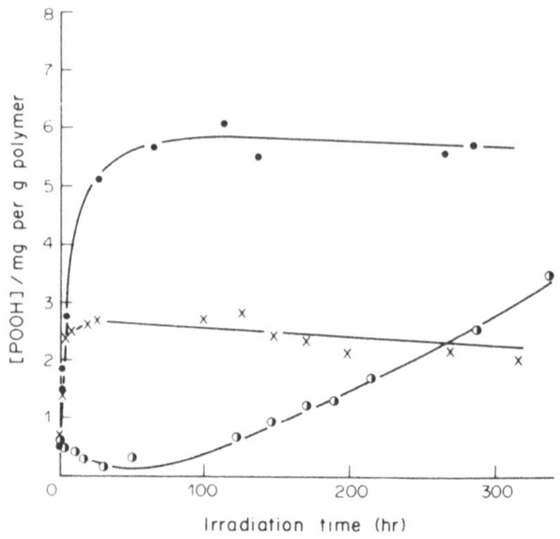

FIG. 21. Hydroperoxide formation in UV-cured resin **XXXI**; (●) 2·5 wt % benzophenone (BZ) + 2·5 wt % MDEA; (×) 5% BZ + 5% MDEA; and in EB-cured resin **XXXI** (◑), during post-cure irradiation in the Microscal unit. (Reprinted with permission from Ref. 26. Copyright 1985, Pergamon Journals Ltd.)

(Figs 23 and 24). This suggests that hydroperoxides are not the key initiating species in the photo-oxidation of these monomers after curing; in the case of UV-cured systems the photoinitiator has the overriding effect whereas aliphatic carbonyl species produced during the EB-curing process, which do not absorb 365 nm light, are the key initiators of photo-oxidation in these systems. The presence of tertiary amine functionality in monomer **XXIX** protects the polymer from photo-oxidation by interacting with residual photoinitiator and being preferentially oxidised.

Allen *et al.* have also investigated the addition of both aromatic epoxy **XXXII** and aliphatic urethane **XXXIII** oligomers to the two monomers **XXIX** and **XXXI**. It was found that formulation with the above oligomers greatly increased the film stability as compared with the monomer **XXXI** alone (Fig. 25). This could possibly be due to either an antioxidant effect produced by the two oligomers and/or the higher crosslink density of the cured film. Interestingly, no difference was observed between the epoxy and urethane oligomers in terms of photo-oxidative stability.

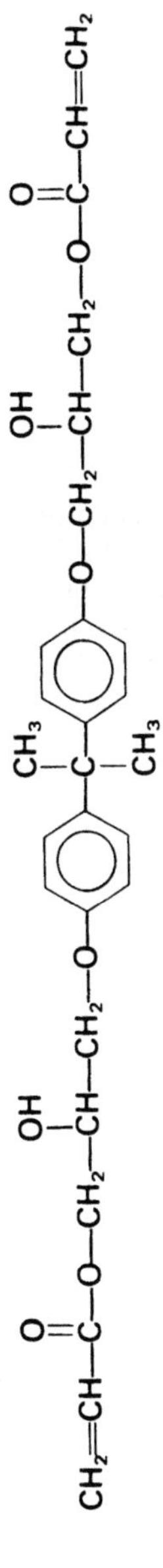

XXXII

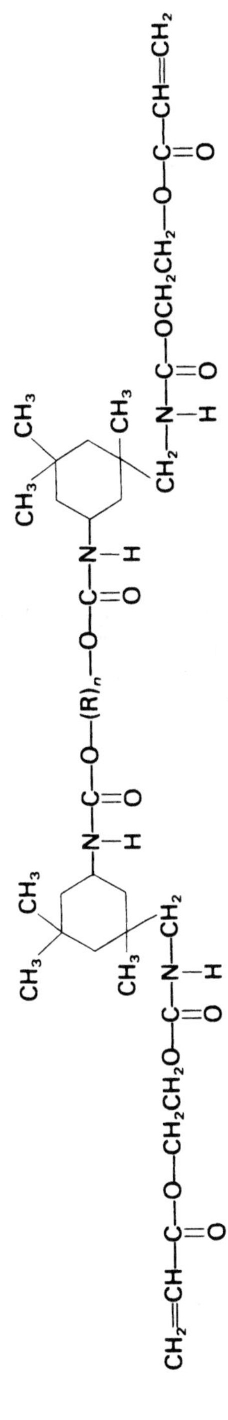

XXXIII

*$HO—(R)_n—OH$ – Polyester of variable composition

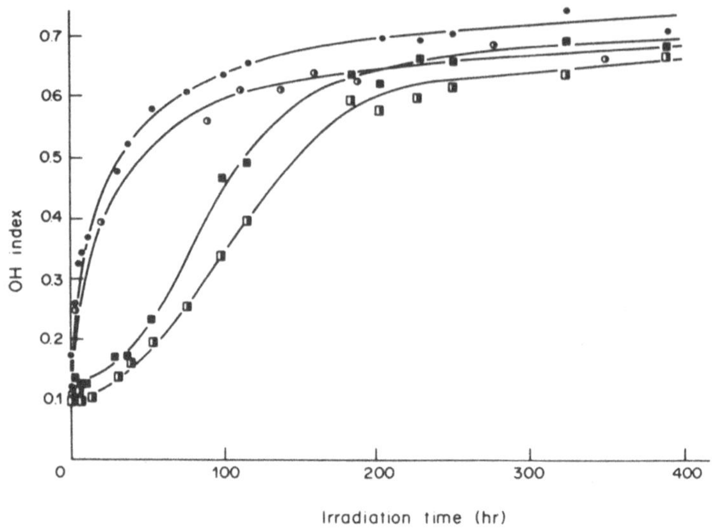

FIG. 22. Hydroxyl production at 3450 cm^{-1} during irradiation in the Microscal unit of UV-cured resin **XXXI** (benzophenone): (●) with SO_2 treatment; (◑) without SO_2 treatment; and of EB-cured resin **XXIX**; (■) with SO_2 treatment; (◨) without SO_2 treatment. (Reprinted with permission from Ref. 26. Copyright 1985, Pergamon Journals Ltd.)

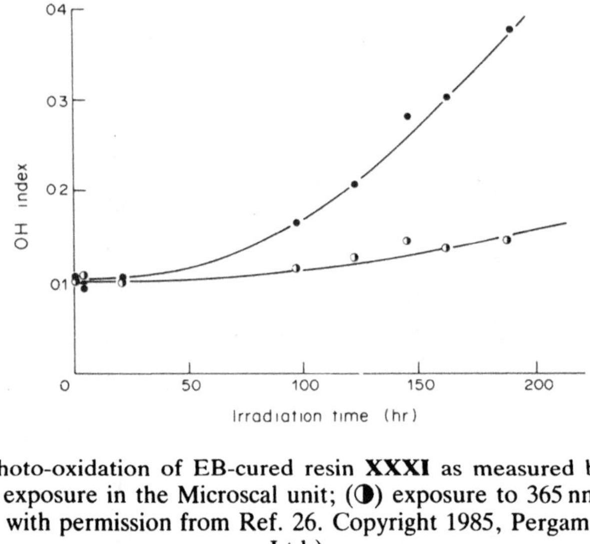

FIG. 23. Photo-oxidation of EB-cured resin **XXXI** as measured by hydroxyl index: (●) exposure in the Microscal unit; (◑) exposure to 365 nm radiation. (Reprinted with permission from Ref. 26. Copyright 1985, Pergamon Journals Ltd.)

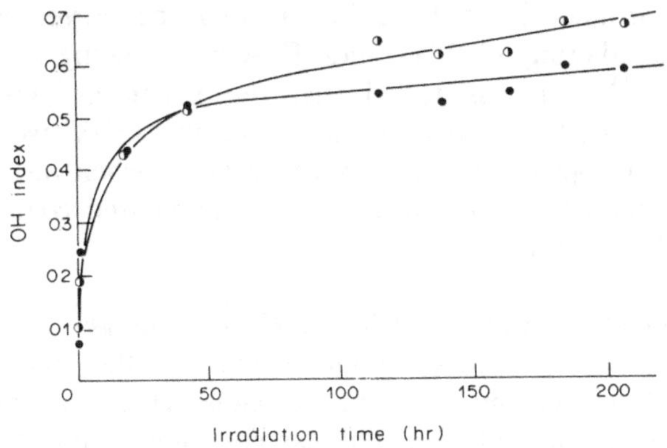

FIG. 24. Photo-oxidation of resin **XXXI** UV-cured with 5 wt % benzophenone as monitored by hydroxyl index: (○) exposure in Microscal unit; (◑) exposure to 365 nm radiation. (Reprinted with permission from Ref. 26. Copyright 1985, Pergamon Journals Ltd.)

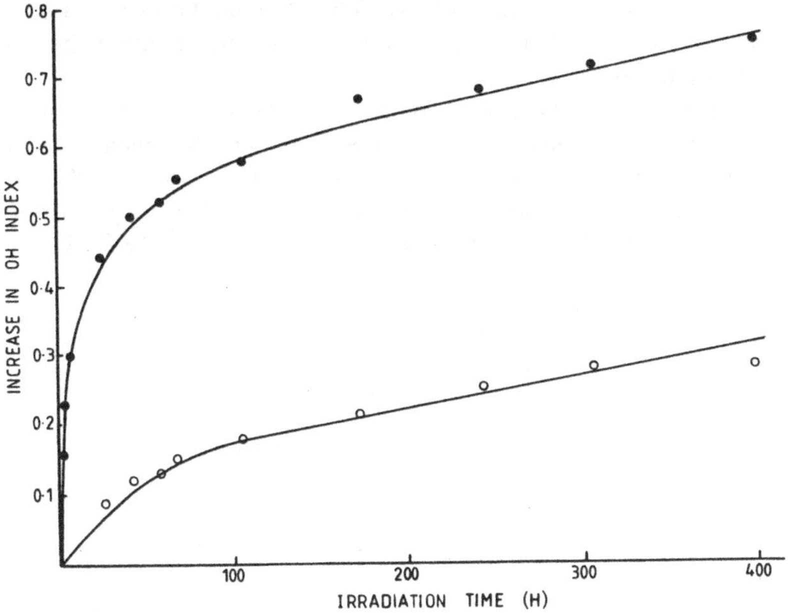

FIG. 25. Hydroxyl index versus irradiation time in a Microscal unit for UV-cured (5 wt % benzophenone) resin **XXXI** (●) and resin **XXXI + XXXIII** (○). (Reprinted courtesy of the Society of Manufacturing Engineers. Copyright 1985; from *Radcure Europe 1985 Proceedings*.)

Gatechair and Evers have investigated the correlation between Florida weathering of UV- and EB-cured coatings, and artificial weathering.[13,32] It was found that the xenon arc weatherometer produced a better correlation than did the QUV weatherometer. However, it appears that care must be taken when investigating the photodegradation characteristics of radiation-cured coatings with artificial light sources.

3.3 Photostabilisation of Radiation-Cured Coatings

In the preceding sections improvement of the photostability of radiation-cured coatings by the judicious choice of photoinitiator diluent monomers and oligomers has been illustrated. An alternative to this approach, where choice is limited, is the use of photostabilisers and/or antioxidants as additives. These additives are already used to improve the photostability of a wide range of polymers.[5–7] To be acceptable for use in radiation-cured coatings, photostabilisers and/or antioxidants must be soluble in the coating formulation and not interfere with the curing process. With the limitations of materials available for formulation, many workers are investigating the use of photostabilisers.

Hult & Ranby have carried out a study of the effect of three common photostabilisers on the photoinitiation effectiveness of some common photoinitiators in the photopolymerisation of methyl methacrylate.[29] The three photostabilisers included a light absorber, 2-(3′,5′-di-isopentyl-2′-hydroxyphenyl)benzotriazole (**XXXIV**); an

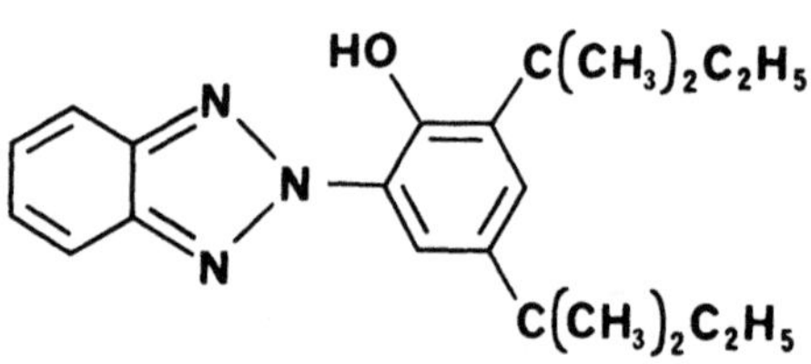

XXXIV

excited-state quencher, [2,2′-thiobis(4-*t*-octylphenolate)]-*n*-butylamine nickel(II) (**XXXV**); and a hindered amine light stabiliser (HALS) bis(1,2,2,6,6-pentamethyl-4-piperidyl) sebacate (**XXXVI**). The photoinitiators studied included **XII**, **XXVII** and **IV** in conjunction with triethylamine.

The benzotriazole UV absorber **XXXIV** was found to reduce the rate of polymerisation and increase the molecular weight. The thioxanthone photoinitiator **IV** was the least affected as it absorbs UV

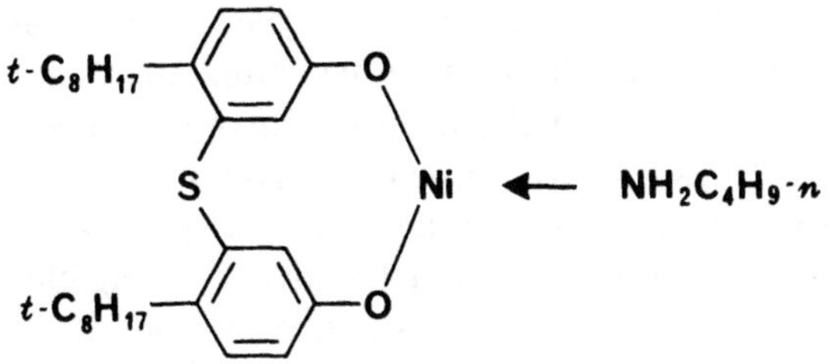

XXXV

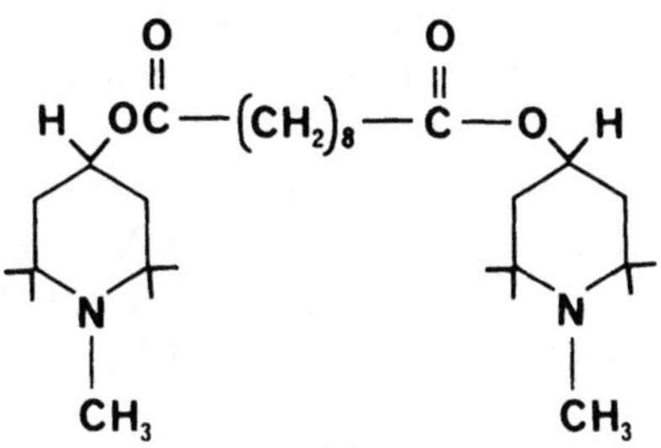

XXXVI

light at longer wavelengths. The excited-state quencher **XXXV** increased the molecular weight because of its light absorption. When used with photofragmenting photoinitiators a narrower polymer molecular weight distribution was produced. It was suggested that this may be due to quenching of secondary reactions of triplet excited acetophenone groups which is difficult to separate from light absorption effects. It was found that this stabiliser could not quench photoexcited photoinitiator **IV**. The HALS **XXXVI** had only a minor effect on the photopolymerisation rate and molecular weight distribution.

Puglisi & Vigeant have reported that the use of UV absorbers such as hydroxytolyl benzotriazole (**XXXVII**) impairs the curing process, as

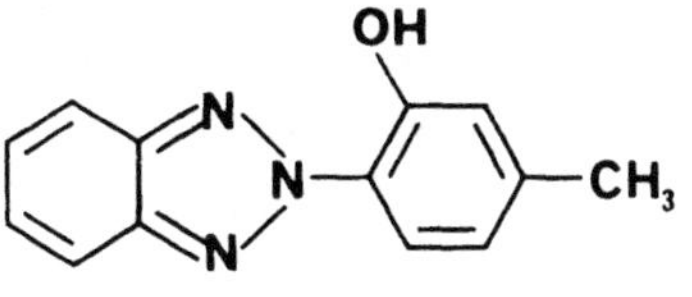

XXXVII

indicated by pendulum hardness measurements.[33] The hardness of the film consistently decreased as more absorber was added. These workers also reported that the use of the HALS **IX** improved the yellowing performance of a urethane acrylate-based coating on Xenotest weathering; this stabiliser had a negligible effect on cure.

O'Hara has investigated the use of a mixed photostabiliser system including HALS **XXXVI** with a benzotriazole absorber to improve the photostability of epoxy acrylate networks on QUV weathering.[10] This photostabiliser mixture was found to have a negligible effect on the cure rate. No improvement in the colour properties of the coatings was observed but photodegradation as measured by film breakdown was greatly improved for the EB-cured coating but only marginally for the UV-cured coating. Decker and Bendaikha have also found that the above UV absorber improved the photostability of epoxy acrylate networks.[23,24]

Decker and Bendaikha have studied methods of reducing the oxygen inhibition which is observed with UV-cured coatings;[34,35] one of these methods used the oxidation of 1,3-diphenylisobenzofuran (**XXXVIII**) with singlet oxygen generated by red-light irradiation of

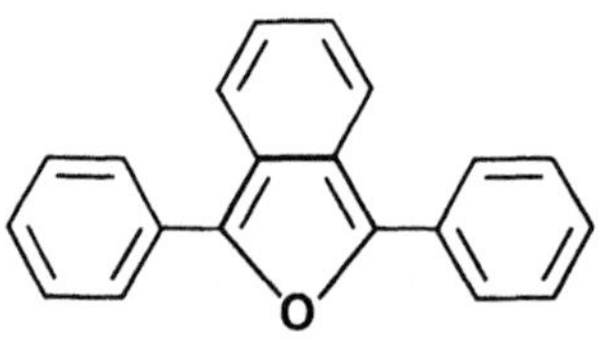

XXXVIII

Methylene Blue (Scheme 9). UV curing is initiated soon after red-light irradiation and proceeds at a faster rate than without the dye-sensitised O_2 removal step. Decker found that coatings prepared by this method had improved stability with less tendency to photoyellowing and photo-oxidative degradation.[34] Presumably this was due to the fact that O_2 was not incorporated into the coating during UV curing to produce prodegradant species such as hydroperoxides.

Tu and coworkers have screened 18 UV absorbers, investigating solubility and compatibility with the UV-curing process.[14] Many of these were eliminated on the grounds of solubility. This study concluded that alkyl cyanodiphenylacrylate **IXL** and dimethyl-*p*-methoxybenzylidenemalonate **XL** were most effective at reducing the photoyellowing observed under xenon-arc weatherometer conditions.

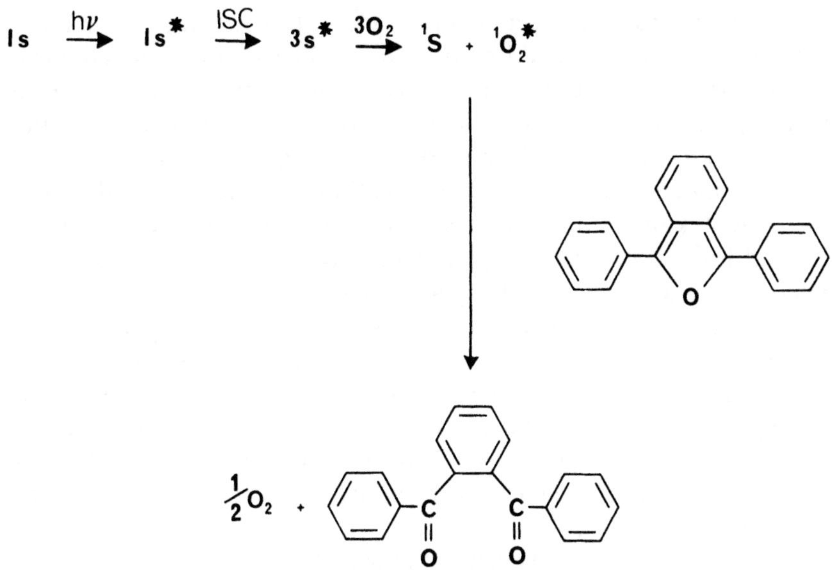

SCHEME 9

Gatechair and coworkers have investigated a number of photostabilisers including UV absorbers, antioxidants and HALS for application in radiation-cured coatings.[13,32] An investigation of the effectiveness of UV absorbers was carried out in a model coating system, with a determination of the maximum practical usable level (MPL) of each UV absorber which results in less than 10% reduction

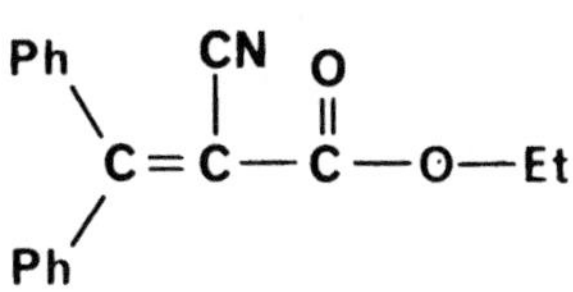

IXL

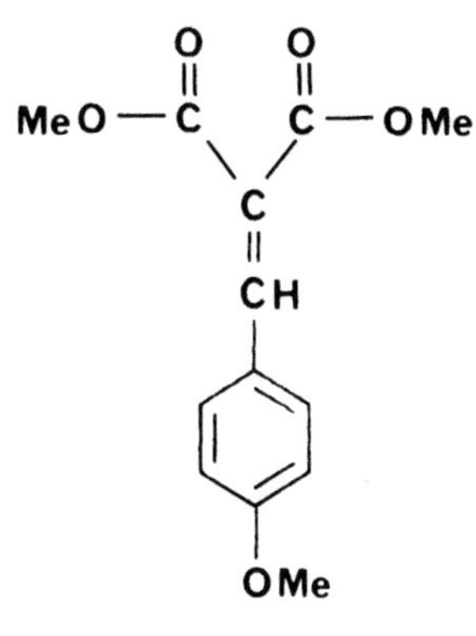

XL

of cure as measured by pendulum hardness. It was found that >1% of octylphenyl salicylate could be used but <0·05% of a hydroxyphenylbenzotriazole can be tolerated. After the cure effects had been investigated, five UV absorbers, representing different structural classes, were tested, at their MPL, with a model aliphatic urethane acrylate under Florida weathering conditions. It was found that after 9–12 months many formulations underwent severe cracking resulting in total loss of physical integrity; 4-octyloxy-2-hydroxybenzophenone, ethyl 2-cyano-3,3-diphenylacrylate, amyl-*p*-dimethylaminobenzoate and two phenolic antioxidants did not protect the test plaques from this catastrophic failure. None of the UV absorbers tested appeared to protect the coatings from photoyellowing; this is in contrast to the findings of Tu and coworkers.[14] In fact, the two antioxidants investigated in this study were found to enhance colour formation by operating as prodegradants and eventually causing premature breakdown of the polymer.

Gatechair has also investigated the use of HALSs in radiation-cured coatings, namely compound **XXXVI**.[13,32,36,37] These materials do not absorb UV light, are catalytic in activity and effective in stabilising radiation-cured coatings. Under Florida ageing conditions the stabilities of radiation-cured resin films were doubled. Under artificial weathering conditions Gatechair noted an initial severe yellowing with HALS stabiliser which is lost on extended ageing but is not observed under natural weathering conditions. Gatechair has rationalised this observation by invoking the probability that a disproportionally high level of nitroxyl radicals are being produced initially by short-wavelength radiation coupled with elevated temperatures which are found in most accelerated weathering devices. Nitroxyl radicals are important intermediates in the HALS stabilisation mechanism and are formed by oxidation of the amine functionality:

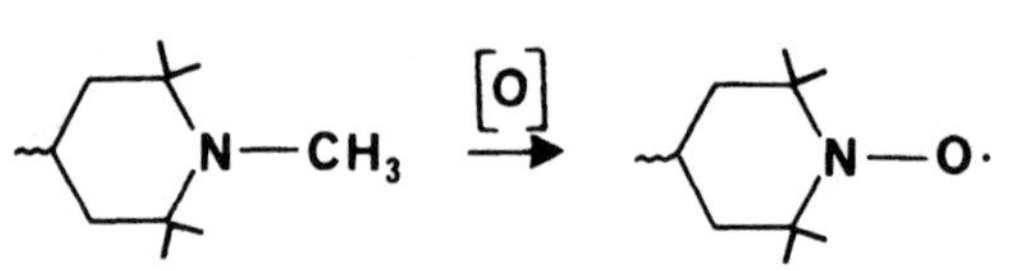

(6)

Even though they stabilise by interaction with the radicals produced during the degradation they do not interfere with the curing process at concentrations below 3%.

Gatechair has also investigated the photostabilisation of EB-cured coatings and UV-cured coatings with a synergistic stabiliser mixture; synergism is a well-known phenomenon in polymer stabilisation. The stabiliser package consisted of a HALS and a hydroxyphenylbenzotriazole UV absorber. It was found that the unstabilised films and those containing the UV absorber alone exhibited a rapid reduction in gloss within the first 12 months of Florida weathering. The HALS on its own was found to improve stability; however, when it was mixed with the hydroxyphenylbenzotriazole UV absorber the stability was further improved.[13,32] It was believed that the UV absorber was protecting the HALS or reactive intermediates from photolysis, thereby increasing the HALS lifetime as a stabilising species.

Olson and coworkers have disclosed a novel approach to the problem of UV absorber addition to UV-cured coatings.[38] In order to remove the internal filter effect of incorporating UV absorbers into a formulation which reduces cure they had the idea of using blocked UV absorbers, either benzotriazole- or benzophenone-type, produced by reaction with a sulphonic acid or an isocyanate. These blocked UV absorbers have the advantage that they exhibit much lower absorptivities at wavelengths greater than 300 nm where there are important photoinitiator absorptions and therefore they can be incorporated at much higher levels in a UV-cured coating formulation without interfering with the curing mechanism. During weathering this blocked UV absorber is released photochemically or thermally, thus producing the active species.

4 CONCLUSIONS

The thermal and photo-oxidative degradation of UV- and EB-cured resin materials is already a difficult subject, often complicated by the variable nature of the formulations used. The structure of the resin is of primary importance in controlling stability, with the initiator, inhibitor and other minor components having a variable controlling influence. These are in turn influenced by the nature and degree of the radiation-curing process used. All these factors control the crosslink density of the resin which has an important influence on oxygen diffusion and hence radiative degradation. Assessment of stability depends, as with all polymers, on the end-use application and whilst experimental studies are useful for assessing certain specific changes, exposure in a real-life environment is essential for long-term predictions to be made. Stabilisation processes are complex but the few studies that have been carried out to date indicate that there is potential for long-term improvements, through the use of hindered piperidine derivatives.

REFERENCES

1. Pappas, S. P., ed., *U.V. Curing: Science and Technology,* Vol. I. Technology Marketing Corp., Norwalk, CT, 1978.
2. Pappas, S. P., ed., *U.V. Curing: Science and Technology,* Vol. II. Technology Marketing Corp., Norwalk, CT, 1985.
3. Roffey, C. G., *Photopolymerization of Surface Coatings.* John Wiley, Chichester, 1982.
4. Phillips, R., *Sources and Applications of Ultraviolet Radiation.* Academic Press, London, 1983.
5. Allen, N. S. & McKellar, J. F., *Photochemistry of Man-Made Polymers.* Applied Science Publishers, London, 1979.
6. Ranby, B. & Rabek, J. F., *Photodegradation, Photooxidation and Photostabilization of Polymers.* John Wiley, London, 1975.
7. Carlsson, D. J. & Wiles, D. M., *J. Radiat. Curing,* **2**(4) (1975) 2.
8. Armbruster, D. C. & Molina, J. F., *J. Radiat. Curing,* **8**(4) (1981) 4.
9. Thalacker, V. P. & Boettcher, T. E., *Proc. Radcure Europe 1985,* SME Technical Paper FC85–450.
10. O'Hara, K. J., *Proc. Radcure Europe 1985,* SME Technical Paper FC85–421.
11. Hult, A. & Ranby, B., *Polym. Deg. Stab.,* **8** (1984) 75.
12. Heyward, I. P., Chan, M. G. & Ludwick, A. G., *Polymer Preprints* **25**(1) (1984) 44.
13. Gatechair, L. R. In Ref. 2, Chapter 7.
14. Tu, R. S., *Proc. Radcure V, 1980,* SME Technical Paper FC80–579.
15. Ohngemach, J., Neisius, K. H., Eichler, J. & Herz, C. P., *Kontakte,* **3** (1980) 15.
16. Schmid, S. T., *J. Radiat. Curing,* **11**(2) (1984) 19.
17. Filipescu, N. & Minn, F. L., *J. Am. Chem. Soc.,* **90** (1968) 1544.
18. Hult, A. & Ranby, B., *Polym. Deg. Stab.,* **8** (1984) 89.
19. Hult, A., Yuan, Y. Y. & Ranby, B., *Polym. Deg. Stab.,* **8** (1984) 241.
20. Gismondi, T. E., *J. Radiat. Curing,* **11**(2) (1984) 14.
21. Allen, N. S. & McKellar, J. F., *J. Appl. Polym. Sci.,* **20** (1976) 1411.
22. Pappas, S. P. & Winslow, F. H. eds, *Photodegradation and Photostabilization of Coatings,* ACS Symposium Series No. 151. American Chemical Society, Washington, DC, 1986.
23. Decker, C. & Bendaikha, T., *Polymer Preprints,* **25**(1) (1984) 42.
24. Decker, C. & Bendaikha, T., *J. Radiat. Curing,* **11**(2) (1984) 6.
25. Allen, N. S., Robinson, P. J., White, N. J. & Skelhorne, G. G., *Eur. Polym. J.,* **20**(1) (1984) 13.

26. Allen, N. S., Robinson, P. J., White, N. J. & Skelhorne, G. G., *Eur. Polym. J.*, **21**(2) (1985) 107.
27. Allen, N. S., White, N. J. & Swale, D. W., *Proc. Radcure Europe 1985*, SME Technical Paper FC85–412.
28. Anon., *Weathering Properties of U.V. Curable Coatings*. Celanese Technical Bulletin (8/80).
29. Hult, A. & Ranby, B., *Polym. Deg. Stab.*, **9** (1984) 1.
30. Sander, M. R. & Osborn, C. L., *Tetrahedron Lett.*, (1974) 415.
31. McGinniss, V. D. In *Developments in Polymer Photochemistry-3*, ed. N. S. Allen. Applied Science, London, 1982, Ch. 1, p. 1.
32. Gatechair, L. R. & Evers, H. J. In *Proc. Water-Borne and Higher-Solids Coatings Symposium, Feb 5–7, 1986, New Orleans, LA USA*.
33. Puglisi, J. & Vigeant, F., *J. Radiat. Curing*, **7**(3) (1980) 31.
34. Decker, C., Fizet, M. & Faure, J., *Org. Coat. Plast. Chem.*, **710** (1980) 42.
35. Decker, C., Bendaikha, T., Fizet, M. & Faure, J., *Proc. Radcure '85*, SME Technical Paper, FC85–432.
36. Gatechair, L. R. & Wostratzky, D., *J. Radiat. Curing*, **10**(3) (1983) 4.
37. Gatechair, L. R. & Wostratzky, D., *Proc. Radiation Curing VI*, 1982, SME Technical Paper FC82–279.
38. Olson, D. R. & Schroeter, S. H., *Polym. Sci.*, **22** (1978) 165.

INDEX

GPSR Compliance

The European Union's (EU) General Product Safety Regulation (GPSR) is a set of rules that requires consumer products to be safe and our obligations to ensure this.

If you have any concerns about our products, you can contact us on ProductSafety@springernature.com

In case Publisher is established outside the EU, the EU authorized representative is:

Springer Nature Customer Service Center GmbH
Europaplatz 3
69115 Heidelberg, Germany

Batch number: 10607198

Printed by Printforce, the Netherlands